Environmental Nanotoxicology

Patrick Omoregie Isibor • Geetha Devi
Alex Ajeh Enuneku
Editors

Environmental Nanotoxicology

Combatting the Minute Contaminants

Springer

Editors
Patrick Omoregie Isibor
Biological Sciences
Covenant University
Ota, Nigeria

Alex Ajeh Enuneku
Environmental Management and Toxicology
University of Benin
Benin City, Nigeria

Geetha Devi
Mechanical and Industrial Engineering
National University of Science and Technology
Sultanate of Oman, Oman

ISBN 978-3-031-54153-7 ISBN 978-3-031-54154-4 (eBook)
https://doi.org/10.1007/978-3-031-54154-4

This Springer imprint is published by the registered company Springer Nature Switzerland AG
The registered company address is: Gewerbestrasse 11, 6330 Cham, Switzerland

Paper in this product is recyclable.

The book is dedicated to God Almighty, the giver of life and inspirations. It is also dedicated to Iwinosa Davina Isibor—the daughter of Dr. Patrick Omoregie Isibor—for being a source of inspiration that envisioned the book.

Preface

In today's age of remarkable technological advancement, the burgeoning field of nanotechnology holds incredible promise across numerous industries, offering groundbreaking solutions to complex challenges. However, this burgeoning promise is accompanied by a profound responsibility—to meticulously examine and comprehend the toxicity potential of nanomaterials on our environment and living organisms.

Environmental Nanotoxicology: Combatting the Minute Contaminants stands at the forefront of this critical discourse, delving deep into the intricate interaction of nanomaterials with the delicate balance of our ecosystems and the well-being of microscopic and macroscopic life forms. As the science and technology advances in harnessing the benefits of nanomaterials for various cutting-edge applications, cutting cross medicine, agriculture, electronics, and environmental remediation, there arises a crucial need to comprehend their potential ecological impact. Comprehending the toxicodynamics and toxicokinetics of nanomaterials requires rigorous scientific inquiries and compendium of perspectives from leading experts in the fields of nanoscience, toxicology, environmental studies, and ecology. The book thus transcends mere exploration; it serves as a guiding light through the complexities surrounding nanotoxicity in our environment. The prelude to this meticulous exposition encompasses an in-depth analysis of the diverse facets of environmental nanotoxicology. From the physicochemical properties of nanoparticles to their interactions with biological systems, their fate in various ecosystems, their potential bioaccumulation and biomagnification through food chains.

Furthermore, the book navigates the intricate landscapes of regulatory policies, offering a critical assessment of existing frameworks while proposing visionary strategies aimed at ensuring the safe and sustainable utilization of nanomaterials. It aims not just to enlighten but to empower policymakers, researchers, and stakeholders with the actionable knowledge necessary to make informed decisions shaping the future of nanotechnology. It serves as an indispensable guide, shedding light on the nuanced interactions between nanomaterials and the environment, propelling us toward a future where innovation coexists harmoniously with ecological stewardship.

The profound commitment to safeguarding our planet's delicate ecosystems thus requires knowledge provided in *Environmental Nanotoxicology: Combatting the Minute Contaminants*, to explore, engage, and contribute to shaping a future where nanotechnology thrives in harmony with nature.

Ota, Nigeria — Patrick Omoregie Isibor
Sultanate of Oman, Oman — Geetha Devi
Benin City, Nigeria — Alex Ajeh Enuneku

Acknowledgement

Covenant University is acknowledged for providing uninterrupted power supply, reliable internet bandwidth, and auspicious environment that enabled the seamless execution of this work.

About the Book

Environmental Nanotoxicology: Combatting the Minute Contaminants stands as a pivotal compendium poised to provide cutting-edge scientific knowledge that meets the critical challenge of safeguarding our environment from the inconspicuous nanoparticles shaping our technological world. This comprehensive material encapsulates a profound exploration into the application of nanomaterials and their impacts on our ecosystems and their lifeforms, from a standpoint of nanotoxicity. This work represents an interdisciplinary collaboration, bringing together the collective knowledge of leading experts from diverse fields—nanoscience, toxicology, environmental studies, ecology, biochemistry and microbiology. It serves as a guide to students, researchers, experts, stake holders, policy makers, and enthusiastic individuals.

The book is rooted in a thorough examination of nanomaterials, dissecting their physicochemical properties, biological interactions, fate in ecosystems, and potential impacts on diverse organisms. From elucidating mechanisms of bioaccumulation and biomagnification to investigating the intricate pathways of nanoparticle-induced toxicity, each chapter offers a scholarly exploration of the complexities surrounding nanotoxicology. It presents the critical aspects of the discourse in sequential chapters in a schematic fashion occasioned by visionary insights and characterized by the pragmatic perspectives. It navigates regulatory frameworks, providing a critical assessment of existing policies while charting visionary strategies aimed at ensuring the safe and sustainable integration of nanomaterial safety in our world. The book's narrative goes beyond academic discourse; it extends an invitation to policymakers, researchers, industry professionals, and stakeholders alike. At the end of this book, the audience is expected to emerge equipped with a profound understanding and practical tools to navigate the landscapes of environmental nanotoxicity.

About the Book

Contents

About the Editors

Patrick Omoregie Isibor, PhD is a specialist in Pollution Studies and Ecotoxicology and a Lecturer/Researcher in the Department of Biological Sciences at Covenant University. He received his PhD and MSc in Environmental Quality Management from the University of Benin and a BSc in Zoology from Ambrose Alli University. Dr. Isibor's research interests include ecotoxicology, hydrobiology, bioaccumulation, biosequestration, biodiversity conservation, and aquatic ecology. He is a member of the Association for Environmental Impact Assessment of Nigeria (AEIAN), the International Association of Risk and Compliance Professionals (IARCP), and the African Society for Toxicological Sciences (ASTS). Dr. Isibor is co-editor of the book *Biotechnological Approaches to Sustainable Development Goals* (Springer, 2023), an Editor for the *African Journal of Health, Safety, and Environment*, and a reviewer for several reputable international journals.

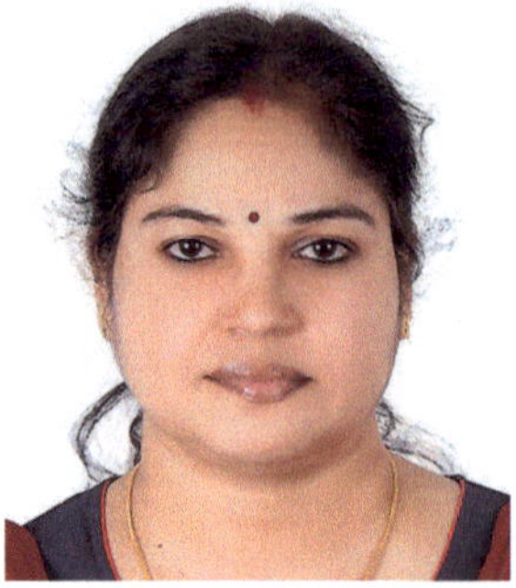

Geetha Devi, PhD is an Associate Professor in Chemical Engineering in the Department of Mechanical and Industrial Engineering, College of Engineering at National University of Science & Technology, Oman. She received her Ph.D. in Nanotechnology from the National Institute of Technology, India. She has more than 80 International peer-reviewed journal and conference publications and 12 funded research projects from various funding agencies, including the Industry Innovation Center (IIC), The Research Council (TRC), Oman, and National University Student and Faculty Research Grants (NUSRG/ NUFRG) in the key research areas of nanotechnology, wastewater treatment, biopharmaceutics, and drug delivery. Dr. Devi has worked on numerous research collaborations with institutions such as the University of South Carolina, USA, Jamia Milia Islamia University, India, the National Institute of Technology (NIT), Durgapur, Khalifa University, Abu Dhabi, Sultan Qaboos University (SQU), Oman, University of Nizwa (UoN), and the Middle East Desalination Research Center (MEDRC), Oman, in the emerging areas of nanotechnology, corrosion control, waste management, and water treatment. She is a Scientific Committee Member for selecting the Omani Young Water Researchers Award for the undergraduate and postgraduate category with the MEDRC. She is an MSc thesis external examiner for Sultan Qaboos University. Dr. Devi received the Best Researcher Award from the National University of Science & Technology for 2021-22. She serves as an editor/editorial board member for several journals and an external reviewer for The Research Council (TRC), Oman.

Alex Ajeh Enuneku, PhD is a specialist in Pollution Studies and Ecotoxicology and Head of the Department of Environmental Management and Toxicology, Faculty of Life Sciences, at the University of Benin. He obtained his PhD and MSc in Pollution Studies and Ecotoxicology and a BSc in Zoology from the University of Benin. He was certified as a chemical hazard control expert in 2016 by Hokkaido University Sapporo. Professor Enuneku's research interests include ecotoxicology, hydrobiology, pollution studies, endocrine disruption, health risk assessment, environmental impact assessment, environmental management, remote sensing, and geographic information systems. He is a member of the Society for Environmental Chemistry and Toxicology (SETAC), the Nigerian Environmental Society, and the Waste Management Association of Nigeria. He is a reviewer for several reputable international journals.

Chapter 1
Introduction to Nanotoxicology

Innocent Ojeba Musa, Patrick Omoregie Isibor, Job Oloruntoba Samuel, Abdulsalam Mustapha, Adams Mustapha, Sikirula Akande, Oluwafemi Adebayo Oyewole, Ifeoluwa Ihotu Kayode-Edwards, and Hephzibah Adeniji

1.1 Introduction

Nanotoxicology is the study of the possible dangers and risks associated with nanoparticles. These nanoengineered and controlled materials (usually range from 1 to 100 nanometers) display specific features that may vary significantly from their bulk counterparts. As a consequence, it is critical to understand how nanomaterials interact with biological systems and their influence on human health and the environment (Yang et al., 2021).

Nanomaterials are a varied group of materials strictly characterized by their small nanoscale dimensions measuring 1–100 nanometers (a nanometer is 1×10^{-9} or one-billionth of a meter). They may be composed of any element, come in different shapes and structures, or be found in nature organically or developed to meet specific purposes (Dghoughi et al., 2023). Environmental nanotoxicology encompasses the examination of the detrimental impacts of nanomaterials on living organisms and ecosystems within the environment, encompassing measures to prevent and alleviate these unfavorable effects, drawing from the Society of Toxicology's established definition of toxicology (Society of Toxicology, 2023).

The unique size range of nanomaterials gives rise to distinct physical, chemical, and biological characteristics which differ from those of bulk materials. These

I. O. Musa (✉) · A. Mustapha
Department of Microbiology, Skyline University Nigeria, Kano, Kano State, Nigeria
e-mail: innocentmusa0011@gmail.com

P. O. Isibor · I. I. Kayode-Edwards · H. Adeniji
Department of Biological Sciences, Covenant University, Ota, Nigeria
e-mail: patrick.isibor@covenantuniversity.edu.ng

J. O. Samuel · A. Mustapha · S. Akande · O. A. Oyewole
Department of Microbiology, Federal University of Technology, Minna, Niger State, Nigeria

P. O. Isibor et al. (eds.), *Environmental Nanotoxicology*,
https://doi.org/10.1007/978-3-031-54154-4_1

characteristics include quantum effects, enhanced reactivity, improved mechanical, optical and magnetic properties, and tunable surface properties. For example, the small size of nanomaterials allows them to be absorbed more readily. Nanomaterials are also present in different forms in the environment. As such, the ubiquitous nature of nanomaterials combined with their unique properties underscores their potential toxicity. Hence, nanomaterials in the environment pose toxicological risks that can be more far-reaching than their bigger-sized counterparts. The potential risks raise a number of safety problems which are within the scope of nanotoxicology (Sengul & Asmatulu, 2020; Yang et al., 2021; Bessa et al., 2023).

Understanding the toxicological effects of nanomaterials is important to assess potential risks and ensure their safe use (Bradford et al., 2022). In size-dependent toxicity, nanomaterials can penetrate cellular structures potentially leading to cellular dysfunction or damage (Fig. 1.1). They can also lead to the generation of reactive oxygen species in living organisms, leading to oxidative stress. Additionally, inflammation, genotoxic DNA damage, and long-term chronic health risks can be attributed to nanotoxicity.

Given the increasing presence of nanotechnology in our everyday lives, the relevance of nanotoxicology cannot be emphasized. Nanomaterials may enter the body through a variety of methods, including inhalation, ingestion, and skin contact. To protect the safety of employees, customers, and patients exposed to nanoproducts, their toxicity must be assessed (Saleh, 2020). Nanoparticles discharged into the environment might have unforeseeable ecological implications. Understanding their possible influence is critical for the long-term viability of nanotechnology and environmental preservation. Around the globe, governments and regulatory organizations are adopting norms and laws to control the use of nanomaterials. Comprehensive nanotoxicological analyses are required for compliance with these standards (Bessa et al., 2023).

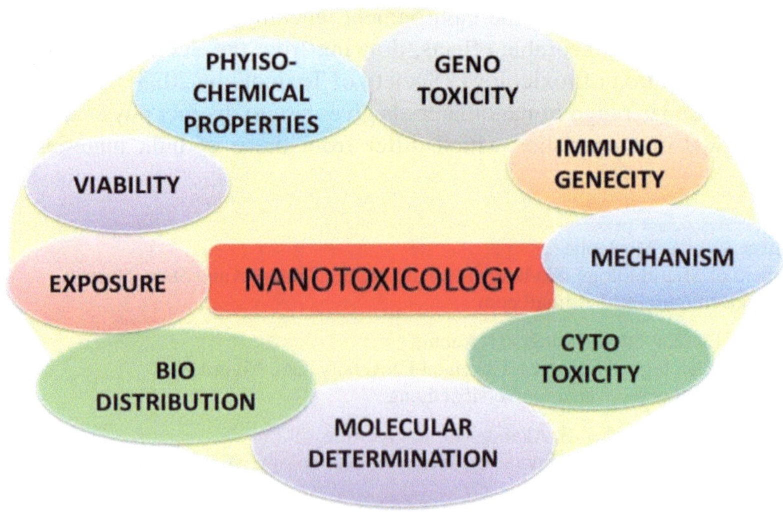

Fig. 1.1 Characteristics of nanotoxicology

Responsible development and use of nanotechnology rely on properly recognizing and reducing safety issues. Nanotoxicology aids in the identification of opportunities for development and innovation in nanomaterial design and applications (Ziglari et al., 2021). While nanotoxicology as a separate profession just recently evolved, safety concerns about nanoparticles date back to the late twentieth century. Significant events, such as the respiratory problems suffered by employees exposed to ultrafine particles at a pigment plant in the 1990s, highlighted the need for a more comprehensive knowledge of nanotoxicity. As a result of these issues, nanotoxicology emerged as an interdisciplinary discipline that drew knowledge from chemistry, biology, toxicology, and materials science. It seeks to solve critical issues concerning the safety of nanomaterials in a constantly evolving technological ecosystem (Naasz et al., 2018; Bessa et al., 2022). This chapter explores the environmental nanotoxicology through a thorough understanding of nanomaterials, the interfaces of nanomaterials in the environment, and the toxicological effects in nanoscale contexts.

1.1.1 Significance of Nanotoxicology in Nanotechnology

Nanotoxicology is a branch of science that focuses on the possible dangers and hazards connected with nanomaterials. Nanomaterials are man-made or naturally occurring materials with nanoscale dimensions ranging from 1 to 100 nanometers. They have distinct features that distinguish them from their bulk counterparts, making them very helpful in a variety of applications such as medicine, electronics, and materials research (Zhang et al., 2022). However, these distinct features raise worries regarding possible toxicity, making nanotoxicology critical. Here are some of the most essential reasons why nanotoxicology is crucial:

(i) Human health and safety: Assessing the possible detrimental effects of nanomaterials on human health is the fundamental issue in nanotoxicology. Nanomaterials may interact with biological systems in ways that bigger particles do not when breathed, eaten, or come into touch with the skin. Understanding these interactions is critical for assuring worker, consumer, and general population safety (Deng et al., 2017).

(ii) Environmental impact: Nanomaterials may enter the environment via a variety of channels, such as industrial waste, consumer items, and medicinal uses. Assessing their environmental effect is critical to preventing ecosystem pollution and possible animal damage (Rosario et al., 2022).

(iii) Regulatory compliance: Before goods incorporating nanoparticles may be launched, regulatory bodies in many countries need safety evaluations. Nanotoxicological investigations contribute to the data needed to set safety criteria and maintain regulatory compliance (Deng et al., 2017).

(iv) Material design and innovation: Nanomaterials offer enormous promise for use in a wide range of applications, including medicine delivery, renewable energy, and water purification. Researchers and engineers may create safer and more effective nanomaterials by understanding their toxicological qualities (Rosario et al., 2022).

(v) Risk assessment and management: Nanotoxicology offers the knowledge required to perform risk assessments and manage possible nanomaterial concerns. This is critical for nanotechnology firms to make educated choices regarding worker safety and product development.

(vi) Public awareness and perception: As the usage of nanotechnology expands, it is critical to address public concerns regarding the safety of nanomaterials. Conducting nanotoxicology research and presenting the results to the public may help establish confidence and guarantee responsible nanotechnology development and usage (Saleh, 2020).

(vii) Multidisciplinary research: Nanotoxicology is an essentially multidisciplinary discipline that includes chemistry, biology, toxicology, physics, and materials science. This partnership promotes a comprehensive knowledge of the dangers and advantages of nanomaterials (Rosario et al., 2022).

(viii) Long-term consequences: Nanomaterials may have subtle or long-term health and environmental consequences that are not immediately evident. Nanotoxicology investigations aid in the discovery of these effects as well as the direction of continuous monitoring and study (Deng et al., 2017).

Nanotoxicology is critical in analyzing and minimizing the possible dangers of nanomaterials, guaranteeing their safe usage in a variety of applications (Renn & Roco, 2020). It is a critical area that promotes responsible nanotechnology development while protecting human health and the environment. Nanotoxicology is a relatively young topic that arose as a result of the increasing usage of nanotechnology in a variety of industries and applications. It is concerned with determining the possible toxicity of nanoparticles and nanomaterials on living beings and the environment (Raftis & Miller, 2019). Concerns regarding the possible hazards connected with nanoparticles and nanomaterials began to emerge as nanotechnology applications extended into new sectors (Chen et al., 2023).

1.1.2 Nanotechnology: A Brief Overview

Nanotechnology is a scientific and technology interdisciplinary topic centered on the manipulation and control of material at the nanoscale, which is commonly described as the size of individual atoms and molecules. It entails creating, producing, and deploying materials, devices, and systems having unique features and functionalities due to specific nanoscale dimensions (Rosario et al., 2022; Bessa et al., 2023).

Nanotechnology is concerned with structures and phenomena at the nanoscale scale, which is generally 1–100 nanometers. A nanometer is one-billionth of a meter or about 100,000 times smaller than the diameter of a human hair. Nanotechnology is a highly multidisciplinary area that draws on physics, chemistry, biology, engineering, and materials science (Cao et al., 2018). Researchers from diverse fields work together to investigate and create nanoscale materials and technology. Materials at the nanoscale often have unique features that distinguish them from their bulk counterparts. These qualities may include higher electrical conductivity, strength, optical characteristics, and reactivity (Yu et al., 2016).

1.2 Applications of Nanotechnology

Nanomedicine utilizes nanoparticles for a variety of purposes, including drug delivery, cancer treatment, imaging, and diagnostics. At the same time, nanoscale transistors and memory devices contribute to the development of faster and smaller electronic components. Materials like carbon nanotubes and graphene are harnessed to craft stronger and lighter nanocomposites. In the field of energy, nanotechnology plays a role in enhancing solar cells, advanced batteries, and fuel cells. Additionally, nanomaterials can purify water and reduce pollution, and when incorporated into polymers, ceramics, and metals, they enhance these materials' mechanical, thermal, and electrical properties.

Nanomaterials such as quantum dots and nanowires boost the efficiency of solar panels, while nanotechnology is instrumental in creating high-capacity, longer-lasting batteries with rapid charging capabilities. Fuel cells utilize nanocatalysts to heighten efficiency and cut costs (Ong et al., 2014). Nanoparticles and nanofilters are applied in water purification processes to eliminate pollutants and heavy metals from water sources. Furthermore, nanomaterials find application in air filtration, effectively absorbing and eliminating contaminants and particulates from the atmosphere. Nanotechnology has the potential to improve the delivery and effectiveness of pesticides, thus minimizing their adverse environmental impacts. Nanomaterials are also poised to revolutionize food packaging by prolonging shelf life and detecting food spoilage. Additionally, they have the capacity to reduce friction and enhance fuel efficiency in automobiles. The automotive and aerospace industries can leverage nanotechnology to create lightweight materials (Fernandez-Bertolez et al., 2021). These diverse applications underscore the versatility and significant impact of nanotechnology across various sectors and domains. However, it is essential to remain cognizant that as nanotechnology progresses, concerns about safety, ethical considerations, and potential environmental consequences must be duly addressed (Asghari et al., 2012).

1.3 Ethical and Safety Considerations

The distinctive attributes of nanomaterials also give rise to concerns regarding their safety and potential impact on the environment. Efforts are underway among researchers and regulatory bodies to comprehend and mitigate these concerns, aiming to ensure the responsible advancement and application of nanotechnology (Klinova et al., 2021). While nanotechnology exhibits significant promise, it encounters challenges in areas like scalability, precision in manufacturing, and the establishment of regulatory frameworks. These obstacles are actively being addressed by researchers to pave the way for the commercialization of innovative nanotechnology solutions (Krug & Nau, 2022).

Nanotechnology stands as a multidisciplinary domain within the realms of science, engineering, and technology. It revolves around the manipulation of matter at the nanoscale, typically characterized by dimensions ranging from 1 to 100 nanometers. To put this in perspective, a nanometer corresponds to one-billionth of a meter or approximately the dimensions of individual atoms and molecules. Nanotechnology encompasses the conceptualization, manipulation, and control of materials and devices at this exceedingly minute scale to engender fresh and unique properties and functionalities (Farre et al., 2009).

Central to nanotechnology is the creation and application of nanomaterials, which encompass materials marked by nanoscale structures. These materials often showcase distinctive traits and behaviors attributable to their diminutive size and remarkable surface-to-volume ratios. Among the most common nanomaterials are nanoparticles, nanotubes, and nanowires. The realm of nanotechnology boasts a broad array of applications spanning across diverse industries, including electronics, medicine, energy, materials science, and more. Illustrative instances of nanotechnology applications encompass nanoelectronics, systems for drug delivery on a nanoscale, nanosensors, and advanced materials endowed with enhanced properties (Oukarroum et al., 2011; Liu et al., 2014).

Nanotechnology offers a dual approach, engaging both bottom-up and top-down methodologies. The bottom-up approaches entail the incremental assembly of nanoscale structures, atom by atom or molecule by molecule. In contrast, top-down strategies involve the size reduction or manipulation of larger structures to reach the nanoscale dimensions (Hong et al., 2011).

1.4 Nanomaterials and Their Properties

Nanomaterials are materials characterized by their exceptional properties and intricate structures at the nanoscale, typically exhibiting dimensions below 100 nanometers (nm). These materials can be systematically engineered and manipulated at this minute scale to manifest distinctive characteristics and functionalities that set them apart from their bulk counterparts. Some frequently encountered nanomaterials encompass nanoparticles, nanowires, nanotubes, and nanosheets, among various others (Zhu et al., 2013). Some fundamental properties and attributes of nanomaterials include the following:

Nanomaterials possess an elevated surface-area-to-volume ratio owing to their diminutive size. This quality renders them valuable in diverse applications, including catalysis, drug delivery, and sensing, where an extensive surface area is of paramount importance (Baun et al., 2008). At the nanoscale, quantum effects emerge prominently. For instance, quantum confinement can induce alterations in the electronic and optical characteristics of nanomaterials. Quantum dots exemplify nanomaterials that exhibit size-dependent quantum effects (Yu et al., 2016).

Nanoparticles may demonstrate heightened chemical reactivity in comparison to bulk materials due to their substantial surface area. This reactivity can be harnessed for catalytic processes and other chemical applications (Cao et al., 2018). Specific nanomaterials, such as carbon nanotubes and graphene, display exceptional mechanical attributes, including remarkable strength and flexibility. These traits render them apt for applications in the realm of materials science and engineering (Yu et al., 2016). Nanomaterials frequently exhibit unique optical properties, encompassing tunable colors and fluorescence. Quantum dots and plasmonic nanoparticles serve as instances of nanomaterials adorned with intriguing optical characteristics (Oukarroum et al., 2011).

Materials like nanowires and certain nanoparticles may manifest augmented electrical conductivity, conferring value in domains like electronics and sensors. Magnetic nanoparticles can be deliberately tailored to attain specific magnetic attributes, rendering them advantageous in applications such as data storage and medical imaging (Zhu et al., 2013). Some nanomaterials, such as carbon nanotubes, exhibit exceptionally high thermal conductivity. This trait proves advantageous in areas like thermal management and the formulation of nanocomposites (Walters et al., 2016). Nanomaterial surfaces are amenable to functionalization with various molecules or ligands. This approach tailors their properties and enhances compatibility with specific applications, such as drug delivery or cancer cell targeting (Heinlaan et al., 2011). Nanomaterials may manifest distinct toxicity profiles owing to their minute size and substantial surface area. Consequently, a comprehensive understanding of their biocompatibility and potential health implications is imperative for applications in the domains of medicine and biology (Walters et al., 2016).

Some nanomaterials possess the capability to autonomously assemble into intricate structures, a feature that can be meticulously controlled and harnessed for fabricating nanodevices and nanoscale patterns (Park & Choi, 2010). Certain nanoparticles, particularly those comprised of iron oxide, have the propensity to exhibit superparamagnetism. This phenomenon renders them invaluable in applications such as magnetic resonance imaging (MRI) and targeted drug delivery (Jayaseelan et al., 2014).

Nanomaterials find a multitude of applications across various sectors, including electronics, energy storage, medicine, environmental remediation, and materials science. Nevertheless, their exceptional attributes simultaneously raise concerns regarding safety, environmental consequences, and ethical considerations.

Therefore, a judicious evaluation and regulatory framework are imperative as nanotechnology progresses (Walters et al., 2016).

1.4.1 Unique Properties of Nanomaterials

Nanomaterials encompass a diverse spectrum of distinctive properties and behaviors attributed to their nanoscale dimensions, typically falling within the range of 1–100 nanometers. These attributes frequently deviate significantly from those exhibited by bulk materials, resulting in a myriad of applications spanning across diverse domains, including electronics, medicine, materials science, and energy (Saleh, 2020; Yang et al., 2021). Here are some of the remarkable properties characterizing nanomaterials:

(i) Size-dependent characteristics: Numerous properties of nanomaterials, spanning optical, electronic, magnetic, and catalytic attributes, manifest strong size-dependent tendencies. As the size of nanoparticles diminishes, their characteristics can undergo dramatic transformations (Yang et al., 2021).

(ii) High surface area-to-volume ratio: Nanoparticles feature a substantially greater surface area relative to their volume when juxtaposed with bulk materials. This elevated surface area proves advantageous in applications like catalysis, where reactions occur predominantly at the surface, and in drug delivery, where nanoparticles can accommodate a substantial payload of therapeutic agents (Yu et al., 2016).

(iii) Quantum effects: Within nanoscale materials, quantum effects emerge as notably significant. For instance, in quantum dots, the confinement of electrons within a confined volume engenders discrete energy levels, giving rise to distinctive optical and electronic attributes, including size-tunable fluorescence (Walters et al., 2016).

(iv) Enhanced mechanical properties: Specific nanomaterials, exemplified by carbon nanotubes and graphene, showcase exceptional mechanical robustness. They exhibit remarkable strength and lightness, rendering them promising candidates for applications within the domains of materials science and engineering.

(v) Improved conductivity: Nanomaterials, such as nanowires and nanotubes, can boast outstanding electrical conductivity. This feature is exploited in the realms of nanoelectronics and conductive coatings (Ziglari et al., 2021).

(vi) Enhanced catalytic activity: Nanoparticles frequently outperform their bulk counterparts in terms of catalytic activity. This superiority emanates from their substantial surface area and distinctive surface properties, making it pivotal in catalysis for chemical reactions, including those found in fuel cells and industrial processes (Dghoughi et al., 2023).

(vii) Size-dependent melting and boiling points: Nanoparticles may exhibit lower melting and boiling points relative to their bulk counterparts, and these points can vary with size. This has implications for materials processing and applications within the sphere of nanomanufacturing (Zhang et al., 2022).

(viii) Magnetic properties: Magnetic nanoparticles, such as iron oxide nanoparticles, can showcase superparamagnetism, wherein they become magnetic when exposed to an external magnetic field and demagnetize when the field is removed. This phenomenon proves valuable in applications like magnetic resonance imaging (MRI) and targeted drug delivery (Deng et al., 2017).

(ix) Optical properties: Quantum dots and other nanomaterials often feature distinctive optical attributes, encompassing size-tunable fluorescence and plasmonic effects. These traits render them invaluable in the domains of sensors, displays, and imaging technologies (Farre et al., 2009).

(x) Enhanced reactivity: The substantial surface area and unique surface chemistry of nanoparticles frequently result in heightened reactivity, endowing them with significance in catalysis, sensors, and environmental remediation.

(xi) Self-assembly: Certain nanomaterials possess the inherent ability to autonomously assemble into ordered structures owing to their size and surface characteristics. This property finds utility in the realm of nanotechnology for constructing intricate structures and devices (Kim et al., 2008).

These remarkable properties of nanomaterials have paved the way for fresh avenues in scientific inquiry and technological progress. They have spurred innovations across an array of industries, culminating in the development of novel materials characterized by customized attributes tailored to specific applications (Baun et al., 2008).

1.5 Types of Nanomaterials

Nanomaterials are materials characterized by their unique properties and structures, existing at the nanoscale, typically defined as having at least one dimension measuring less than 100 nanometers (nm). These materials can be categorized into distinct classes based on their composition, configuration, and attributes (Yang et al., 2021).

Nanoparticles exhibit unique size-dependent properties, affecting their interaction with biological systems. Smaller nanoparticles can penetrate cellular structures more easily, potentially causing greater harm (Zhao et al., 2011). Additionally, coatings and functionalization can alter the surface chemistry of nanoparticles, influencing their reactivity and interactions with biological molecules (Albanese et al., 2012). The surface charge of nanoparticles (zeta potential) also affects their stability and cellular uptake as positively or negatively charged nanoparticles can interact differently with cell membranes and proteins (Fröhlich, 2012). Nanoparticles come in various shapes such as nanorod, spherical nanoparticles, nanobar, nanostar, nanocube, polyhedral, dendritic, nanoneedles, pyramid-shaped, nanoprism, nanohusks, and tetrapod (Fig. 1.2), impacting their biological interactions (Tzitzios et al., 2008; Zhang et al., 2008; Al-Sherbini, 2010; Paul et al., 2010; Sun et al., 2012, 2014; Woo

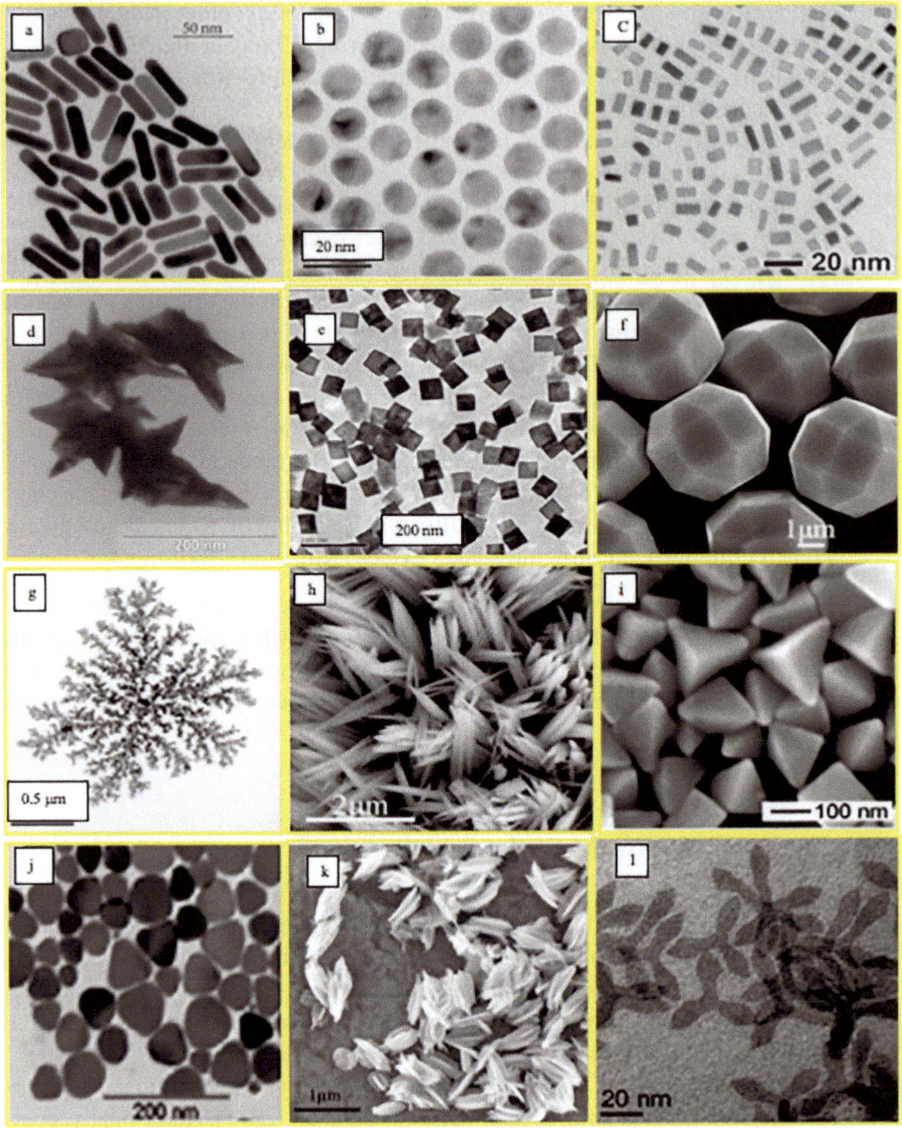

Fig. 1.2 (**a**) Nanorod (Al-Sherbini, 2010), (**b**) spherical nanoparticles (Tzitzios et al., 2008), (**c**) nanobar (Li et al., 2014), (**d**) nanostar (Melnikau et al., 2014), (**e**) nanocube (Woo et al., 2012), (**f**) polyhedral (Sun et al., 2012), (**g**) dendritic (Zhang et al., 2008), (**h**) nanoneedles (Liu et al., 2020), (**i**) pyramid-shaped (Khodashenas & Ghorbani, 2019), (**j**) nanoprism (Castillo-Henríquez et al., 2020), (**k**) nanohusks, and (**l**) tetrapod (Demille et al., 2018).

et al., 2012; Li et al., 2014; Melnikau et al., 2014; Gayathri et al., 2015; Sayed & Polshettiwar, 2015; Demille et al., 2018; Khodashenas & Ghorbani, 2019; Castillo-Henríquez et al., 2020; Liu et al., 2020). Needle-like or high-aspect-ratio nanoparticles may cause more damage due to their sharp edges (Lin et al., 2018).

Nanomaterials can further be categorized according to the materials they are made of:

(i) Nanoparticles: Metal nanoparticles (e.g., gold, silver, platinum): frequently harnessed in catalysis, electronics, and medical applications.
(ii) Semiconductor nanoparticles (e.g., quantum dots): Exhibiting tunable optical features, they serve roles in displays, imaging, and solar cells (Hughes & Asmatulu, 2021).
(iii) Polymer nanoparticles: Applied in drug delivery, coatings, and nanocomposite materials.
(iv) Carbon nanoparticles (e.g., carbon nanotubes, fullerenes): Distinguished by remarkable mechanical, electrical, and thermal characteristics, they find utility in various domains, encompassing electronics, materials science, and medicine (Zhang et al., 2022).
(v) Nanowires and nanorods: Elongated nanomaterials characterized by nanoscale diameters and variable lengths. They are instrumental in applications within electronics and sensors and as foundational building components for nanodevices (Saleh, 2020).
(vi) Nanosheets and nanofilms: Two-dimensional nanomaterials, exemplified by graphene and transition metal dichalcogenides, boast exceptional electronic, thermal, and mechanical properties. They are prevalent in the domain of nanoelectronics and composite materials (Deng et al., 2017).
(vii) Nanocomposites: These materials unite nanoparticles or nanofillers with bulk materials to enhance their characteristics. Instances encompass polymer nanocomposites and metal matrix nanocomposites.
(viii) Nanotubes: Carbon nanotubes (CNTs) present cylindrical structures composed of carbon atoms. Renowned for their remarkable mechanical robustness, they serve diverse purposes, including nanoelectronics, nanocomposites, and drug delivery (Huang et al., 2021).
(ix) Nanoporous materials: These materials incorporate nanoscale pores or voids, typified by zeolites and metal-organic frameworks (MOFs). They are integral in applications such as gas storage, catalysis, and drug delivery.
(x) Nanoparticles for drug delivery: Utilizing carriers like liposomes and micelles, these nanomaterials facilitate drug delivery with improved targeting precision and controlled release (Rosario et al., 2022).
(xi) Nanosensors: Nanomaterials are routinely employed in sensors designed for the detection of various analytes, including gases, biomolecules, and environmental pollutants.
(xii) Nanomagnets: Magnetic nanoparticles and nanomaterials endowed with unique magnetic characteristics play pivotal roles in data storage, medical imaging, and magnetic resonance imaging (MRI).

(xiii) Nanoporous membranes: These membranes incorporate nanoscale pores and are integral in processes such as filtration and separation, serving applications like water purification and gas separation (Naasz et al., 2018).

(xiv) Nanoparticles for imaging: Quantum dots and superparamagnetic nanoparticles are deployed as contrast agents in medical imaging, encompassing MRI and fluorescence imaging. These categories encompass a wide array of nanomaterials, each possessing diverse properties and applications. As nanotechnology advances, it continues to yield novel and innovative nanomaterials for an array of industries and research domains (Cao et al., 2018).

1.6 Nanomaterials and Environmental Concerns

Nanomaterials are materials that have been meticulously crafted and manipulated at the nanoscale, typically exhibiting dimensions ranging from 1 to 100 nanometers. Their uniqueness often stems from their diminutive size, expansive surface area, and the intriguing quantum effects they display, rendering them invaluable for a myriad of applications spanning the domains of electronics, medicine, energy, and materials science. Nevertheless, nanomaterials are a source of concern for several compelling reasons (Walters et al., 2016). Certain nanomaterials have the potential to exhibit toxicity to living organisms, including humans, in ways that are distinct from their bulk counterparts. The minute size and expansive surface area of nanoparticles can heighten reactivity and the possibility of cellular absorption, leading to potential adverse health effects. Research endeavors continue to explore the toxicity of various nanomaterials and their interactions within biological systems.

When nanomaterials are released into the environment, unintended consequences may arise. They could accumulate within ecosystems, impacting plants and animals, potentially entering the food chain. The long-term environmental ramifications of nanomaterials remain not fully comprehended (Yu et al., 2016). Regulatory agencies in many nations have grappled with the rapid evolution of nanotechnology. Consequently, regulatory gaps may exist, governing the production, utilization, and disposal of nanomaterials, resulting in insufficient oversight and potential risks (Bessa et al., 2022).

The consequences of prolonged exposure to nanomaterials remain insufficiently understood. Ongoing research endeavors aim to uncover potential health risks, necessitating comprehensive studies to evaluate the safety of diverse nanomaterial types and exposure scenarios. Individuals involved in the production and handling of nanomaterials may face potential exposure to nanoparticles. Without proper safety measures and guidelines, occupational exposure can give rise to health-related concerns (Bundsachuh et al., 2018). The utilization of nanomaterials across diverse applications, such as surveillance, military technology, and the augmentation of human capabilities, prompts ethical inquiries encompassing issues of privacy, security, and potential misuse. Traditional risk assessment methods may not be readily applicable to nanomaterials, given their unique attributes and behaviors.

The development of suitable risk assessment frameworks for nanomaterials remains an ongoing challenge (Sengul & Asmatulu, 2020).

Nanoparticles can manifest divergent behaviors within biological and environmental systems when compared to their larger counterparts. A profound understanding of how nanoparticles interact with cells, tissues, and the environment is indispensable for assessing their potential risks (Park & Choi, 2010). Nanoparticles possess the potential to migrate within the human body and traverse biological barriers, engendering concerns about bioaccumulation and long-term effects. Public attitudes and acceptance of nanomaterials can be swayed by apprehensions regarding safety, environmental impact, and ethical considerations. Addressing these concerns is vital for the responsible advancement and utilization of nanotechnology (Krug & Nau, 2022).

Exposure to nanomaterials can transpire through various pathways, and it is imperative to comprehend these avenues to evaluate and mitigate the potential health and environmental hazards associated with nanotechnology. Nanomaterials are meticulously engineered at the nanoscale, endowing them with unique properties while also introducing potential risks (Klinova et al., 2021).

Inhalation stands as one of the most prevalent means of exposure to nanomaterials, especially in occupational settings where workers may inhale airborne nanoparticles. These nanoparticles can become suspended in the air in the form of aerosols or dust during the manufacturing, handling, or utilization of nanomaterials. Once inhaled, nanoparticles have the potential to reach deep within the respiratory system, potentially posing health risks (Liu et al., 2014).

Skin contact with nanomaterials can transpire during the manipulation of products containing nanomaterials or through direct contact with substances containing nanomaterials. While the skin serves as a protective barrier, some nanoparticles may possess dimensions small enough to penetrate the skin and enter the bloodstream, potentially resulting in localized or systemic effects (Carvalhais et al., 2021). The ingestion of nanomaterials can manifest when they are present in consumables like food, beverages, or other consumer products. Contaminated food, for instance, could expose individuals to nanoparticles. Furthermore, in occupational contexts, inadvertent ingestion can occur when workers handle nanomaterials without adhering to proper hygiene practices (Raftis & Miller, 2019).

In medical and research applications, nanomaterials might be intentionally injected into the body, such as for drug delivery or diagnostic objectives. While this pathway is deliberate and under control, it carries its own array of considerations and prospective risks, including immune responses and distribution within the body (Zhang et al., 2022). In medical applications, nanomaterials may be intentionally administered directly into the bloodstream through intravenous injection or other parenteral routes. While this route ensures precise delivery to specific target tissues, it mandates careful consideration of potential adverse effects (Rosario et al., 2022).

Accidental eye contact with nanomaterials can occur through splashes and aerosol exposure or when dealing with products containing nanomaterials. Nanoparticles entering the eyes can lead to potential irritation or other eye-related health concerns. Nanomaterials can be discharged into the environment during the manufacturing, use, or disposal of nanoproducts. This release can contaminate the air, water, and soil. Once within the environment, nanoparticles can impact ecosystems and potentially infiltrate the food chain, resulting in indirect human exposure (Walters et al., 2016).

It is crucial to recognize that the toxicity and health implications of nanomaterials can fluctuate depending on factors such as their composition, dimensions, morphology, surface chemistry, and the route of exposure. To guarantee the safe utilization of nanomaterials, risk assessments and suitable safety precautions should be instituted, guided by the particular characteristics of the nanomaterials in question and the potential exposure scenarios. Regulatory agencies in numerous nations have established protocols and regulations to address the safe management and use of nanomaterials in a range of applications (Huang et al., 2021). Nanoparticle toxicity pertains to the potential adverse impacts of nanoparticles on living organisms, encompassing humans, animals, and the environment. Nanoparticles are minuscule particles, typically characterized by at least one dimension being less than 100 nanometers (nm). Owing to their diminutive size and distinct attributes, nanoparticles can engage with biological systems in ways that are distinct from larger particles composed of the same material (Yang et al., 2021).

Nanoparticles possess a substantial surface area-to-volume ratio, which augments their reactivity and their potential to interact with biological molecules. This elevated reactivity can render them more inclined to traverse biological barriers, such as cell membranes (Walters et al., 2016). The toxicity of nanoparticles can vary significantly contingent on their chemical composition. For example, certain nanoparticles constructed from metals like silver or titanium dioxide have exhibited toxic effects, whereas others, like those composed of gold, are generally considered to be less toxic (Asghari et al., 2012). Nanoparticles can infiltrate the organism through multiple routes, including inhalation, ingestion, dermal contact, and injection. The specific route of exposure can impact the magnitude of toxicity and the organs or tissues that are affected (Krug & Nau, 2022). Upon entrance into a biological system, nanoparticles can disperse to different organs and tissues, potentially inducing toxicity at particular sites. The capacity of nanoparticles to traverse biological barriers, such as the blood-brain barrier, can be a matter of concern (Asghari et al., 2012; Domingues et al., 2022).

Nanoparticle toxicity can ensue through diverse mechanisms, encompassing oxidative stress, inflammation, genotoxicity (DNA damage), and disruption of cellular functions. The precise mechanisms hinge on the properties of the nanoparticle and the specific biological context (Bessa et al., 2023). Analogous to many toxic substances, nanoparticle toxicity frequently adheres to a dose-response correlation, where the gravity of the toxic effects amplifies with the dose or level of exposure (Ong et al., 2014). The long-term repercussions of nanoparticle exposure represent an ongoing area of exploration. Prolonged exposure to specific nanoparticles may lead to cumulative effects and chronic conditions, such as respiratory ailments or cancer (Hong et al., 2011). Owing to apprehensions regarding nanoparticle toxicity, regulatory bodies in numerous nations have initiated the formulation of guidelines and safety assessment protocols for nanoparticles employed in consumer products and medical applications.

This scientific field is dedicated to scrutinizing the toxicological repercussions of nanoparticles. Nanotoxicology researchers aim to apprehend the risks associated with nanoparticles and devise strategies to mitigate these risks (Bessa et al., 2023). Strategies to mitigate nanoparticle toxicity might encompass the engineering of safer nanoparticles, the utilization of protective coatings, the implementation of appropriate workplace safety measures, and the execution of comprehensive risk assessments (Naasz et al., 2018).

It is paramount to acknowledge that not all nanoparticles inherently possess toxicity, and their potential for harm is contingent upon numerous factors, inclusive of their particular properties, levels of exposure, and individual susceptibilities. Scientists persist in investigating nanoparticle toxicity to enhance comprehension of the risks related to nanomaterials and to create safety directives for their utilization across various applications (Walters et al., 2016).

1.6.1 Toxicodynamics of Nanomaterials

Nanomaterials are introduced into water and soil ecosystems through various pathways. Dissolved nanoparticles in the atmosphere may get into soil and water via precipitation. Other nanoparticles may be introduced into water and soil bodies through wastewater discharge and surface runoff or through the direct applications of nanoparticle-containing products in water treatment and agricultural practices. When present in these environmental matrices, nanoparticles in soil and water interact with both media and resident organisms influencing ecosystem dynamics (Chethipuzha et al., 2021).

1.6.2 Nanomaterials in Living Organisms

Nanomaterials can work their way from the environment into living organisms through the process of bioaccumulation and trophic transfer. Nanoparticles in water and soil are absorbed by plants which in turn are consumed by animals. When consumed by plants or animals, assimilated nanomaterials may be altered within those organisms and deposited in their cells and tissues. For more toxic nanoparticles such as metal oxides, this results in toxicity in plants and bioaccumulation in food chains. As nanomaterials bioaccumulate, they increase toxicity exposure for consumers higher up in the food chains and across food webs (Montes et al., 2012; Gottardo et al., 2021). Engineered nanoparticles can also make their way directly into the human body. In the food industry, for example, nanoparticles are used in food packaging. These nanoparticles may leach into food itself and can then be easily ingested, inhaled, or absorbed by humans and other living organisms (Chen et al., 2023).

1.6.3 Toxicokinetics of Nanomaterials

To evaluate if and to what extent nanoparticles elicit toxicological responses, it is necessary to understand dose-response relationships. There is limited research on the dose-dependent toxicological impact of environmental nanoparticles. However, some studies have shown the toxic effects (such as liver inflammation) of NPs based on increasing doses and particle size (Lu et al., 2012). Traditional mass-based measurements for dosing are insufficient for comprehending the nanotoxicology of a nanoparticle (NP), as these particles demonstrate distinct properties at the nanoscale, diverging from their bulk forms (Hull et al., 2012). Utilizing alternative dosimetry methods and characterizing NPs by particle number, particle number density, and suspension surface area might offer a more comprehensive understanding of dose-response connections. Additionally, hormetic-like biphasic dose responses can better help characterize the toxic effects of NPs at low levels of exposure (Iavicoli et al., 2014).

1.7 Factors Influencing Nanotoxicity

Nanotoxicity pertains to the potential adverse effects of nanoparticles on biological systems, encompassing cells, tissues, and organisms. Grasping nanotoxicity is imperative for the prudent advancement and secure application of nanotechnology. Numerous factors can influence nanotoxicity; these include:

(i) Size of nanoparticles: Nanoparticle size is a pivotal determinant of nanotoxicity. Smaller nanoparticles typically exhibit a larger surface area relative to their volume, rendering them more reactive and potentially more toxic. Their reduced size facilitates penetration of cells and tissues, potentially leading to harm (Krug & Nau, 2022).
(ii) Surface chemistry: The surface chemistry of nanoparticles is a noteworthy contributor to their toxicity. Functional groups and coatings on nanoparticle surfaces can either amplify or diminish their toxicity. Surface alterations can enhance biocompatibility or enable targeted delivery, for instance (Cao et al., 2018).
(iii) Shape: Nanoparticles' shape can exert an influence on their toxicity. Certain shapes, such as nanorods or nanowires, may present distinct toxicological profiles compared to spherical nanoparticles, as suggested by some studies.
(iv) Aggregation and dispersion: The aggregation or dispersion of nanoparticles in a medium can influence their toxicity. Aggregated nanoparticles may exhibit distinct biological interactions and behaviors in comparison to well-dispersed nanoparticles (Bessa et al., 2022).
(v) Composition: The chemical composition of nanoparticles wields substantial influence over their toxicity. Various materials exhibit different levels of

biocompatibility. For instance, some metals (e.g., silver, cadmium) can be more toxic than others.

(vi) Dose and concentration: The quantity of nanoparticles administered or present in the environment can affect toxicity. Elevated doses or concentrations are more likely to induce adverse effects (Heinlaan et al., 2011).

(vii) Exposure route: The mode of exposure to nanoparticles holds significance. Inhalation, ingestion, dermal contact, and injection can lead to differing toxicological responses.

(viii) Duration of exposure: The length of time during which exposure to nanoparticles occurs can impact toxicity. Short-term and long-term exposure may yield varying effects on biological systems (Park & Choi, 2010).

(ix) Biological variability: Variations in individual susceptibility, genetic factors, and pre-existing health conditions can affect how individuals respond to exposure to nanomaterials.

(x) Environmental factors: Environmental conditions, such as temperature, humidity, and the presence of other chemicals or pollutants, can modify the behavior and toxicity of nanoparticles in the environment (Saleh, 2020).

(xi) Biological interactions: Nanoparticles can interact with a variety of biomolecules, such as proteins, lipids, and DNA. These interactions can influence their toxicity, potentially affecting cellular processes and triggering toxic responses (Sengul & Asmatulu, 2020).

(xii) Biodistribution and clearance: The capacity of the body to distribute, metabolize, and eliminate nanoparticles can impact their toxicity. Nanoparticles that accumulate in specific tissues may lead to localized toxicity (Ziglari et al., 2021).

(xiii) Physicochemical changes: Some nanoparticles may undergo physicochemical transformations in biological environments, which can alter their toxicological properties.

(xiv) Aging of nanoparticles: The stability and reactivity of nanoparticles can evolve over time, potentially influencing their toxicity when they are stored or used for extended durations (Bessa et al., 2023).

Apprehending these factors and conducting comprehensive toxicity assessments are essential for the responsible development and secured utilization of nanomaterials across diverse applications, encompassing medicine, electronics, and environmental remediation. Researchers and regulatory agencies persist in their endeavors to explore and formulate guidelines for assessing and managing nanotoxicity (Walters et al., 2016; Bessa et al., 2023).

1.8 Conclusion

Nanotoxicology assumes a pivotal role in ensuring the prudent development and utilization of nanomaterials. As the field continues to evolve, it will be indispensable in addressing emerging challenges and preserving human health and the environment from potential hazards associated with nanotoxicology. Nanomaterials' environmental impacts on living organisms and ecosystems, as well as efforts to prevent these impacts, constitute the field of environmental nano-ecotoxicology. The size, shape, and surface charge of nanomaterials affect how they interact with biological and environmental systems to cause toxicity. Nanotoxicology is a field that has evolved over centuries as nanotechnology and its applications have advanced. As a result, with increasing advancements in nanotechnology in the modern day, there is an increased need to understand and mitigate the potential impacts of nanomaterials. As nanotechnology continues to evolve, research efforts must ensure its sustainable integration into human society, with adequate considerations of ethical and safety concerns.

The subsequent chapters focus on specific characteristics of nanomaterials, their interactions with living organisms, and emerging technologies in environmental nanotoxicology.

References

Albanese, A., Tang, P. S., & Chan, W. C. (2012). The effect of nanoparticle size, shape, and surface chemistry on biological systems. *Annual Review of Biomedical Engineering, 14*, 1–16.

Al-Sherbini, A. S. A. (2010). Thermal reshaping of gold nanorods in micellar solution of water/glycerol mixtures. *Journal of Nanomaterials, 2010*, 1–6.

Asghari, S., Johari, S. A., Lee, J. H., Kim, Y. S., Jeon, Y. B., & Choi, H. J. (2012). Toxicity of various silver nanoparticles compared to silver ions in Daphnia magna. *Journal of Nanobiotechnology, 10*, 14. https://doi.org/10.1186/1477-3155-10-14

Baun, A., Sørensen, S. N., Rasmussen, R. F., Hartmann, N. B., & Koch, C. B. (2008). Toxicity and bioaccumulation of xenobiotic organic compounds in the presence of aqueous suspensions of aggregates of nano-C(60). *Aquatic Toxicology, 86*(3), 379387. https://doi.org/10.1016/j.aquatox.2007.11.019

Bessa, M. J., Brandão, F., Fokkens, P. H. B., Leseman, D. L. A. C., Boere, A. J. F., Cassee, F. R., Salmatonidis, A., Viana, M., Monfort, E., Fraga, S., et al. (2022). Unveiling the toxicity of fine and nano-sized airborne particles generated from industrial thermal spraying processes in human alveolar epithelial cells. *International Journal of Molecular Sciences, 23*, 4278.

Bessa, M. J., Brandão, F., Rosário, F., Moreira, L., Reis, A. T., Valdiglesias, V., Laffon, B., Fraga, S., & Teixeira, J. P. (2023). Assessing the in vitro toxicity of airborne (nano)particles to the human respiratory system: From basic to advanced models. *Journal of Toxicology and Environmental Health. Part B, Critical Reviews, 26*, 67–96.

Bradford, S. A., Shen, C., Kim, H., Letcher, R. J., Rinklebe, J., Ok, Y. S., & Ma, L. (2022). Environmental applications and risks of nanomaterials: An introduction to CREST publications during 2018–2021. *Critical Reviews in Environmental Science and Technology, 52*(21), 3753–3762.

Bundschuh, M., Filser, J., Lüderwald, S., McKee, M. S., Metreveli, G., Schaumann, G. E., Schulz, R., & Wagner, S. (2018). Nanoparticles in the environment: Where do we come from, where do we go to? *Environmental Sciences Europe, 30*, 6.

Cao, Y., Gong, Y., Liao, W., Luo, Y., Wu, C., Wang, M., & Yang, Q. A. (2018). Review of cardiovascular toxicity of TiO2, ZnO and Ag nanoparticles (NPs). *Biometals, 31*, 457–476.

Carvalhais, A., Pereira, B., Sabato, M., Seixas, R., Dolbeth, M., Marques Guilherme, A. S., Pereira, P., Pacheco, M., & Mieiro, C. (2021). Mild effects of sunscreen agents on a marine flatfish: Oxidative stress, energetic profiles, neurotoxicity and behaviour in response to titanium dioxide nanoparticles and oxybenzone. *International Journal of Molecular Sciences, 22*, 1567.

Castillo-Henríquez, L., Alfaro-Aguilar, K., Ugalde-Álvarez, J., Vega-Fernández, L., Montes de Oca-Vásquez, G., & Vega-Baudrit, J. R. (2020). Green synthesis of gold and silver nanoparticles from plant extracts and their possible applications as antimicrobial agents in the agricultural area. *Nanomaterials, 10*(9), 1763.

Chen, J., Guo, Y., Zhang, X., Liu, J., Gong, P., Su, Z., Fan, L., & Li, G. (2023). Emerging nanoparticles in food: Sources, application, and safety. *Journal of Agricultural and Food Chemistry, 71*(8), 3564–3582. https://doi.org/10.1021/acs.jafc.2c06740

Chethipuzha, M., Abraham, A. R., Kalarikkal, N., Thomas, S., & Sreeja, S. (2021). Embracing nanotechnology concepts in the electronics industry. In *Fundamentals and properties of multifunctional nanomaterials* (pp. 405–421). Elsevier.

Demille, T. B., Hughes, R. A., Preston, A. S., Adelung, R., Mishra, Y. K., & Neretina, S. (2018). Light-mediated growth of noble metal nanostructures (Au, Ag, Cu, Pt, Pd, Ru, Ir, Rh) from micro-and nanoscale ZnO tetrapodal backbones. *Frontiers in Chemistry, 6*, 411.

Deng, R., Lin, D., Zhu, L., Majumdar, S., White, J. C., Gardea-Torresdey, J. L., & Xing, B. (2017). Nanoparticle interactions with co-existing contaminants: Joint toxicity, bioaccumulation and risk. *Nanotoxicology, 11*, 591–612.

Dghoughi, A., Raji, M., Nekhlaoui, S., Essabir, H., Bouhfid, R., & Qaiss, A. E. K. (2023). Introduction of nanomaterials and polymer nanocomposites. In *Mechanics of nanomaterials and polymer nanocomposites* (pp. 1–17). Springer. https://doi.org/10.1007/978-981-99-2352-6_1

Domingues, C., Santos, A., Alvarez-Lorenzo, C., Concheiro, A., Jarak, I., Veiga, F., Barbosa, I., Dourado, M., & Figueiras, A. (2022). Where is nano today and where is it headed? A review of nanomedicine and the dilemma of nanotoxicology. *ACS Nano, 16*(7), 9994–10041.

Farré, M., Gajda-Schrantz, K., Kantiani, L., & Barceló, D. (2009). Ecotoxicity and analysis of nanomaterials in the aquatic environment. *Analytical and Bioanalytical Chemistry, 393*(1), 81–95. https://doi.org/10.1007/s00216-008-2458-1

Fernández-Bertólez, N., Brandão, F., Costa, C., Pásaro, E., Teixeira, J. P., Laffon, B., & Valdiglesias, V. (2021). Suitability of the in vitro cytokinesis-block micronucleus test for genotoxicity assessment of TiO2 nanoparticles on SH-SY5Y cells. *International Journal of Molecular Sciences, 22*, 8558.

Fröhlich, E. (2012). The role of surface charge in cellular uptake and cytotoxicity of medical nanoparticles. *International Journal of Nanomedicine*, 5577–5591.

Gayathri, T., Sundaram, N. M., & Kumar, R. A. (2015). Gadolinium oxide nanoparticles for magnetic resonance imaging and cancer theranostics. *Journal of Bionanoscience, 9*(6), 409–423.

Gottardo, S., Mech, A., Drbohlavová, J., Małyska, A., Bøwadt, S., Sintes, J. R., & Rauscher, H. (2021). Towards safe and sustainable innovation in nanotechnology: State-of-play for smart nanomaterials. *NanoImpact, 21*, 100297.

Heinlaan, M., Kahru, A., Kasemets, K., Arbeille, B., Prensier, G., & Dubourguier, H. C. (2011). Changes in the Daphnia magna midgut upon ingestion of copper oxide nanoparticles: A transmission electron microscopy study. *Water Research, 45*(1), 179–190. https://doi.org/10.1016/j.watres.2010.08.026

Hong, S. C., Lee, J. H., Lee, J., Kim, H. Y., Park, J. Y., & Cho, J. (2011). Subtle cytotoxicity and genotoxicity differences in superparamagnetic iron oxide nanoparticles coated with various functional groups. *International Journal of Nanomedicine, 6*, 3219–3231. https://doi.org/10.2147/IJN.S26355

Huang, H. J., Lee, Y. H., Hsu, Y. H., Liao, C. T., Lin, Y. F., & Chiu, H. W. (2021). Current strategies in assessment of nanotoxicity: Alternatives to in vivo animal testing. *International Journal of Molecular Sciences, 22*, 4216.

Hughes, S. M., & Asmatulu, E. (2021). Nanotoxicity and nanoecotoxicity: Introduction, principles, and concepts. In *Environmental chemistry for a sustainable world*. Springer. https://doi.org/10.1007/978-3-030-63241-0_1

Hull, M. S., Kennedy, A. J., Detzel, C. J., Vikesland, P. J., & Chappell, M. A. (2012). Moving beyond mass: The unmet need to consider dose metrics in environmental nanotoxicology studies. *Environmental Science & Technology, 46*(20), 10881–10882. https://doi.org/10.1021/es3035285

Iavicoli, I., Fontana, L., Leso, V., & Calabrese, E. J. (2014). Hormetic dose–responses in nanotechnology studies. *Science of the Total Environment, 487*, 361–374. https://doi.org/10.1016/j.scitotenv.2014.04.023

Jayaseelan, C., Abdul Rahuman, A., Ramkumar, R., Perumal, P., Rajakumar, G., Vishnu Kirthi, A., et al. (2014). Effect of sub-acute exposure to nickel nanoparticles on oxidative stress and histopathological changes in Mozambique tilapia, Oreochromis mossambicus. *Ecotoxicology and Environmental Safety, 107*, 220–228. https://doi.org/10.1016/j.ecoenv.06.012

Khodashenas, B., & Ghorbani, H. R. (2019). Synthesis of silver nanoparticles with different shapes. *Arabian Journal of Chemistry, 12*(8), 1823–1838.

Kim, Y. S., Kim, J. S., Cho, H. S., Rha, D. S., Kim, J. M., & Park, J. D. (2008). Twenty-eight-day oral toxicity, genotoxicity, and gender-related tissue distribution of silver nanoparticles in SpragueDawley rats. *Inhalation Toxicology, 20*(6), 575–583. https://doi.org/10.1080/08958370701874663

Klinova, S. V., Katsnelson, B. A., Minigalieva, I. A., Gerzen, O. P., Balakin, A. A., Lisin, R. V., Butova, K. A., Nabiev, S. R., Lookin, O. N., Katsnelson, L. B., et al. (2021). Cardioinotropic effects in subchronic intoxication of rats with lead and/or cadmium oxide nanoparticles. *International Journal of Molecular Sciences, 22*, 3466. [CrossRef] [PubMed].

Krug, H. F., & Nau, K. (2022). Editorial: Methods and protocols in nanotoxicology. *Frontiers in Toxicology, 4*, 1093765.

Li, Y., Bian, T., Du, J., Xiong, Y., Zhan, F., Zhang, H., & Yang, D. (2014). Facile synthesis of high-quality Pt nanostructures with a controlled aspect ratio for methanol electro-oxidation. *CrystEngComm, 16*(36), 8340–8343.

Lin, N., Berton, P., Moraes, C., Rogers, R. D., & Tufenkji, N. (2018). Nanodarts, nanoblades, and nanospikes: Mechano-bactericidal nanostructures and where to find them. *Advances in Colloid and Interface Science, 252*, 55–68.

Liu, Y., Xia, Q., Liu, Y., Zhang, S., Cheng, F., & Zhong, Z. (2014). Genotoxicity assessment of magnetic iron oxide nanoparticles with different particle sizes and surface coatings. *Nanotechnology, 25*(42), 425101. https://doi.org/10.1088/0957-4484/25/42/425101

Liu, N., Tang, M., & Ding, J. (2020). The interaction between nanoparticles-protein corona complex and cells and its toxic effect on cells. *Chemosphere, 245*, 125624.

Lu, X., Jin, T., Jin, Y., Wu, L., Hu, B., Tian, Y., & Fan, X. (2012). Toxicogenomic analysis of the particle dose- and size-response relationship of silica particles-induced toxicity in mice. *Nanotechnology, 24*(1), 015106. https://doi.org/10.1088/0957-4484/24/1/015106

Melnikau, D., Savateeva, D., & Rakovich, Y. P. (2014). Coupling effect in hybrid system of plasmonic nanoparticles and J-aggregates yields double Rabi splitting. In *2014 16th International Conference on Transparent Optical Networks (ICTON)* (pp. 1–4). IEEE.

Montes, M. O., Hanna, S. K., Lenihan, H. S., & Keller, A. A. (2012). Uptake, accumulation, and biotransformation of metal oxide nanoparticles by a marine suspension-feeder. *Journal of Hazardous Materials, 225–226*, 139–145. https://doi.org/10.1016/j.jhazmat.2012.05.009

Naasz, S., Altenburger, R., & Kühnel, D. (2018). Environmental mixtures of nanomaterials and chemicals: The Trojan-Horse phenomenon and its relevance for ecotoxicity. *Science of the Total Environment, 635*, 1170–1181.

Ong, K. J., MacCormack, T. J., Clark, R. J., Ede, J. D., Ortega, V. A., Felix, L. C., Dang, M. K. M., Ma, G., Fenniri, H., Veinot, J. G. C., et al. (2014). Widespread nanoparticle-assay interference: Implications for nanotoxicity testing. *PLoS One, 9*, e90650. [CrossRef].

Oukarroum, A., Bras, S., Perreault, F., & Popovic, R. (2011). Inhibitory effects of silver nanoparticles in two green algae, Chlorella vulgaris and Dunaliella tertiolecta. *Ecotoxicology and Environmental Safety, 2012*(78), 80–85. https://doi.org/10.1016/j.ecoenv.11.012

Park, S. Y., & Choi, J. (2010). Geno- and ecotoxicity evaluation of silver nanoparticles in freshwater crustacean Daphnia magna. *Environmental Engineering Research, 15*(1), 23–27. https://doi.org/10.4491/eer.2010.15.1.428

Paul, K. B., Athanasopoulos, G. I., Doumanidis, C. C., & Rebholz, C. (2010). Growth and characterization of nanowires. *Advances in Condensed Matter Physics*, 2010: 1–6.

Raftis, J. B., & Miller, M. R. (2019). Nanoparticle translocation and multi-organ toxicity: A particularly small problem. *Nano Today, 26*, 8–12.

Renn, O., & Roco, M. C. (2020). Nanotechnology and the need for risk governance. In *Emerging technologies* (pp. 321–359). Routledge.

Rosário, F., Costa, C., Lopes, C. B., Estrada, A. C., Tavares, D. S., Pereira, E., Teixeira, J. P., & Reis, A. T. (2022). In vitro hepatotoxic and neurotoxic effects of titanium and cerium dioxide nanoparticles, arsenic and mercury co-exposure. *International Journal of Molecular Sciences, 23*, 2737. [CrossRef] [PubMed].

Saleh, T. A. (2020). Nanomaterials: Classification, properties, and environmental toxicities. *Environmental Technology and Innovation, 20*, 101067.

Sayed, F. N., & Polshettiwar, V. (2015). Facile and sustainable synthesis of shaped iron oxide nanoparticles: Effect of iron precursor salts on the shapes of iron oxides. *Scientific Reports, 5*(1), 9733.

Sengul, A. B., & Asmatulu, E. (2020). Toxicity of metal and metal oxide nanoparticles: A review. *Environmental Chemistry Letters, 18*, 1659–1683.

Society of Toxicology. (2023). Relevance of toxicology to public health. *Society of Toxicology (SOT)*. https://www.toxicology.org/about/relevance.asp

Sun, S., Kong, C., You, H., Song, X., Ding, B., & Yang, Z. (2012). Facet-selective growth of Cu–Cu 2 O heterogeneous architectures. *CrystEngComm, 14*(1), 40–43.

Sun, J., Ge, J., Liu, W., Lan, M., Zhang, H., Wang, P., Wang, Y., & Niu, Z. (2014). Multi-enzyme co-embedded organic–inorganic hybrid nanoflowers: synthesis and application as a colorimetric sensor. *Nanoscale, 6*(1), 255–262.

Tzitzios, V., Georgakilas, V., Zafiropoulou, I., Boukos, N., Basina, G., Niarchos, D., & Petridis, D. (2008). A general chemical route for the synthesis of capped nanocrystalline materials. *Journal of Nanoscience and Nanotechnology, 8*(6), 3117–3122.

Walters, C., Pool, E., & Somerset, V. (2016). Nanotoxicity in aquatic invertebrates. In M. L. Larramendy & S. Soloneski (Eds.), *Invertebrates—experimental models in toxicity screening* (1st ed., pp. 13–34). InTech. Available from: https://www.intechopen.com/books/invertebratesexperimental-models-in-toxicity-screening/nanotoxicity-inaquatic-invertebrates

Woo, H., Kang, H., Kim, A., Jang, S., Park, J. C., Park, S., Kim, B. S., Song, H., & Park, K. H. (2012). Azide-alkyne huisgen [3+ 2] cycloaddition using CuO nanoparticles. *Molecules, 17*(11), 13235–13252.

Yang, W., Wang, L., Mettenbrink, E. M., DeAngelis, P. L., & Wilhelm, S. (2021). Nanoparticle toxicology. *Annual Review of Pharmacology and Toxicology, 61*, 269–289.

Yu, X., Hong, F., & Zhang, Y. Q. (2016). Bio-effect of nanoparticles in the cardiovascular system. *Journal of Biomedical Materials Research. Part A, 104*, 2881–2897.

Zhang, J., Meng, L., Zhao, D., Fei, Z., Lu, Q., & Dyson, P. J. (2008). Fabrication of dendritic gold nanoparticles by use of an ionic polymer template. *Langmuir, 24*(6), 2699–2704.

Zhang, F., Wang, Z., Peijnenburg, W. J. G. M., & Vijver, M. G. (2022). Review and prospects on the ecotoxicity of mixtures of nanoparticles and hybrid nanomaterials. *Environmental Science & Technology, 56*, 15238–15250.

Zhao, F., Zhao, Y., Liu, Y., Chang, X., Chen, C., & Zhao, Y. (2011). Cellular uptake, intracellular trafficking, and cytotoxicity of nanomaterials. *Small, 7*(10), 1322–1337.

Zhu, X., Hondroulis, E., Liu, W., & Li, C. Z. (2013). Biosensing approaches for rapid genotoxicity and cytotoxicity assays upon nanomaterial exposure. *Small, 9*(9–10), 1821–1830. https://doi.org/10.1002/smll.201201593

Ziglari, T., Wang, Z., & Holian, A. (2021). Contribution of particle-induced lysosomal membrane hyperpolarization to lysosomal membrane permeabilization. *International Journal of Molecular Sciences, 22*, 2277. [CrossRef] [PubMed].

Chapter 2
Nanoparticle Properties and Characterization

Onwaeze Oritseweyinmi Ogochukwu, Micheal Bolarinwa Fabiyi, Oluwadurotimi Samuel Aworunse, Oluwafemi Adebayo Oyewole, and Patrick Omoregie Isibor

2.1 Introduction

A host of ecological processes are intertwined with nanoscale particles allowing them to be influential in various fields of science and technology (Tan et al., 2021; Shoukat & Khan, 2021). Nanoparticles exhibit remarkable versatility, manifesting in various forms, including nanoparticles, nanowires, and nanotubes, each with unique properties and applications (Huang et al., 2021; Khan & Hossain, 2022). They are also frequently seen as complex frameworks with key elements such as a surface layer, which may be functional in engineering for various biocompatibility objectives, a shell layer, and a core layer (Petrini et al., 2023; Akbarzadeh et al., 2023). These qualities and dimensions endow them with particular characteristics, such as large surface area-to-volume proportions and quantum size effects, which can have a significant impact on their behaviour and chemistry (Karak, 2019; Khan & Hossain, 2022; Joudeh & Linke, 2022).

The dimension of the particles is responsible for the majority of the attributes related to nanostructures (Coetzee et al., 2020; Khan & Hossain, 2022). According to the prevalent hypothesis around this phenomenon, when particle size decreases, the proportion of particle area relative to volume rises exponentially (Coetzee et al., 2020; Chen et al., 2020a, b; Ağbulut, 2022). The influence of the bonds binding the particles decreases exponentially as the dimensions of the particles decrease (Jiang

O. O. Ogochukwu (✉) · O. S. Aworunse · P. O. Isibor
Department of Biological Sciences, Covenant University, Ota, Nigeria
e-mail: ogochukwu.onwaeze@stu.cu.edu.ng; patrick.isibor@covenantuniversity.edu.ng

M. B. Fabiyi
Universidade Federal do Pará, Belem, Brazil

O. A. Oyewole
Department of Biochemistry, Federal University of Technology, Gidan Kwano campus, Minna, Nigeria

P. O. Isibor et al. (eds.), *Environmental Nanotoxicology*,
https://doi.org/10.1007/978-3-031-54154-4_2

et al., 2022). The electrical and temperature conductivity, polymer phase behaviour and strength, physical characteristics, weight, and optic, magnetic, mechanical, or electric attributes of the matrix are all affected by nanoparticle size (Zhang et al., 2020; Chen et al., 2020a, b; Coetzee et al., 2020; Ağbulut, 2022). This phenomenon results in characteristics that differ dramatically from those of bulk materials (Chen et al., 2020a, b).

Coetzee et al. (2020) characterise this as a result of various quantum features including spatially constraint and minimised defects not found in bulk substances and bigger surface elements and energy. Nanomaterials' extremely small dimensions emphasise area-volume ratio, culminating in more "surface" contingent and thus size contingent characteristics of the material (Zhang, 2020). Nanotechnology has dominated biomedical and environmental remediation research, particularly in recent years, where we have witnessed an increase in the utilisation of nanocomposite structures and tools, which has led to exceptional and frequently unexpectedly groundbreaking studies (Nguyen et al., 2021a, b; Danish et al., 2022; Rohilla & Chaudhary, 2023). The size-dependent characteristics of these particles are pivotal in modern research and innovation, impacting fields as diverse as materials science, medicine, and electronics (Zhang, 2020); therefore a detailed understanding of their properties and in vivo interactions is vital for continued development. This chapter gives an overview of size-dependent properties of nanoparticles, addressing how attributes such as optical, electronic, and magnetic properties change with nanoparticle size. It also discusses surface area and surface chemistry, delving into the influence of high surface area and unique surface chemistry on nanoparticle reactivity and interactions.

2.2 Size-Dependent Properties of Nanoparticles

2.2.1 Optical Properties

Nanomaterial optical properties like the absorbing, scattering, reflecting, and emitting of light are fluid and may vary dramatically from bulk substance characteristics (Andhere et al., 2020; Wang et al., 2020; Said et al., 2022). These size-dependent optical properties are primarily a result of the optical energy bandgap and surface plasmon resonance that occur as particles reach nanoscale dimensions (Chen et al., 2020a, b; Lee & Link, 2021). The optical bandgap comes from disparities in the electron energy in valence and conduction bands and amplifies as nanoparticles shrink and experience quantum confinement effects (Aziz et al., 2019). The bandgap amplification results in a discrete energy level structure, causing smaller nanoparticles to have larger bandgaps. It also allows them to absorb and emit light at higher energies, shifting towards the blue end of the electromagnetic spectrum (Aziz et al., 2019). Electrons located inside the particles jointly vibrate while resisting stabilising force generated by the nucleus (Wang et al., 2020). They orient one another with the direction of the arriving electromagnetic wave (Aziz et al., 2019). The combined movement of electrons is analogous to a dipole oscillation source (Fig. 2.1). Semiconductors and several metals exhibit these major shifts in optical characteristics as particle size increases (Chen et al., 2020a, b; Wang et al., 2020).

The surface plasmon resonance causes colouring in nanoparticles (Lee et al., 2008; Shivras et al., 2015). This phenomenon is a resonance of the particles' outer electron bands with light wavelengths. When nanoparticles are exposed to light, they preferentially capture photons with energy differences equal to the amount of energy gap within their restricted electron states (Verger et al., 2021). The energy values grow more defined as the nanoparticle size reduces, which leads to discrete peak absorption levels in the spectrum (Khurgin & Levy, 2020). This effect is known as "quantisation of energy levels" and is a feature of quantum confinement (Abid & Al-Rashid, 2020; Khurgin & Levy, 2020). The size-colour relationship is an established model and has been quantitatively measured using prominent scattering models (Wrigglesworth & Johnston, 2021).

Emission spectra, which describe the colours released by nanoparticles after stimulation, vary greatly with dimension as well (Khattab et al., 2021; Gao et al., 2021). Bigger nanoparticles produce near-IR light with longer wavelengths. In contrast, nanoparticles with smaller dimensions demonstrate "blue shift", as they emit light that is nearer the blue region of the electromagnetic spectrum (Gao et al., 2021). The increasing deficits in energy across quantum states drives this

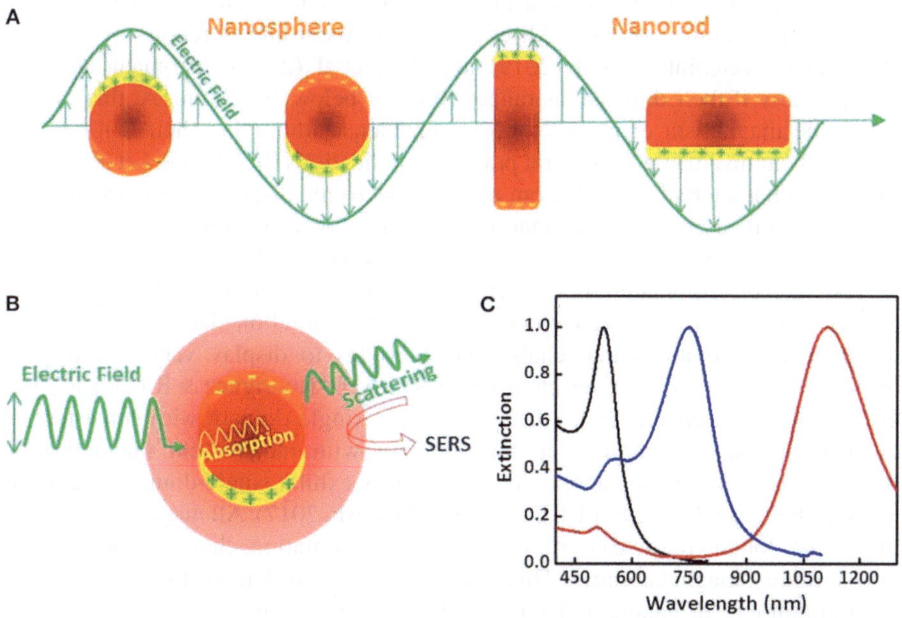

Fig. 2.1 Surface plasmon resonance and optical properties of nanoparticles: (**a**) nanoparticle to electromagnetic radiation interaction, (**b**) nanoparticle interaction with lights, and (**c**) scattering and absorption by gold nanoparticles (Bodelon et al., 2018)

phenomenon. For instance, Zhao et al. (2003) and Zayadi and Bakar (2020) demonstrate that the colour of the colloidal gold solution may be orange or shift from red to yellow as its dimension grows. Round three-dimensional silver nanomaterials appear orange at 100 nm and green at 25 nm. This also differs greatly depending on the type of nanoparticles as spherical gold nanomaterials may display orange hues at 100 nm but a comparable size spherical silver particle will display yellow (Mendis et al., 2016; Wrigglesworth & Johnston, 2021).

By merely modifying the structure, dimensions, and surface functioning, an extensive number of optical effects can be achieved from nanomaterials and nanocomposites for an array of purposes (Huang et al., 2020; Paredes et al., 2022). Based on the substance type, size, and perspective of the nanoparticle, this modification can be accomplished using techniques like quantum dot Blinking (Nguyen et al., 2020), photodoping (Ghini et al., 2021), electrochromism (Ezhilmaran & Bhat, 2022), or plasmonic resonance tuning (Yakovkin & Reshetnyak, 2023).

2.2.2 Magnetic Properties

Magnetism in nanotechnology is invaluable with multifaceted relevance in biomedicine (Materón et al., 2021), water and environmental remediation (Shaumbwa et al., 2021; Vallinayagam et al., 2021), waste control (Shukla et al., 2021), magnetic fluid production (Soetaert et al., 2020), magnetic resonance imaging (Jeon et al., 2021), and data storage (Ali et al., 2021). According to reports by Popović et al. (2010), Vetchinkina et al. (2019), and Kang et al. (2020), nanomaterials operate optimally if their dimensions range within or below 10–20 nm. At this critical value, the magnetism of particles takes precedence improving their abilities and making them invaluable for several practical applications (Vetchinkina et al., 2019). Magnetic characteristics result from unequal electronic dispersion and are affected by the methods used to prepare nanoparticles, such as co-precipitation, microemulsion, thermal breakdown, and so on (Joudeh & Linke, 2022).

The size of nanomaterials is a critical factor that determines their magnetic behaviour (Caizer, 2016; Nguyen et al., 2021a, b). Based on size, we can observe magnetic configurations that enable nanoparticles to display varying magnetic responses in different magnetic environments (Rivera-Rodriguez & Rinaldi-Ramos, 2021; Coene & Leliaert, 2022). They may be a single domain with consistent or inconsistent magnetisation or a multi-domain with inconsistent magnetisation (Caizer, 2016). Magnetic nanoparticles frequently exhibit single-domain characteristics as their size decreases (Till et al., 2015; Daoush, 2017). All magnetic moments in a single-domain particle coordinate in a single orientation, changing the particles into a unitary magnetic entity. This transition to single-domain function happens most commonly in nanoparticles lower than a certain critical size (<10–20 nm), which differs based on the substance (Daoush, 2017; Kianfar, 2021). Single-domain nanoparticles have intriguing magnetic properties, making them extremely desirable in a variety of fields (Dong et al., 2021). For example, they have a high susceptibility to magnetic and can produce significant magnetic moments, which is important in magnetic storage media (Nejad et al., 2021) and MRI contrast agents (Avasthi et al., 2020). Furthermore, the lack of domain boundaries in single-domain particles minimises energy dissipation while increasing stability (Daoush, 2017; Kianfar, 2021).

Nanoparticles are also typically classified as magnetic or nonmagnetic based on their inherent magnetic properties and how they respond to magnetic fields, both in the presence and absence of a magnetic environment (Kumari et al., 2020). They may be classified as diamagnetic, ferrimagnetic or antiferromagnetic, ferromagnetic, and paramagnetic (Fig. 2.2) (Li & Fortner, 2020; Rubel & Hossain, 2022). Diamagnetic substances are characterised by the absence of unpaired electrons that allow the paramagnetic particles to create individual magnetic moments (Xu et al., 2021). Hence, without a magnetic field, they display no magnetism, and within a magnetic field, they possess dipoles that demonstrate weak repulsion to the direction of the force (Sagayaraj et al., 2021). As illustrated in Fig. 2.2, in paramagnetic substances, magnetic dipoles are present; therefore in the presence of a magnetic field, they demonstrate weak attractions that persist when the field is lost (Sagayaraj et al., 2021; Li & Fortner, 2020). Similarly, in the absence of an external magnetic field, ferromagnetic nanomaterials have magnetic dipole expressions; however, they demonstrate magnetism independently from the magnetic field (Sagayaraj et al., 2021). Antiferromagnetic and ferrimagnetic materials are identical to this; however, they are anti-aligned such that the net magnetic moments of the nanoparticles cancel out or produce a weak magnetic force (Xu et al., 2021). Furthermore, there is an intriguing magnetic effect in nanoparticles known as "exchange bias". When a ferromagnetic material contacts an antiferromagnetic material, the ferromagnetic

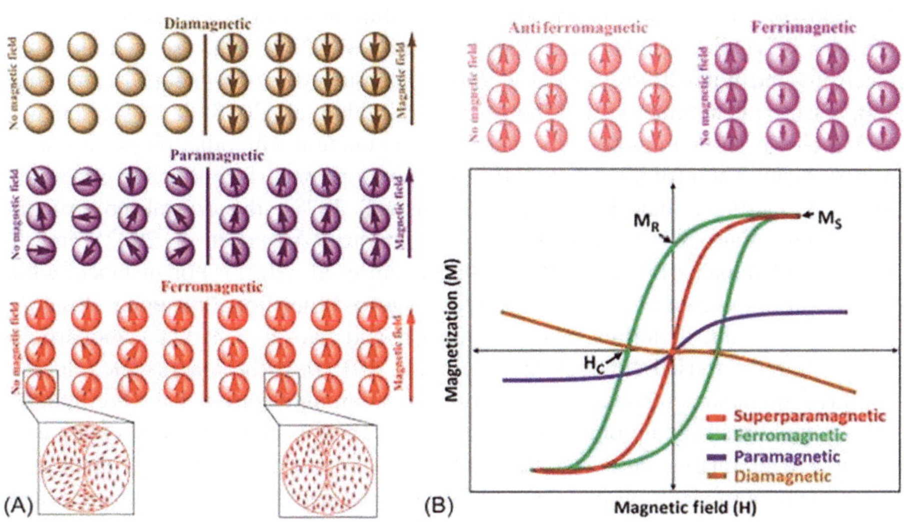

Fig. 2.2 Magnetism in nanoparticles (Ciofani, 2018)

material's hysteresis loop shifts (Yüksel, 2019). This effect has practical applications, such as magnetic sensors and hard disk drive read heads.

Nanoparticles with dimensions greater than 100 nm typically exhibit ferrimagnetic behaviour consistent with their bulk material origins (Karak, 2019). Nanoparticles with dimensions less than 100 nm and 10 nm typically show no oscillation and exhibit superparamagnetic behaviour, respectively (Karak, 2019; Avval et al., 2020). For instance, Kolen'ko et al. (2014) demonstrate that though iron oxide is a ferromagnetic at the macro-scale and larger nanoparticles, at below 10 nm, they display a single-domain structure and demonstrate superparamagnetism. Superparamagnetism is a prominent size-dependent magnetic phenomena observed in nanoparticles (Karak, 2019; Mohapatra et al., 2020). Magnetic characteristics in macro-scale materials are caused by the cohesiveness of distinct magnetic moments (Zufelato et al., 2022). However, as a nanoparticle's size shrinks, its thermal energy increases significantly to overcome the energy barrier that preserves magnetic alignment. This creates the moments of magnetic attraction within the particle to vary arbitrarily. Where a magnetic field is absent, superparamagnetism in nanoparticles is temporary (Karak, 2019). The particular scale at which superparamagnetism emerges is governed by the balance of thermal energy and net orientation of magnetic moments, and it varies greatly depending on the bulk material (Zufelato et al., 2022).

2.2.3 *Electric Properties*

Due to quantum size effects, the capacity for electric conductivity of different sized nanoparticles may vary substantially and act differently to bulk materials (Khan & Hossain, 2022). By virtue of the electrical confinement effect, electrical properties are particularly restricted by size (Pandey & Jaiswal, 2022). When scale diminishes, electrons and holes are restricted within the semiconductor's energy band, hence suppressing the conductivity of the particles (Singh et al., 2018; Pandey & Jaiswal, 2021). However, quantum effects of size exist in tandem with other properties of the nanoparticle. Therefore, nanoparticles may be more or less conductive depending on the constituent material (Pandey & Jaiswal, 2022), the crystalline structure (Tatarchuk et al., 2017), the surface area chemistry (Ansari et al., 2022), and the presence of dopants (Hong et al., 2006; Saleem et al., 2022). For instance, while they observe lowered electric conductivity in zero-dimensional nanoparticles compared to the bulk materials, one-dimensional and two-dimensional nanoparticles demonstrate increased conductivity (Fu et al., 2019). At the macroscale, electron conduction is characterised by delocalisation, meaning electrons have the freedom to migrate unrestrictedly in multiple directions (Thomas et al., 2014). In nanoparticles where all dimensions exist on the nanoscale (zero-dimensional), we can observe a confinement of electrons and restricted movement within a three-dimensional spatial domain (Premaratne & Agrawal, 2021; Pandey, 2022). However, in nanoparticles where only one or more dimensions exist on the nanoscale (one-dimensional, two-dimensional), we can observe a partial delocalisation of electrons. This creates discrete energy levels which diversify the electrical properties of nanoparticles, often minimising or eradicating electrical conductivity (Premaratne & Agrawal, 2021).

Nanoparticles also display surface-related electrical behaviour which becomes a critical factor when surface area-to-volume ratio increases with reducing sizes. There are greater amounts of disorganised or unsaturated bonds on the exterior of the particle with a higher number of atoms, and unsatisfied bonds cause unique electronic states (Bhushan & Maini, 2019; Sanap et al., 2020). The state of the surface area, which is greatly depending on their size, can cause a retention or liberation of electrons, influencing electrical behaviour dramatically (Vertegel et al., 2004). As a result of the increased curvature of their surface, smaller nanoparticles possess an especially noticeable charged surface and increased charge density (Abbas et al., 2008; Shi et al., 2018). The stability of nanoparticle suspensions and their interactions with different materials in various applications, including coatings and films (Woeppel et al., 2019), drug delivery (Mirvakili & Langer, 2021), targeted therapy (Woeppel et al., 2019), and energy storage (Ghosh et al., 2022), are significantly impacted by these phenomena. In catalysis and sensing applications (Dippong et al., 2021), the surface characteristics of nanoparticles are critical as they drive reactivity and electron migration processes.

2.3 Surface Area and Surface Chemistry

The analysis of surface chemistry and surface area constitutes a critical facet in nanoparticle characterisation, thereby shedding light on their distinctive properties and potential applications. Surface chemistry is often evaluated through zeta potential measurements. Zeta potential is the difference between the dispersion medium and the stationary layer of fluid attached to the dispersed particle's surface. It is typically measured in millivolts (mV) (Du et al., 2023). When particles have a high zeta potential (either positive or negative), they tend to repel each other, resulting in a stable dispersion as they are less likely to aggregate or flocculate; the zeta potential therefore gauges the electric charge at the nanoparticle's surface (Jiang et al., 2022). This is an important process because of its influence on particle's stability, interactions with surrounding media, and biological responses. Similarly, assessing surface area is paramount in evaluating the nanoparticle's reactivity and adsorption capabilities. Nanoparticles possess an expansive surface area compared to their bulk counterparts due to their small size, which significantly impacts their chemical and physical behaviour. This characteristic makes them highly efficient in various applications such as catalysis, adsorption, and drug delivery (Fujii et al., 2022; Stordy et al., 2022).

The crucial aspect of surface chemistry lies in establishing concrete connections between surface structures and their chemical properties, crucial for technological advancements. Early endeavours, such as ammonia synthesis studies on various Fe crystal surfaces, utilised specialised experimental setups capable of isolating single crystals and characterising their surfaces with advanced techniques (Gabor & Somorjai, 2011). By correlating catalytic activity measurements of these isolated crystal surfaces with specific atomic arrangements, it was revealed that certain Fe crystal faces exhibit significantly higher catalytic activity for ammonia synthesis. This knowledge, coupled with investigations into elementary reaction steps and the influence of catalyst supports, underscores the pivotal role of well-controlled model systems in distilling essential insights from complex surface chemical processes, ultimately driving enhancements in industrial processes (Gabor & Somorjai, 2011).

2.4 Common Techniques Used for Surface and Surface Area Characterisation

Currently, an extensive array of methods has been meticulously developed to drive surface and surface chemistry research. These techniques, as outlined in Table 2.1, are becoming indispensable for nanoparticle research and quality control, as they possess remarkable abilities (Almomani et al., 2021).

2.4.1 Nuclear Magnetic Resonance (NMR) Spectroscopy

Although nuclear magnetic resonance (NMR) spectroscopy is widely acknowledged as the primary technique for structural characterisation of organic molecules (Zhang & Yan, 2010), it can also be used for the characterisation of nanoparticles that has good solubility. In this method, the surface-bound molecules are examined through the application of solution 1H NMR methodology. Several studies, such as the analysis of thiophenol-coated cadmium sulphide (CdS) quantum dots (Sachleben et al., 1992), 2-carboxyethanephosphonic acid-coated SnO_2 nanoparticles (Holland

Table 2.1 Techniques for characterising surface properties of nanoparticles

Techniques	Properties characterised	References
Nuclear magnetic Resonance (NMR) spectroscopy	Surface/interface structure, the size, shape, and crystallinity of nanocrystals	Gabor and Somorjai (2011)
X-ray diffraction (XRD)	Structure of micro- and mesoporous materials	Gabor and Somorjai (2011)
Ambient pressure X-ray photoelectron spectroscopy (AP-XPS)	Surface chemical composition	Almomani et al. (2021)
Sum frequency generation spectroscopy (SFG)	Surface reaction intermediates	Almomani et al. (2021)
Energy-dispersive X-ray spectroscopy (EDX)	Morphological and chemical properties	Sharma et al. (2022a, b)

et al., 2007), and oleic acid-coated iron-oxide nanoparticles (Willis et al., 2005), demonstrated this method. However, this method is often limited by the low solubility and heterogeneity exhibited by nanomaterials. In instances where the molecules are attached to solid particles, their inability to undergo rapid tumbling precludes the averaging of signals, resulting in less distinct NMR signals compared to those observed in solution (Wang et al., 2007). Attaching molecules to nanoparticles using an extended, flexible, and soluble linker can augment the limitation associated with low solubility of nanoparticles, consequently leading to an overall enhancement of their NMR spectra. Nonetheless, the practical incorporation of such configurations into targeted molecules presents a synthetic hurdle. Even though this revealed that conventional NMR spectroscopy is valuable for monitoring reactions and product formation in characterisation of nanoparticles, it is not optimal for complete structural elucidation of organic molecules situated on the surface of nanoparticles (Huang et al., 2022).

2.4.2 X-Ray Diffraction Analysis (XRD)

X-ray diffraction analysis (XRD) stands as a cornerstone technique for material characterisation. Its fundamental principle revolves around irradiating a material with X-rays and then measuring the intensities and scattering angles of the X-rays that emerge from the material (Noreen et al., 2022). This method finds widespread application in scrutinising the phase and crystalline nature of nanoparticles. In an experiment conducted by Hazrat et al., nanostructure of silver nanoparticles was characterised using crystallographic X-ray diffraction, and the measurements were performed with Bragg-Brentano geometry. When the Ag nanoparticle powder samples were subjected to analysis using the XRD equipment, the resulting XRD pattern, also known as a diffractogram, of the Ag nanoparticles was indexed using a software. In the XRD pattern, certain sharp peaks at different positions (38.070, 44.290, 64.440, and 77.340) were observed. These positions correspond to specific arrangements of atoms within the crystal structure, and it is referred to as lattice planes (Ali et al., 2023). Nonetheless, it is imperative to acknowledge that the effectiveness of XRD can be diminished under specific circumstances. For instance, samples exhibiting high amorphous characteristics with varying interatomic distances or nanoparticles smaller than a few hundred atoms can present challenges that impact result precision (Kamminga et al., 2023).

2.4.3 Sum Frequency Generation (SFG) Spectroscopy

Sum frequency generation (SFG) is a type of optical spectroscopy that operates at the second-order nonlinear level, offering detailed vibrational spectra of surfaces and interfaces with remarkable sensitivity even for very thin layers (Cherednichenko et al., 2022). These spectra essentially serve as unique "fingerprints" for molecules. Consequently, SFG is used to assess the chemical make-up and molecular arrangements at surfaces and concealed interfaces (Du et al., 2023). The SFG technique's selection rule ensures that the acquired data specifically originates from the targeted surfaces and interfaces, thereby providing highly selective information. To better understand how this works, we can imagine shining two types of light on a surface at the same time – one with a wide range of colours in the mid-infrared range and another with a narrow range of colours in the visible range. These lights combine to create a new light signal with a frequency that is the sum of the original frequencies (Sengottiyan et al., 2023). This process occurs only in specific situations, like on surfaces or at interfaces, where the arrangement of molecules is not symmetrical. Unlike regular infrared (IR) spectroscopy, which looks at all molecules, SFG only focuses on the molecules attached to the surface and interface. Even if the same molecules are in the air or gas around, SFG helps us understand only the ones on the surface and the interface due to the selection rule of the technique. It can also study how molecules behave at the meeting point of air and liquid or between different liquids. Surprisingly, it can even see what is happening at interfaces hidden inside solids. The light it produces changes depending on how the molecules are arranged and how many of them are present. This helps us figure out the structure and number of molecules on the surface. By tweaking the types of light, we use and how they are oriented, we can also figure out how the molecules are lined up. Scientists have used SFG in many areas of research, like understanding how chemical reactions happen on surfaces (Zou et al., 2022) making better materials for various uses, studying how light and matter interact, and learning about plastics and how things come together on a tiny scale and even in the field of medicine involving very small particles (nanomedicine) (Kanamaru et al., 2022).

2.4.4 Ambient Pressure X-Ray Photoelectron Spectroscopy (AP-XPS)

Ambient pressure X-ray spectroscopy operates within a total reactant pressure of up to 10 torr. What sets this technique apart from conventional XPS is the utilisation of a differentially pumped electrostatic lens system. This system effectively refocuses photoelectrons emitted from the sample surface into the object plane of a high-vacuum electron energy analyser. The kinetic energy of these photoelectrons is adjustable through modifications to the X-ray source's energy (Gabor & Somorjai, 2011). Fine-tuning the kinetic energy to an appropriate level enables the minimisation of the electron mean-free path, facilitating determination of the oxidation state and composition of the surface layer with a thickness of approximately 1 nm (Tan et al., 2022).

2.4.5 Energy-Dispersive X-Ray Spectroscopy (EDX)

Energy-dispersive X-ray spectroscopy (EDX) is a valuable technique that involves irradiating the sample with an electron beam. When the electron beam interacts with the sample surface, inner shell electrons are ejected, and subsequent transition of outer shell electrons to fill these vacancies generates X-rays (Kumari et al., 2020). Each element yields a characteristic X-ray emission pattern based on its unique atomic structure, allowing for compositional analysis (Sharma et al., 2022a, b). However, it is important to note that EDX provides qualitative compositional information, showing the chemical elements present in the sample without quantification. Nevertheless, the intensity of peaks can offer some insight into the relative abundance of elements in the sample. Another advantage of EDX is its practicality, as it can be easily integrated with existing SEM or TEM setups, providing a synergistic approach for morphological and chemical analysis (Jiang et al., 2022).

In a study conducted by El-Hagary et al. (2019), the structural and optical characteristics of pure CeO_2 and Fe-doped CeO_2 nanoparticles were investigated, produced using a precipitation method. The investigation encompassed varying Fe concentrations ranging from 0.02% to 0.08%. X-ray diffraction (XRD) analyses of the nanoparticles at different Fe levels consistently revealed a cubic zinc-blend-type structure. Notably, nanoparticle sizes were observed to diminish from 20 nm (0%) to 14 nm (8%) with increasing Fe content, a finding corroborated by TEM micrographs. The optical absorption spectra exhibited a discernible blue-shift from 550.2 nm to 336.4 nm as Fe concentration increased, attributed to the reduction in particle size. Additionally, a marginal rise in the bandgap energy was observed, from 3.55 eV ($x = 0$) to 3.69 eV ($x = 0.08$), linked to the quantum confinement effect. Interestingly, room temperature of ferromagnetism was observed in Fe-doped CeO_2, with the saturated magnetisation (Ms) augmenting with higher Fe content. These compelling findings emphasise the potential of Fe-doped CeO_2 as a promising candidate for the design of optoelectronic and spintronic devices. Similarly, in another work done by He et al. (2015), precise control over cobalt monoxide (CoO) nanoparticle structure, shape, and size via thermal decomposition of cobalt (III) acetylacetonate (Co $(acac)_3$) in oleylamine was achieved. The characterisation methods employed involved various techniques like EDX, SFG, XRD, and more. Particle properties were tailored by adjusting precursor concentration, reaction time, and temperature. CoO nanoparticles displayed stable behaviour with distinct optical properties, exhibiting different crystal structures. Notably, hexagonal CoO nanoparticles exhibited size-dependent optical band gaps. These nanoparticles showed potential for applications in pigments, sensors, information storage, batteries, and magnetic recording. Gabor and Somorjai (2011) argued that surface chemistry has a major impact on industries like chemicals, electronics, and medicine. For example, in commercial production of ammonia, it was suggested that the surface atoms with high coordination number (C7 and C8) are the active site for ammonia synthesis.

2.5 Conclusion

Nanoparticles possess unique properties such as increased surface functionality and altered physical characteristics compared to bulk materials. Their higher surface-to-mass ratios and abundant surface atoms significantly impact material properties, including catalytic activity. Size characterisation is essential, considering various dimensions and aspects such as atomic structure-based size and effective size within a matrix. Surface area and surface chemistry analysis are crucial for understanding reactivity and adsorption capabilities. Techniques like NMR, XRD, SFG, AP-XPS, and EDX provide vital insights into surface structure, composition, and chemical properties of nanoparticles. These insights are key to harnessing nanoparticles for a wide range of applications, including catalysis, energy storage, and medical technologies, making them pivotal in both research and consumer product domains.

References

Abbas, Z., Labbez, C., Nordholm, S., & Ahlberg, E. (2008). Size-dependent surface charging of nanoparticles. *The Journal of Physical Chemistry C, 112*(15), 5715–5723.

Abid, H. A., & Al-Rashid, S. N. T. (2020). Study of the effect of nanoparticle size on the dielectric constant and concentration of charge carriers of Si and CdS materials. *Chalcogenide Letters, 17*, 623–629.

Ağbulut, Ü. (2022). Understanding the role of nanoparticle size on energy, exergy, thermoeconomic, exergoeconomic, and sustainability analyses of an IC engine: A thermodynamic approach. *Fuel Processing Technology, 225*, 107060.

Akbarzadeh, H., Abareshi, N., & Kamrani, M. (2023). Role of the middle-shell in the stability of three-shell nanoparticles: A molecular dynamics study. *Colloids and Surfaces A: Physicochemical and Engineering Aspects, 676*, 132163.

Ali, A., Shah, T., Ullah, R., Zhou, P., Guo, M., Ovais, M., et al. (2021). Review on recent progress in magnetic nanoparticles: Synthesis, characterization, and diverse applications. *Frontiers in Chemistry, 9*, 629054.

Ali, M. H., Azad, M. A. K., Khan, K. A., Rahman, M. O., Chakma, U., & Kumer, A. (2023). Analysis of crystallographic structures and properties of silver nanoparticles synthesized using PKL extract and nanoscale characterization techniques. *ACS Omega, 8*(31), 28133–28142.

Almomani, F., Al-Jaml, K. L., & Bhosale, R. R. (2021). Solar photo-catalytic production of hydrogen by irradiation of cobalt co-doped TiO2. *International Journal of Hydrogen Energy, 46*(22), 12068–12081.

Andhare, D. D., Patade, S. R., Kounsalye, J. S., & Jadhav, K. M. (2020). Effect of Zn doping on structural, magnetic and optical properties of cobalt ferrite nanoparticles synthesized via. Co-precipitation method. *Physica B: Condensed Matter, 583*, 412051.

Ansari, S. M., Sinha, B. B., Sen, D., Sastry, P. U., Kolekar, Y. D., & Ramana, C. V. (2022). Effect of oleylamine on the surface chemistry, morphology, electronic structure, and magnetic properties of cobalt ferrite nanoparticles. *Nanomaterials, 12*(17), 3015.

Avasthi, A., Caro, C., Pozo-Torres, E., Leal, M. P., & García-Martín, M. L. (2020). Magnetic nanoparticles as MRI contrast agents. In *Surface-modified nanobiomaterials for electrochemical and biomedicine applications* (pp. 49–91). Springer.

Avval, Z. M., Malekpour, L., Raeisi, F., Babapoor, A., Mousavi, S. M., Hashemi, S. A., & Salari, M. (2020). Introduction of magnetic and supermagnetic nanoparticles in new approach of targeting drug delivery and cancer therapy application. *Drug Metabolism Reviews, 52*(1), 157–184.

Aziz, S. B., Marif, R. B., Brza, M. A., Hassan, A. N., Ahmad, H. A., Faidhalla, Y. A., & Kadir, M. F. Z. (2019). Structural, thermal, morphological and optical properties of PEO filled with biosynthesized Ag nanoparticles: New insights to band gap study. *Results in Physics, 13*, 102220.

Bhushan, J., & Maini, C. (2019). Nanoparticles: A promising novel adjunct for dentistry. *Indian Journal of Dental Sciences, 11*(3), 167–173.

Bodelon, G., Montes-García, V., Perez-Juste, J., & Pastoriza-Santos, I. (2018). Surface-enhanced Raman scattering spectroscopy for label-free analysis of P. aeruginosa quorum sensing. *Frontiers in Cellular and Infection Microbiology, 8*, 143.

Caizer, C. (2016). Nanoparticle size effect on some magnetic properties. In *Handbook of nanoparticles* (p. 475). Springer.

Chen, Y. C., Huang, Y. S., Huang, H., Su, P. J., Perng, T. P., & Chen, L. J. (2020a). Photocatalytic enhancement of hydrogen production in water splitting under simulated solar light by band gap engineering and localized surface plasmon resonance of ZnxCd1-xS nanowires decorated by Au nanoparticles. *Nano Energy, 67*, 104225.

Chen, Y., Lai, Z., Zhang, X., Fan, Z., He, Q., Tan, C., & Zhang, H. (2020b). Phase engineering of nanomaterials. *Nature Reviews Chemistry, 4*(5), 243–256.

Cherednichenko, Y. V., Konnova, S. A., & Fakhrullin, R. F. (2022). Self-assembly of halloysite nanotubes as a tool for the formation of 3D structures. *Colloid Journal, 84*(3), 344–352.

Ciofani, G. (Ed.). (2018). *Smart nanoparticles for biomedicine*. Elsevier.

Coene, A., & Leliaert, J. (2022). Magnetic nanoparticles in theranostic applications. *Journal of Applied Physics, 131*(16).

Coetzee, D., Venkataraman, M., Militky, J., & Petru, M. (2020). Influence of nanoparticles on thermal and electrical conductivity of composites. *Polymers, 12*(4), 742.

Danish, M. S. S., Estrella-Pajulas, L. L., Alemaida, I. M., Grilli, M. L., Mikhaylov, A., & Senjyu, T. (2022). Green synthesis of silver oxide nanoparticles for photocatalytic environmental remediation and biomedical applications. *Metals, 12*(5), 769.

Daoush, W. M. (2017). Co-precipitation and magnetic properties of magnetite nanoparticles for potential biomedical applications. *Journal of Nanomedicine Research, 5*(3), 00118.

Dippong, T., Levei, E. A., & Cadar, O. (2021). Recent advances in synthesis and applications of MFe2O4 (M= Co, Cu, Mn, Ni, Zn) nanoparticles. *Nanomaterials, 11*(6), 1560.

Dong, Y., Zhang, T. L., Wang, H., Liu, X., & Jiang, C. B. (2021). Chemical synthesis and characterization of SmCo 5/Co magnetic nanocomposite particles. *Rare Metals, 40*, 1224–1231.

Du, M., Guo, L., Ren, H., Tao, X., Li, Y., Nan, B., et al. (2023). Non-noble FeCrO x bimetallic nanoparticles for efficient NH3 decomposition. *Nanomaterials, 13*(7), 1280.

El-Hagary, M., Shaaban, E. R., Moustafa, S. H., & Gad, G. M. A. (2019). The particle size-dependent optical band gap and magnetic properties of Fe-doped CeO2 nanoparticles. *Solid State Sciences, 91*, 15–22.

Ezhilmaran, B., & Bhat, S. V. (2022). Enhanced charge transfer in TiO_2 nanoparticles/MoO3 nanostructures bilayer heterojunction electrode for efficient electrochromism. *Materials Today Communications, 31*, 103497.

Fu, S., Sun, Z., Huang, P., Li, Y., & Hu, N. (2019). Some basic aspects of polymer nanocomposites: A critical review. *Nano Materials Science, 1*(1), 2–30.

Fujii, S., Takano, S., Nakazawa, K., & Sakurai, K. (2022). Impact of zwitterionic polymers on the tumor permeability of molecular bottlebrush-based nanoparticles. *Biomacromolecules, 23*(7), 2846–2855.

Gabor, A., & Somorjai, Y. (2011). Impact of surface chemistry. *Proceedings of the National Academy of Sciences, 108*, 3.

Gao, D., Gao, J., Gao, F., Kuang, Q., Pan, Y., Chen, Y., & Pan, Z. (2021). Quintuple-mode dynamic anti-counterfeiting using multi-mode persistent phosphors. *Journal of Materials Chemistry C, 9*(46), 16634–16644.

Ghini, M., Curreli, N., Camellini, A., Wang, M., Asaithambi, A., & Kriegel, I. (2021). Photodoping of metal oxide nanocrystals for multi-charge accumulation and light-driven energy storage. *Nanoscale, 13*(19), 8773–8783.

Ghosh, S., Polaki, S. R., Macrelli, A., Casari, C. S., Barg, S., Jeong, S. M., & Ostrikov, K. K. (2022). Nanoparticle-enhanced multifunctional nanocarbons—recent advances on electrochemical energy storage applications. *Journal of Physics D: Applied Physics, 55*(41), 413001.

He, X., Song, X., Qiao, W., Li, Z., Zhang, X., Yan, S., et al. (2015). Phase-and size-dependent optical and magnetic properties of CoO nanoparticles. *The Journal of Physical Chemistry C, 119*(17), 9550–9559.

Holland, G. P., Sharma, R., Agola, J. O., Amin, S., Solomon, V. C., Singh, P., et al. (2007). NMR characterization of phosphonic acid capped SnO2 nanoparticles. *Chemistry of Materials, 19*(10), 2519–2526.

Hong, C. S., Park, H. H., Moon, J., & Park, H. H. (2006). Effect of metal (Al, Ga, and In)-dopants and/or Ag-nanoparticles on the optical and electrical properties of ZnO thin films. *Thin Solid Films, 515*(3), 957–960.

Huang, J., Liu, J., & Wang, J. (2020). Optical properties of biomass-derived nanomaterials for sensing, catalytic, biomedical and environmental applications. *TrAC Trends in Analytical Chemistry, 124*, 115800.

Huang, W., Zhu, J., Wang, M., Hu, L., Tang, Y., Shu, Y., et al. (2021). Emerging mono-elemental bismuth nanostructures: Controlled synthesis and their versatile applications. *Advanced Functional Materials, 31*(10), 2007584.

Huang, Z., Callmann, C. E., Wang, S., Vasher, M. K., Evangelopoulos, M., Petrosko, S. H., & Mirkin, C. A. (2022). Rational vaccinology: Harnessing nanoscale chemical design for cancer immunotherapy. *ACS Central Science, 8*(6), 692–704.

Jeon, M., Halbert, M. V., Stephen, Z. R., & Zhang, M. (2021). Iron oxide nanoparticles as T1 contrast agents for magnetic resonance imaging: Fundamentals, challenges, applications, and prospectives. *Advanced Materials, 33*(23), 1906539.

Jiang, B., Lu, D., Jiang, N., Jiang, H., Qi, L., Shi, S., et al. (2022). Synthesis of non-precious N-doped mesoporous Fe3O4/CoO@ NC materials towards efficient oxygen reduction reaction for microbial fuel cells. *Electroanalysis, 34*(8), 1245–1255.

Joudeh, N., & Linke, D. (2022). Nanoparticle classification, physicochemical properties, characterization, and applications: A comprehensive review for biologists. *Journal of Nanobiotechnology, 20*(1), 262.

Kamminga, M. E., Birk, J. O., í Hjøllum, J., Jacobsen, H., Lass, J., Koch, T. L., et al. (2023). Nanocriticality in the magnetic phase transition of CoO nanoparticles. *Physical Review B, 107*(6), 064424.

Kanamaru, T., Sakurai, K., & Fujii, S. (2022). Impact of polyethylene glycol (peg) conformations on the in vivo fate and drug release behavior of pegylated core-cross-linked polymeric nanoparticles. *Biomacromolecules, 23*(9), 3909–3918.

Kang, H., Rho, S., Stiles, W. R., Hu, S., Baek, Y., Hwang, D. W., et al. (2020). Size-dependent EPR effect of polymeric nanoparticles on tumor targeting. *Advanced Healthcare Materials, 9*(1), 1901223.

Karak, N. (2019). Fundamentals of nanomaterials and polymer nanocomposites. In *Nanomaterials and polymer nanocomposites* (pp. 1–45). Elsevier.

Khan, S., & Hossain, M. K. (2022). Classification and properties of nanoparticles. In *Nanoparticle-based polymer composites* (pp. 15–54). Woodhead Publishing.

Khattab, T. A., El-Naggar, M. E., Abdelrahman, M. S., Aldalbahi, A., & Hatshan, M. R. (2021). Facile development of photochromic cellulose acetate transparent nanocomposite film immobilized with lanthanide-doped pigment: Ultraviolet blocking, superhydrophobic, and antimicrobial activity. *Luminescence, 36*(2), 543–555.

Khurgin, J. B., & Levy, U. (2020). Generating hot carriers in plasmonic nanoparticles: When quantization does matter? *ACS Photonics, 7*(3), 547–553.

Kianfar, E. (2021). Magnetic nanoparticles in targeted drug delivery: A review. *Journal of Superconductivity and Novel Magnetism, 34*(7), 1709–1735.

Kolen'ko, Y. V., Bañobre-López, M., Rodríguez-Abreu, C., Carbó-Argibay, E., Sailsman, A., Piñeiro-Redondo, Y., et al. (2014). Large-scale synthesis of colloidal Fe3O4 nanoparticles exhibiting high heating efficiency in magnetic hyperthermia. *The Journal of Physical Chemistry C, 118*(16), 8691–8701.

Kumari, K., Aljawfi, R. N., Chawla, A. K., Kumar, R., Alvi, P. A., Alshoaibi, A., et al. (2020). Engineering the optical properties of Cu doped CeO2 NCs for application in white LED. *Ceramics International, 46*(6), 7482–7488.

Lee, S. A., & Link, S. (2021). Chemical interface damping of surface plasmon resonances. *Accounts of Chemical Research, 54*(8), 1950–1960.

Lee, K. C., Lin, S. J., Lin, C. H., Tsai, C. S., & Lu, Y. J. (2008). Size effect of Ag nanoparticles on surface plasmon resonance. *Surface and Coatings Technology, 202*(22–23), 5339–5342.

Li, W., & Fortner, J. D. (2020). (Super) paramagnetic nanoparticles as platform materials for environmental applications: From synthesis to demonstration. *Frontiers of Environmental Science & Engineering, 14*, 1–9.

Materón, E. M., Miyazaki, C. M., Carr, O., Joshi, N., Picciani, P. H., Dalmaschio, C. J., et al. (2021). Magnetic nanoparticles in biomedical applications: A review. *Applied Surface Science Advances, 6*, 100163.

Mendis, P., de Silva, R. M., de Silva, K. N., Wijenayaka, L. A., Jayawardana, K., & Yan, M. (2016). Nanosilver rainbow: A rapid and facile method to tune different colours of nanosilver through the controlled synthesis of stable spherical silver nanoparticles. *RSC Advances, 6*(54), 48792–48799.

Mirvakili, S. M., & Langer, R. (2021). Wireless on-demand drug delivery. *Nature Electronics, 4*(7), 464–477.

Mohapatra, J., Xing, M., Elkins, J., Beatty, J., & Liu, J. P. (2020). Size-dependent magnetic hardening in CoFe2O4 nanoparticles: Effects of surface spin canting. *Journal of Physics D: Applied Physics, 53*(50), 504004.

Nejad, F. G., Tajik, S., Beitollahi, H., & Sheikhshoaie, I. (2021). Magnetic nanomaterials based electrochemical (bio) sensors for food analysis. *Talanta, 228*, 122075.

Nguyen, A. T., Gao, F., Baucom, D., & Heyes, C. D. (2020). CuInS2-Doped ZnS quantum dots obtained via non-injection cation exchange show reduced but heterogeneous blinking and provide insights into their structure–optical property relationships. *The Journal of Physical Chemistry C, 124*(19), 10744–10754.

Nguyen, D. T. C., Nguyen, T. T., Le, H. T., Nguyen, T. T. T., Bach, L. G., Nguyen, T. D., et al. (2021a). The sunflower plant family for bioenergy, environmental remediation, nanotechnology, medicine, food and agriculture: A review. *Environmental Chemistry Letters, 19*, 3701–3726.

Nguyen, M. D., Tran, H. V., Xu, S., & Lee, T. R. (2021b). Fe3O4 Nanoparticles: Structures, synthesis, magnetic properties, surface functionalization, and emerging applications. *Applied Sciences, 11*(23), 11301.

Noreen, S., Tahir, M. B., Hussain, A., Nawaz, T., Rehman, J. U., Dahshan, A., et al. (2022). Emerging 2D-Nanostructured materials for electrochemical and sensing application-a review. *International Journal of Hydrogen Energy, 47*(2), 1371–1389.

Pandey, P. (2022). Role of nanotechnology in electronics: A review of recent developments and patents. *Recent Patents on Nanotechnology, 16*(1), 45–66.

Pandey, B. K., & Jaiswal, R. L. (2021). Theoretical prediction for band gap of semiconducting nanoparticles. *Advanced Materials Letters, 12*(10), 1–4.

Pandey, B. K., & Jaiswal, R. L. (2022). Electrical conductivity of semiconducting nanoparticles. *Physica B: Condensed Matter, 646*, 414279.

Paredes, P., Rauwel, E., & Rauwel, P. (2022). Surveying the synthesis, optical properties and photocatalytic activity of Cu3N nanomaterials. *Nanomaterials, 12*(13), 2218.

Petrini, N., Ghini, M., Curreli, N., & Kriegel, I. (2023). Optical modelling of plasmonic nanoparticles with electronically depleted layers. *The Journal of Physical Chemistry C, 127*(3), 1576–1587.

Popović, Z., Liu, W., Chauhan, V. P., Lee, J., Wong, C., Greytak, A. B., et al. (2010). A nanoparticle size series for in vivo fluorescence imaging. *Angewandte Chemie, 122*(46), 8831–8834.

Premaratne, M., & Agrawal, G. P. (2021). *Theoretical foundations of nanoscale quantum devices*. Cambridge University Press.

Rivera-Rodriguez, A., & Rinaldi-Ramos, C. M. (2021). Emerging biomedical applications based on the response of magnetic nanoparticles to time-varying magnetic fields. *Annual Review of Chemical and Biomolecular Engineering, 12*, 163–185.

Rohilla, D., & Chaudhary, S. (2023). Advanced nanoparticles: A boon or a bane for environmental remediation applications. In *Advanced functional nanoparticles "boon or bane" for environment remediation applications: Combating environmental issues* (pp. 27–74). Springer International Publishing.

Rubel, M. H., & Hossain, M. K. (2022). Crystal structures and properties of nanomagnetic materials. In *Fundamentals of low dimensional magnets* (1st ed.). CRC Press, Taylor & Francis Group.

Sachleben, J. R., Wooten, E. W., Emsley, L., Pines, A., Colvin, V. L., & Alivisatos, A. P. (1992). NMR studies of the surface structure and dynamics of semiconductor nanocrystals. *Chemical Physics Letters, 198*(5), 431–436.

Sagayaraj, R., Aravazhi, S., & Chandrasekaran, G. (2021). Review on structural and magnetic properties of (Co–Zn) ferrite nanoparticles. *International Nano Letters, 11*(4), 307–319.

Said, Z., Arora, S., Farooq, S., Sundar, L. S., Li, C., & Allouhi, A. (2022). Recent advances on improved optical, thermal, and radiative characteristics of plasmonic nanofluids: Academic insights and perspectives. *Solar Energy Materials and Solar Cells, 236*, 111504.

Saleem, S., Jameel, M. H., Akhtar, N., Nazir, N., Ali, A., Zaman, A., et al. (2022). Modification in structural, optical, morphological, and electrical properties of zinc oxide (ZnO) nanoparticles (NPs) by metal (Ni, Co) dopants for electronic device applications. *Arabian Journal of Chemistry, 15*(1), 103518.

Sanap, P., Hegde, V., Ghunawat, D., Patil, M., Nagaonkar, N., & Jagtap, V. (2020). Current applications of chitosan nanoparticles in dentistry: A review. *International Journal of Applied Dental Science, 6*(4), 81–84.

Sengottiyan, S., Mikolajczyk, A., Jagiełło, K., Swirog, M., & Puzyn, T. (2023). Core, coating, or corona? The importance of considering protein coronas in nano-QSPR modeling of zeta potential. *ACS Nano, 17*(3), 1989–1997.

Sharma, A., Varshney, M., Gautam, T., Sharma, A., Vij, A., Sharma, R. K., et al. (2022a). XANES investigations on electronic structure and magnetic properties of GaFeO3 nanocrystals. *Journal of Electronic Materials, 51*(8), 4133–4138.

Sharma, R. K., Dey, G., Banerjee, P., Maity, J. P., Lu, C. M., Siddique, J. A., et al. (2022b). A new aspect of lipopeptide incorporated nanoparticles synthesis and recent advancement in biomedical and environmental sciences: A review. *Journal of Materials Chemistry B, 11*, 10–32.

Shaumbwa, V. R., Liu, D., Archer, B., Li, J., & Su, F. (2021). Preparation and application of magnetic chitosan in environmental remediation and other fields: A review. *Journal of Applied Polymer Science, 138*(42), 51241.

Shi, Y. R., Ye, M. P., Du, L. C., & Weng, Y. X. (2018). Experimental determination of particle size-dependent surface charge density for silica nanospheres. *The Journal of Physical Chemistry C, 122*(41), 23764–23771.

Shoukat, R., & Khan, M. I. (2021). Carbon nanotubes: A review on properties, synthesis methods and applications in micro and nanotechnology. *Microsystem Technologies, 303*, 1–10.

Shrivas, K., Shankar, R., & Dewangan, K. (2015). Gold nanoparticles as a localized surface plasmon resonance based chemical sensor for on-site colorimetric detection of arsenic in water samples. *Sensors and Actuators B: Chemical, 220*, 1376–1383.

Shukla, S., Khan, R., & Daverey, A. (2021). Synthesis and characterization of magnetic nanoparticles, and their applications in wastewater treatment: A review. *Environmental Technology & Innovation, 24*, 101924.

Singh, M., Goyal, M., & Devlal, K. (2018). Size and shape effects on the band gap of semiconductor compound nanomaterials. *Journal of Taibah University for Science, 12*(4), 470–475.

Soetaert, F., Korangath, P., Serantes, D., Fiering, S., & Ivkov, R. (2020). Cancer therapy with iron oxide nanoparticles: Agents of thermal and immune therapies. *Advanced Drug Delivery Reviews, 163*, 65–83.

Stordy, B., Zhang, Y., Sepahi, Z., Khatami, M. H., Kim, P. M., & Chan, W. C. (2022). Conjugating ligands to an equilibrated nanoparticle protein corona enables cell targeting in serum. *Chemistry of Materials, 34*(15), 6868–6882.

Tan, P., Fu, H., & Ma, X. (2021). Design, optimization, and nanotechnology of antimicrobial peptides: From exploration to applications. *Nano Today, 39*, 101229.

Tan, Y., Cai, W., Luo, C., Tang, J., Kwok, R. T., Lam, J. W., et al. (2022). Rapid biotransformation of luminescent bimetallic nanoparticles in hepatic sinusoids. *Journal of the American Chemical Society, 144*(45), 20653–20660.

Tatarchuk, T., Bououdina, M., Judith Vijaya, J., & John Kennedy, L. (2017). Spinel ferrite nanoparticles: Synthesis, crystal structure, properties, and perspective applications. In *Nanophysics, nanomaterials, interface studies, and applications: Selected proceedings of the 4th international conference Nanotechnology and Nanomaterials (NANO2016)* (pp. 305–325). Springer International Publishing.

Thomas, S., Rafiei, S., Maghsoodlou, S., & Afzali, A. (2014). *Foundations of nanotechnology, volume two: Nanoelements formation and interaction*. CRC Press.

Till, J. L., Guyodo, Y., Lagroix, F., Morin, G., & Ona-Nguema, G. (2015). Goethite as a potential source of magnetic nanoparticles in sediments. *Geology, 43*(1), 75–78.

Vallinayagam, S., Rajendran, K., Lakkaboyana, S. K., Soontarapa, K., Remya, R. R., Sharma, V. K., et al. (2021). Recent developments in magnetic nanoparticles and nano-composites for wastewater treatment. *Journal of Environmental Chemical Engineering, 9*(6), 106553.

Verger, A., Brandhonneur, N., Molard, Y., Cordier, S., Kowouvi, K., Amela-Cortes, M., & Dollo, G. (2021). From molecules to nanovectors: Current state of the art and applications of photosensitizers in photodynamic therapy. *International Journal of Pharmaceutics, 604*, 120763.

Vertegel, A. A., Siegel, R. W., & Dordick, J. S. (2004). Silica nanoparticle size influences the structure and enzymatic activity of adsorbed lysozyme. *Langmuir, 20*(16), 6800–6807.

Vetchinkina, E., Loshchinina, E., Kupryashina, M., Burov, A., & Nikitina, V. (2019). Shape and size diversity of gold, silver, selenium, and silica nanoparticles prepared by green synthesis using fungi and bacteria. *Industrial & Engineering Chemistry Research, 58*(37), 17207–17218.

Wang, Z., Liu, Q., Zhu, H., Liu, H., Chen, Y., & Yang, M. (2007). Dispersing multi-walled carbon nanotubes with water–soluble block copolymers and their use as supports for metal nanoparticles. *Carbon, 45*(2), 285–292.

Wang, L., Hasanzadeh Kafshgari, M., & Meunier, M. (2020). Optical properties and applications of plasmonic-metal nanoparticles. *Advanced Functional Materials, 30*(51), 2005400.

Willis, A. L., Turro, N. J., & O'Brien, S. (2005). Spectroscopic characterization of the surface of iron oxide nanocrystals. *Chemistry of Materials, 17*(24), 5970–5975.

Woeppel, K. M., Zheng, X. S., Schulte, Z. M., Rosi, N. L., & Cui, X. T. (2019). Nanoparticle doped PEDOT for enhanced electrode coatings and drug delivery. *Advanced Healthcare Materials, 8*(21), 1900622.

Wrigglesworth, E. G., & Johnston, J. H. (2021). Mie theory and the dichroic effect for spherical gold nanoparticles: An experimental approach. *Nanoscale Advances, 3*(12), 3530–3536.

Xu, K., Xu, S., & Wei, F. (2021). Recent progress in magnetic applications for micro-and nanorobots. *Beilstein Journal of Nanotechnology, 12*(1), 744–755.

Yakovkin, I., & Reshetnyak, V. (2023). Controlling plasmon resonance of gold and silver nanoparticle arrays with help of liquid crystal. *Photonics, 10*(10), 1088.

Yüksel, Y. (2019). Effects of the particle size and shape of the magnetic nanoparticles on the magnetic hyperthermia and exchange bias properties☆. *Physica B: Condensed Matter, 575*, 411689.
Zayadi, R. A., & Bakar, F. A. (2020). Comparative study on stability, antioxidant and catalytic activities of bio-stabilized colloidal gold nanoparticles using microalgae and cyanobacteria. *Journal of Environmental Chemical Engineering, 8*(4), 103843.
Zhang, H. (2020). Molecularly imprinted nanoparticles for biomedical applications. *Advanced Materials, 32*(3), 1806328.
Zhang, B., & Yan, B. (2010). Analytical strategies for characterizing the surface chemistry of nanoparticles. *Analytical and Bioanalytical Chemistry, 396*, 973–982.
Zhang, Z., Ouyang, Y., Cheng, Y., Chen, J., Li, N., & Zhang, G. (2020). Size-dependent phononic thermal transport in low-dimensional nanomaterials. *Physics Reports, 860*, 1–26.
Zhao, C., Qu, S., Qiu, J., & Zhu, C. (2003). Photoinduced formation of colloidal Au by a near-infrared femtosecond laser. *Journal of Materials Research, 18*(7), 1710–1714.
Zou, W., Zhao, C., Zhang, X., Jin, C., Jiang, K., & Zhou, Q. (2022). Mitigation effects and associated mechanisms of environmentally relevant thiols on the phytotoxicity of molybdenum disulfide nanosheets. *Environmental Science & Technology, 56*(13), 9556–9568.
Zufelato, N., Aquino, V. R., Shrivastava, N., Mendanha, S., Miotto, R., & Bakuzis, A. F. (2022). Heat generation in magnetic hyperthermia by manganese ferrite-based nanoparticles arises from Néel collective magnetic relaxation. *ACS Applied Nano Materials, 5*(5), 7521–7539.

Chapter 3
Sources of Nanoparticles

Konjerimam Ishaku Chimbekujwo, Aishat Rabiu Sani, Oluwafemi Adebayo Oyewole, and Patrick Omoregie Isibor

3.1 Introduction

Nano is a prefix that comes from the Greek word *nanos*, which means "short" according to Buzea et al. (2007). In numerous articles, the term nano has become a common way to describe tiny existence and procedures in modern science (Buzea et al., 2007). Nano is a term used to refer to small entities or procedures, typically actual micrometer scale. The study of matter at the microscale is known as nanoscience which is concerned with the distinct dimension based on characteristics of solid-state matter (Mulvaney, 2015). The term nanomaterials, which encompass materials of size 1–100 nm, is used in nanotechnology (Hasan, 2015). Nanoscience deals with tiny materials which are utilized and manipulated at a small level. Tiny particles function distinctly at this dimension and offer diverse and intriguing applications. A wide variety of product areas have been extensively studied in the area of nanotechnology and nanoscience. The advancement of materials like those of medicinal purposes will be possible beyond the limitations of traditional methods. The use of nanotechnology involves the creation, utilization, and manufacturing of

K. I. Chimbekujwo (✉)
Department of Microbiology, Modibbo Adama University, Yola, Sangerei, Yola, Nigeria
e-mail: chimbekonje@gmail.com

A. R. Sani
Department of Microbiology, Ibrahim Badamasi Babangida University, Lapai, Niger State, Nigeria

O. A. Oyewole
Department of Microbiology, Federal University of Technology, Minna, Nigeria

P. O. Isibor
Department of Biological Sciences, Covenant University, Ota, Nigeria
e-mail: patrick.isibor@covenantuniversity.edu.ng

P. O. Isibor et al. (eds.), *Environmental Nanotoxicology*,
https://doi.org/10.1007/978-3-031-54154-4_3

materials at tiny, molecular levels to create novel nano-dimension substances (Hahens et al., 2007).

Drug carriers that are biodegradable and solid, such as pharmaceutical nanomaterials, are defined by their size of less than 100 nm. The term nanoparticle is used to both describe microsphere and micro-capsules. The drugs are distributed through microsphere and micro-capsules, which are a type of model approach that surrounds specialized membrane polymer. This analytical method centers on the preparation process, classifying, utilization, categorization, healthcare perspective, and nanoparticle area of pharmaceutics (Couvreur et al., 1995).

Nano-object with its outward size in the microscale are used to characterize nanoparticles, where the distance between each axis is relatively constant. If the dimensions are often thrice more different, then NPs may be better defined by using terms like microfibers/microplates (Mulvaney, 2015). NP structures and dimensions are diverse. The shapes that can be used include round, concave, cylindric, tube, hollowed out, corkscrew, or nonuniform (Ealia & Saravanakumar, 2017). The dimension of nanoparticle is from 1 nm to 100 nm. Atom cluster is a better term for NPs that are smaller than 1 nm in size. Nanoparticles maybe crystal with one or more solidified crystal or amorphous and also loosely fitted or cluster (Machado et al., 2015). Nanoparticles can be unvarying or many coatings. Typically, the last-mentioned layer consists of small particles, surface-active agents, metallic ion, or polymeric. The shell layer is composed of a chemically different material from the core coatings. The core layer is the center layer of nanoparticle (Khan et al., 2019).

3.2 Sources of Nanoparticles

3.2.1 Natural Sources

Many natural procedures, such as volcanic explosion, reactions from photochemistry, bush fire, and weathering, can produce NPs including living creatures and plants. The global impact of air pollution can be attributed to activity of human, including industries and motors, and burning of coal, which can result in the production of nanoparticulate matter. The energy equilibrium of the planet is influenced by small particles suspended in the atmosphere, called aerosols, which take up sunlight and disperse back into space. The decrease in abundance of mineral suspended particulate along with reduction, wind deflation, marine salt, and volcanic outburst as a result of human activities is the most significant component of global atmospheric aero-pollution (Muneeb, 2021).

Harmattan are the most significant sources of NPs of the surroundings. Minerals from deserts make up roughly half of the aerosol particles in the troposphere (Shi et al., 2005). Harmattan dust particle dimension formed can range from 100 nm to a number of microns, with one-third to half being tiny than 2.5 mm (Shi et al., 2005). The emission of NPs by crucial surrounding occurrence in a specific region can have an impact on locality thousandfold distance far away. In the interplanetary space, NPs are prevalent, for example, data gathered from space, the moon and Mars. Unlike most terrestrial dust, lunar dust is extremely fine and has a nanometer density that makes up over 50% of particles (Buzea et al., 2007). The magnetic NPs found in lunar dust particles are capable of adhering to electrostatic energized external surface (Buzea et al., 2007). The astronaut's space is so large that it is almost not possible to get rid of.

Marine salt droplets are a significant emission from oceans and ocean waters due to water evaporation (Buseck & Pósfai, 1999). Water drops that are formed by waves and released into the air result in the formation of aerosols. Their dimension ranges from 100 nm to a number of microns. Precipitation and evaporation of water can lead to the formation of NPs in these substances. $CaCO_3$ is a strong component of water that stays soluble in chilly water, but its temperature becomes greater. The precipitation of $CaCO_3$ from the water can create nanodimension of particles that form brightly swirls (Buzea et al., 2007).

The formation of NPs during industrial activities, natural decay events, and mineral mining and components can occur spontaneously. NPs that are present in the environment are typically composed of oxides of metals, for example, sulfides. In a distinct state of matter, colloids exist amidst aqueous solution and suspension. Chemical degradation and biological decomposition are potential causes of colloids (Hartland et al., 2013). The particles in colloids are typically tiny, ranging from a small number of nm to 1 μm; suspensions are diverse mixtures containing larger than 1 μm particles and are susceptible to sedimentation. NPs and suspended particles exist in diverse forms. The colloids' tiny particle dimension and vast exterior area and charges make them highly reactive against pollutant component, making them distinct from materials with greater particle dimension and tiny molecules (Wigginton et al., 2007). Nonliving particle suspensions like iron and Mn are present in both oxygenized and deepwater environment (Sigg, 1994).

3.2.1.1 Desert Sources of Nanoparticles

Silicon is the primary component of desert sand, along with some aluminum and calcium minerals and elements. Desert dust storms may contain N_2 and C as well as Al or added ions. Hamattan dust is subjected to crushing to the micrometer scale due to high temperatures and turbulence in the deserts. Desert dust turbulence particles can be transported through air as NPs. About 50% of the minerals in the troposphere's particles in air are derived from desert sources (Muneeb, 2021).

3.2.1.2 Humic-Assisted Synthesis (HS) of Nanoparticles

Hyperbranched self-gathering of electrolyte of high molecular weight with unstructured molecule forms and essential biosphere function are known as humus material (Xu et al., 2013). Humic lacks a clear form, structure, or standard. Amorphous is black and has a gluey texture and spongy characteristics. Microscope analysis reveals that it contains mechanical and biochemically decay living organisms and remains of microbes (Bernier & Ponge, 1994). Humus act as a suspended particle in the environment, and when charged static stabilized, they cluster to form larger molecular forms in the range of nanoscale. Humic-assisted synthesis interplay with metallic oxides on a microscale is utilized in advanced equipment. HS is employed as a stabilizer in medicine to avert the cluster of magnetic nanoparticles of Fe_2O_3 in mixture (Illes & Tombacz, 2006). Dubas and Pimpan (2008) reported the utilization of humus fabric to synthesize manipulated gold, silver, zinc sulfide, and other NPs. The silver NPs produced from silver ion sources in water samples are not purely due to human actions as nature as also played a role in them (Mukherjee et al., 2001).

3.2.1.3 Anthropogenic Nanomaterials

Microparticles have been formed by humans for thousands of years, resulting from simple burning processes like cooking of food, production of chemical substances, binding of materials, and purifying of ore and burning of fuel in cars and aeroplane engine (Rogers et al., 2005). Linak et al. (2000) discussed the use of fuel and charcoal to generate power. The public market has been offering modified NPs that are widely used in various fields such as beauty products, tires, and sportwear. A small fraction of environmental microparticles including these NPs and newly created or artificial nanoparticles like micro-carbon tubes are present. The major part of nano- and microparticles in cities are generated by petrol and exhaust valve, as reported by EPA (2002). The size of particles released from automobile exhaust valve can be between 20 nm and 60 nm, with gasoil having a much larger size. They are usually roughly round (Sioutas et al., 2005). Recent toxicology research has revealed the presence of carbon microtubes and fibrous materials in engine valve as a waste of burning diesel and in the vicinity of gas burning origin.

3.2.1.4 Volcano Eruption and Ash

The erupting volcano causes different particle sizes from ashes and gas to rise far in the aerosphere and may get as far higher than 18,000 meters. A significant number of tiny particles are discharged into the aerosphere, leading to adverse effects on ecological community. Volcano explosion and erosion are sources of NPs. The ashes are produced at a temperature that exceeds 1400 °C during the explosion. Various NPs (100–200 nm) are discharged into the aerosphere as ash from the air current is transported, ultimately being dispersed in rocks, on plants, in soil, and in sources of water (Griffin et al., 2018). The lava and residue are hazardous to ecological community and man's health. The upper respiratory systems like alveolar, trachea, and bronchi areas are affected by air droplets, which could cause significant sickness of the respiratory tract. The number of natural NPs in heavy materials was found to be 10–500% above than the total contents (Dubas & Pimpan, 2008). It was shown that these NPs of residual volcano are essential conveyors on the worldwide scale for poisonous metals and metalloids. The residual materials from the Pichincha volcano's explosion contains elements such as magnesium, sodium, aluminum, phosphorus, potassium, iron, and calcium (Akaighe et al., 2011).

3.2.1.5 Biological Nanoparticles

In comparison to NPs produced chemically, those made from microorganisms and plants are both inexpensive and environmentally friendly. In addition, inanimate NPs within and outside of the cell are produced by a single-cell organisms and cultural tolerable organism metabolism. In external cellular process, metallic ions are decreased for nanoparticle by microbe catalytic activity and polypeptides or by living particles present in medium for culture. Processes within the cell employ the transport of metallic ion via the cell by attractive static electricity using carboxylic group on the walls of the cell. They also undergo a decrease of polypeptides found within cell to form nanoparticles (Siddiqi et al., 2018).

3.2.1.6 Nanoparticle Synthesis by Plant Extracts

The absence of coating agent and poison make plant-based NPs produce a more efficient option. Au and Ag NPs can be produced from geranium distillate (Shankar et al., 2004). A leaf distillate of *Azadirachta indica* and *Cinnamomum camphora* has been described by Patil et al. (2012). The production of bimetallic Au and Ag with plant leaf distillate is often inexpensive. The gum of the plant, *Plumeria rubra* (Patil et al., 2012), has been identified as an ideal source for creating Ag NPs. *Mirabilis jalapa* distillate was used to explain the biological production of Au NPs (Vankar & Bajpai, 2010).

3.2.1.7 Nanoparticle Synthesis by Bacteria

Metallic NPs can be produced by species like *Bacillus*; scientists have demonstrated their ability to reduce Ag and create uniform extracellular nanostructures that are 10–20 nm in diameter (Sunkar & Nachiyar, 2012). Bacteria that produce silver from the mine are known to contain Ag NPs buildup in the periplasmic space of *Pseudomonas stutzeri* AG259. Also, the production of Au NPs involves bacteria. Sharma et al. (2012) reported that *Marinobacter pelagius*, a new type of microorganism, was found to be suitable for producing a stable uniform particle sizes of Au. Prasad et al. (2007) reported species of *Lactobacillus* utilized to create titanium NPs. Comprehension essential procedures will supposedly aid in the identification of an undiscovered method for synthesizing metallic NPs.

3.2.1.8 Nanoparticle Synthesis by Fungi/Yeast

The reception of toxic, increased biological concentration, relatively inexpensive production, and easy procedure for downstream and handling of waste are the factors that determine the biological production of NPs by fungi. Ag NPs are synthesized from *Aspergillus niger* and *Fusarium solani* outside of an organism cell (Binupriya et al., 2010). The spheroidal NPs are also produced by *Trichoderma viride* (Thakkar et al., 2010). Silver hydrosols are produced by a precursor of Ag ions by *Fusarium oxysporum* (Ahmad et al., 2003). Ag NP produced in *Penicillium* genus is an alternative that has been found to involve system outside the cell organism. Yeast species obtained from extreme environment of acid mine was examined by Mourato et al. (2011) for the biological synthesis of Ag and Au NPs. The synthesis of PbS NPs internally from the sea yeast *Rhodosporidium dichloride* has been studied (Seshadri et al., 2011).

3.2.2 *Synthetic Nanoparticles*

Two main methods for creating nanoparticles are top-down approaches and bottom-up processes.

3.2.2.1 Top-Down Approaches

The process where mass materials are disintegrated materials into a range of smaller forms or tiny form is termed the top-down process, and this includes laser ablation, sputtering, mechanical milling, lithography, etc.

Mechanical Milling

The production of large materials at the nano size can be achieved through inexpensive granulation using machinery. The production of nanoparticles achievable through grinding mechanically is a productive process for diverse blending aspect. Resilient coatings such as oxides and carbide-reinforced Al alloys and spray or micro-alloys made from Al-, Mg-, and Cu-based substances are all produced through machinery grinding. A new type of micro-materials and carbon nanotubes that are ball-milled can be used to address cleanup of the environment, power storage, and power change requirements (Lyu et al., 2017).

Laser Ablation

The production of NPs during laser ablation formation is by using a strong laser ray that hits the material regretted. When a particle is spotlight illuminated during ablation, it can be used to create worthy metal NPs without the use of balanced agents or additional chemicals, making it a biological procedure. The procedure can create a diverse range of microparticles including carbon and metallic NPs and oxide compound. By using pulsed ray ablation in water, it is possible to create particles of uniform size in a dispersed phase without the use of surface-active agent or ligands. Modifying the emittance, wavelength, and incorporating ray salt can change NPs' characteristics including its distribution and dimension. The dimension of lead NPs can be significantly altered by the pulsed beam wavelength and emission (Park et al., 2017).

Electrospinning

Electrical spinning is a straightforward technique of creating nanocomposites from the top-down process. Microfibers are typically made from a range of substances, with polymer being the primary component (Ostermann et al., 2011).

The advancement of coaxial electrospinning was a significant milestone. Two coaxial tubes of small internal diameter are incorporated into the spinneret in coaxial electrical spin. Core-sheath nanostructure can be created by using narrow tube with an adhesive liquid as the sheath and refracting liquid as the element in the electrical field. To accomplish slim fibers on a massive scale, an easy and efficient top-down method for producing core sheath is used. Diverse nanometers are the maximum diameter of these thin nanomaterials. Coaxial electrical spin, core-sheath, cupped polymer, nonorganic, organic, and blended materials have been created using this technique (Kumar et al., 2014).

Sputtering

Robust particles like gas are bombarded onto a solid exterior to create nanoparticles through sputtering. This technique is believed to be an efficient for creating scales of nanoparticles. Fragments are physically expelled from the aimed exterior during this sputtering displacement procedure, which is influenced by energy from gas ionization (Wender et al., 2013). Sputtering from DC diode and radio frequency diode, magnetron utilization methods are used in achieving sputtering (Wender et al., 2013). The displaced cubicle is where sputtering takes place when a gas sputter is injected. Free electrons bombard with the gas at high-powered energy to form gaseous ions. Atoms are constantly ejected from the cathode as a result of speed caused by cation in the electrical field toward the aimed exterior (Garcıa-Munoz et al., 2009). WSe2-stratified nanomaterials are created by means of electron tube sputtering on silicon dioxide and substrate of paper carbon. This method is noteworthy for its ability to maintain a similar nanoparticle content with lower contaminant than electron ray lithography (Nie et al., 2009).

Lithography

Nano-construction can be created through lithography using a target ray of light or electrons. There are two main types of lithography: mask and unmask lithography (Pimpin & Srituravanich, 2012). A specific template is utilized in masked micro-lithograph to transfer micro-model over a huge exterior. Examples of mask lithography are photo, micro-imprint, and soft lithography (Yin et al., 2000). The use of scanning examination lithography, EBL (electron beam lithography), and FIB (focused ion beam) are all techniques used in unmasked lithography. Unmasked techniques permit the use of nano-model writing in the absence of masking. Ion implantation, combined with the use of a targeted ion ray and wet chemically etch, allows for the creation of three-dimensional free morpheme micro-nano-production (Garg et al., 2020).

3.2.2.2 The Bottom-Up Approach

A bottom-up technique to creating great nanoparticles such as atom molecules and its subunit employs the use of tiny constituents. In this approach, atom particles and molecules constitute structure that unifies themselves to meet specific requirements. Various techniques such as sol-gel process, solvothermal and hydrothermal methods, CVD (chemical vapor deposition), and reverse micelle are employed (Baig et al., 2021).

Chemical Vapor Deposition (CVD)
Carbon-based nanoparticles are heavily dependent on chemical vapor deposition technique. A slim film in CDV on substrate exterior is produced by precursor at vaporization stage in a chemical reaction (Jones & Hitchman, 2008). CVD can be obtained from a precursor that exhibits high chemical clarity, steadiness when evaporating, inexpensiveness, and long life span as well as moderate volatilization. Its decay should not result in any contaminant remains (Shah & Tali, 2016). A mixture is made hot to an extreme temperature in oven when producing graphite nanofibers using CVD. A precursor that contains carbon, such as hydrocarbons, is slowly introduced into the system. When the gas is decomposed at temperature that is high, carbon particles are released and fused to form graphene microfibers on substrate (Shah & Tali, 2016). The morphology and type of nanoparticle obtained are heavily influenced by the stimulant chosen. The preparation of carbon using CVD involves the use of nickel and cobalt stimulant, while single-coated coating is produced by a copper stimulant (Ago, 2015). CVD is a highly effective method for creating a superior nanoparticle, and it is widely known for the formation of multidimensional nanoparticles (Wu et al., 2009).

Reverse Micelle Methods
Nanoparticle can be created in the preferred structure and dimension using the reverse micelle technique. An oil and water mixture produce ordinary micelles, which contain water-resistant tails that are directed at an internal core with cutoff drip of oil. In the case of immiscible liquid, reverse micelles are generated by aligning strong water affinity heads with moisten nucleus (Malik et al., 2012). The reverse micelle nucleus is used as a nanoscale reactor producing nanomaterials. Nanoparticles can be created using pool of H_2O. By adjusting surface-active agent and water proportion to different dimensions, the nanoscale reactor can be made smaller, eventually leading to the creation of nanofibers. The formation of small-scale nanomaterial is caused by reduced drops of water (Shiri et al., 2019). The reverse micelle technique supplies a convenient means of producing uniform nanoparticles with precisely controlled particle size. Nanoparticles created using the reverse micelle technique exhibit exceptional fineness and are particles of uniform sizes (Nguyen, 2013).

Solvothermal and Hydrothermal Methods
The widely known and used technique for the production of nanoparticles is known as hydrothermal technique (Cao et al., 2016). Nanoparticles are obtained via diverse response in an essentially liquid medium at temperature and pressure that are extreme. Solvent thermal technique is used to describe its process. It is performed in a fluid state, which is distinct. Closed systems are typically subjected to heated water and heated solvent techniques (Chen & Holt-Hindle, 2010). Hydrothermal process that involves electromagnetic wave has gained significant concentration in the field of nanoscale construction, as it merges as the advantages of both methods. Various nanoforms of materials like rods, sheets, and wires along with nanostructure are exciting new uses for water and solvent thermal techniques (Dong et al., 2020).

The Sol-Gel Method
A moist chemical process known as the sol-gel techniques is employed to create a range of superior metallic oxide nanoparticles. The process of synthesizing metallic oxide nanomaterials into the resulting fluid precursor is known as the sol-gel method (Danks et al., 2016). Metal alkyl oxides are commonly used as precursor in the sol-gel process for creating nanoparticles, by hydrolyzing the metallic oxide in liquid or with alcohol resulting to sol; nanomaterials are produced and then subjected to condensing to rise the adhesiveness of the solvent and create permeable forms that are kept to old. During condensing process, hydroxo- (M–OH–M) or oxo- (M–O–M) viaduct are formed in the solution of metallic hydroxo polymerization (Tseng et al., 2010). As maturing progresses, condensation polymerization persists and alters its form, characteristics, and permeability. The colloid suspension spacing rises as the permeability reduces. After maturity procedure, the gel is dried and treated to get rid of liquid and solvents. The process of heating is used to create nanoparticles (Parashar et al., 2020). The sol-gel process is utilized to exhibit the development of feature film and fine particles (Znaidi, 2010). Besides being cost-effective, the sol-gel technique has several benefits, including homogeneity of the material formed, low temperatures, and the easy manufacturing of compounds and nanocomposites (Parashar et al., 2020).

3.3 Classification of Nanomaterials

The fundamental components of nano-innovation are nanoparticles. Tiny materials have a nanoscale size at most lower than 100 nm (Kolahalam et al., 2019). Four classifications are used to categorize nanocomposites, including the following:

(i) Zero-dimensional (0D) nanoparticles: The three entire sizes of nanostructure limits are present here. Nanoscale, quantum dots, and fullerenes are all examples (Kolahalam et al., 2019).
(ii) One-dimensional (1D) nanoparticles: This category of micro-materials containing a single side at the outer particle like encircled microtube, micro-tubulin,

micro-horn, and quantum wires. Unidimensional or linear approach slime layer or produced exteriors have been utilized in hardware, chemistry, and bioengineering for long period of time. Nowadays, it is common to create single layers or slim layer (ranging from 1 to 100 nm) for catalytic process. Information storage systems, biogenic and chemical detector, optical fiber method, magneto-optical disks, and optical instrument are among the innovations of this slim film (Kolahalam et al., 2019).

(iii) Two-dimensional (2D) nanoparticles: Nanoparticles exhibit two-dimensional nanostructure like carbon microtubes, slim layer, and film coating. A cylinder-shaped many-sided mesh of carbon particles, measuring 1 nm width and 100 nm length, is made up of graphene coating. Carbon nanotubes are classified into two; single-walled carbon nanotubes (SWCNTs) and multiwalled carbon nanotubes (MWCNTs). Micro-carbon tubes are exceptional matter due to their small size and outstanding characteristics like mechanical, electrical, and physiological properties (Kohler et al., 2004). Their characteristic may be semiconductor or metal. It is electrical conductor, with very elevated electric current quantity that can reach one billion amperes. Micro-carbon tube is higher than finest steel 60 times in tensile robustness; it also possesses a remarkable ability to absorb molecular components and offer symmetry. Furthermore, they exhibit high chemical stability.

(iv) Three-dimensional (3D) nanoparticles: It is a massive particle that is not constrain to nanostructure in size. It involves large particle, discharged of microparticles, and arrangements of microtubes and wires.

Other types include the following:

1. Fullerenes

   Fullerenes are circularly enclosed with carbon particles between 28 and 100. It is a hollowed-out sphere made of 60 particles of carbon, which looks like a football. The class of fullerenes possesses physically distinct characteristics. They can be put through intense force and then recover their form upon release. The particles are not repelled by each other, making them ideal as lubricator. Electrically fascinating characteristics can be utilized for electrical applications such as packed information and solar panel manufacturing. The utilization of fullerenes in microelectronics is a promising area due to their empty shape and closely related size to that of biogenic particle. Fullerenes can be filled with various particles and could be used for medical purposes (Tomalia, 2004).

2. Dendrimers

   Dendrimers are a new category of nano-sized restrained-shaped polymer. Drug transporter and image use dendrimers are with diameter between 10 and 100 nm which are carriers of numerous hydroxyl group on the exterior. The dendrimers used in modern dentary can be highly specialized and contain useful particles. The basic of these compounds contains curative or medicine (Li et al., 2007). The components that are treated essential for the creation of biotic and abiotic micro-form with size from 1 to 100 nm are treated on an extensive extent (Tomalia, 2004). They can be adapted to biotic form like DNA and made in metallic micro-form or with an enveloped competence (Fu et al., 2007). Dendrimers possess diverse micro-form and are well suited to biotic form, such as DNA making them a popular choice in medical and biomedical settings. The cation of dendrimers can cause damage to plasma membrane and become poisonous (Mecke et al., 2004).

3. Quantum Dots (QDs)

   These are gadgets that have small number of conductive electrons. Quantum dots are dispersed particles and semiconductive nanoparticles that have radius between 2 nm and 10 nm. Electrical chemistry and colloid production are both methods used to produce QDs from material of semi-conduction. The frequently used quantum dots are indium arsenide, cadmium selenide, indium phosphide, and cadmium telluride. QDs range from just one electron to diverse numbers. Electronic structure, dimension, and number can be managed with precision. Semiconductive materials, metallic oxide, and magnetizable substance have been created as the result. QDs are the quantum custody of electrical charged element and hole conveyors at size lesser than radius of Bohr. QD nanoparticles are typically made up of particles at their center. To enhance the stable light and obtained quantum emission, it is possible to incorporate a layer of zinc sulfide and cadmium sulfide that stops the exterior put off of electron excitement in the emission center (Goldberg et al., 2007). QDs possess sufficient exterior area to bind curative agents for concurrent drug conveyor and images within cell as well as for constructing tissues (Larson et al., 2003).

Nanoparticles are typically classified into groups, namely, carbon-based, organic, and nonorganic (Ealia & Saravanakumar, 2017).

1. *Organic NPs*

   Lipids, polypeptides, sugars, and complex material as well as biotic component (Pan & Zhong, 2016) are included in this class. It is primarily composed of dendron branches, liposomes, micelles, and polypeptide compound like ferritin. Nanoparticles are usually not poisonous, can be broken down, and may have the ability to produce a cylindrical center of liposomes. Heat and electrical magnetic ray, including thermal and photo, are favorable elements for biological NPs (Ealia & Saravanakumar, 2017). Moreover, they are frequently generated by nonbonding existing between molecules, which brings about less susceptibility in nature and provides an exit route from the body (Ng & Zheng, 2015). The scope of this NPs can be explored in diverse ways, depending on various parameters. Various factors include exterior shape, composition, stable state, carrying capability, and more. Targeted drug conveyance in the field of medicine is primarily facilitated by NPs (Ealia & Saravanakumar, 2017). Cancer treatment has been suggested by Gujarat et al. (2014).

2. *Carbon-Based NPs*

   The NP category is composed of particles of carbon alone (Ealia & Saravanakumar, 2017). QD, fullerenes, and carbon black are known examples. Particles of carbon have an enclosed shape known as fullerenes. C_{60} are composed of 60 particles of carbon that shape them like a football (Long et al., 2013),

Fullerenes of various types, including C_{70} and C_{540} fullerenes, was narrated. Highly fused round particles that are similar to grapes make up nanoparticles of carbon black (Yuan et al., 2019). QDs are made up of distinct, round nanoparticles that are smaller than 10 nm in diameter (Lu et al., 2016). Nanoparticles combine the unique characteristics of sp^2 blended covalent bonds with the non-nanoparticle with physical and chemical traits and their outstanding electric conduction, tensile durability, and attraction of electron as well as their optic, heat, and absorption trait properties (Mauter & Elimelech, 2008; Khan et al., 2019). NPs have diverse uses including drug conveyor (Oh et al., 2010), energy storage (Liu et al., 2018), and biological imaging (Chandra et al., 2011); the use of light voltage and surrounding sensing utilization was noted to be used to observe microbiological habitat (Mauter & Elimelech, 2008). NPs are more intricate and less poisonous and biologically compatible of this composite, which makes them ideal for use in drug conveyor and construction of tissue (Ahlawat et al., 2021).

3. *Inorganic NPs*

   Non-carbon and organic compounds are included in this category. Examples are usually metallic, ceramic, and semiconducting material. The stimulants of metallic oxide in metallic NPs can be single-metals, two-metals (Toshima & Yonezawa, 1998), or numerous metals (Nascimento et al., 2018). Based on the precursor characteristics of the metallic oxide, the formation of double metal nanoparticles can involve the use of alloys for its production (Toshima & Yonezawa, 1998). These NPs have special visible and electric characteristic (Khan et al., 2019) due to their confine exterior plasmon resonance; moreover, certain nanoparticulate metals exhibit exceptional magnetic, heat, and biogenic traits (Ealia & Saravanakumar, 2017). NPs are now highly essential for the creation of nanomachine that can be used in a variety of usage (Mody et al., 2010). In recent times, the control of metallic nanoparticle dimension and structure is crucial for the creation of advanced materials (Dreaden et al., 2012). NPs made from semi-conducive substance exhibit traits that lie between metallic and non-metallic oxides. NPs are of great importance in light-activated catalyst and optical and electrical gadgets (Gupta & Tripathi, 2012). The abiotic materials that make up NPs of ceramic are CO_3^{2-}, PO_4^{3-}, carbides, and metallic oxides and metalloids, such as Ca and Ti (Thomas et al., 2015). NPs are commonly produced through thermal and cool consecutively and exhibit structureless polycrystals and thick, permeable, or cupped structure (Khan et al., 2019). Their excessive stable state and great ability for load make them a popular choice in medicine (Moreno-Vega et al., 2012).

Based on the source materials, nanoparticle types are specifically grouped as follows:

(i) *Metal NPs*

   All metallic oxides are utilized in the composition of nanoparticles. Confine exteriors' plasmon resonance (LSPR) is responsible for the unique electrical traits of the NPs. A broad band for sorption is present in the light part of the electrical solar magnetic scope, which includes nanoparticles like copper and silver. The enhanced traits of metallic NPs, such as facet, dimension, and structure, have made them useful in various areas of science (Khan et al., 2019).

(ii) *Ceramic NPs*

Nanoparticles of ceramic are tiny, non-metal, thermal-treated, and cool to obtain distinct characteristics. The structure is varied, ranging from structureless to polycrystal, thick, permeable, and cupped. They possess both heat aversion and durability. Utilization of nanoparticles of ceramic spans from coverings to cells (Sigmund et al., 2006).

(iii) *Semiconductor NPs*

Nanoparticles made from semiconductive material share close with metals and nonmetals. Chemical and chemical characteristics of semiconductive nanoparticles make them valuable for various utilizations. The ability of nano-semiconductive particles to take up and discharge light can be utilized to manufacture more effective photovoltaic cells. The creation of small and speedy electrical gadget, like transistors, is possible, which can also be utilized in medicine and cancer treatment (Biju et al., 2008).

(iv) *Polymeric NPs (PNPs)*

The exterior sorption of energetic materials on the center of polymer or body can occur in nanopolymer particles with dimension ranging from 1 to 1000 nm. Organic compound is found frequently as nanoparticles and they are referred to as PNP. The majority of their features are reminiscent of micro-spherical and micro-capsulated particles (Khan et al., 2019).

(v) *Lipid-Based NPs*

Nanoparticles are useful in many biogenic areas since they contain glycolipids. Spherical lipid microparticles possess width which ranges from 10 nm to 1000 nm. The nanoparticles of the lipids that are NPs are compound and have a strong lipid center, while their form comprises of dissolvable lipid affinity substances (Khan et al., 2019).

3.4 Conclusion

Nanomaterials offer an incredibly attractive environment for a variety of biogenic uses. Nanoparticles are typical substances that fall within the limit of 1–100 nm in size. The scope of nanotechnology in various scientific fields has broadened in diverse ways to note materials from micro- to tiny-scale dimension. There are numerous organic and biogenic origins that are both ecofriendly and economically viable. Two main processes are used to produce nanomaterials, namely, top-down and bottom-up methods. The most prevalent origins of carbon are carbon-based nano-substances including nanotubes and fullerenes which are acceptable alternatives for industrial manufacturing and other fields. Nanoparticles' small size and wide exterior area make them a potential ecofriendly choice for use.

References

Ago, H. (2015). CVD growth of high-quality single-layer graphene. In K. Matsumoto (Ed.), *Frontiers of graphene and carbon nanotubes* (pp. 3–20). Springer. https://doi.org/10.1007/978-4-431-55372-4_1

Ahlawat, J., Asil, S. M., Barroso, G. G., Nurunnabi, M., & Narayan, M. (2021). Application of carbon nano onions in the biomedical field: Recent advances and challenges. *Biomaterials Science, 10*, 1039.

Ahmad, A., Mukherjee, P., Senapati, S., Mandal, D., Khan, M. I., Rajiv, K., & Murali, S. (2003). Extracellular biosynthesis of silver nanoparticles using the fungus *Fusarium oxysporum. Colloids Surfaces Biosynthesis Biointerfaces, 28*, 313–318.

Akaighe, N., MacCuspie, R. I., Navarro, D. A., Aga, D. S., Banerjee, S., Sohn, M., & Sharma, V. K. (2011). Humic acid-induced silver nanoparticle formation under environmentally relevant conditions. *Environmental of Science and Technology, 45*(9), 3895–3901.

Baig, N., Kammakakam, I., & Falath, W. (2021). Nanomaterials: A review of synthesis methods, properties, recent progress, and challenges. *Materials Advances, 2*, 1821–1871.

Bernier, N., & Ponge, J. F. (1994). Humus form dynamics during the sylvogenetic cycle in a mountain spruce forest. *Soil Biology and Biochemistry, 26*(2), 183–220.

Biju, V., Itoh, T., Anas, A., Sujith, A., & Ishikawa, M. (2008). Semiconductor quantum dots and metal nanoparticles: Syntheses, optical properties, and biological applications. *Analytical and Bioanalytical Chemistry, 391*, 2469–2495.

Binupriya, A. R., Sathishkumar, M., & Yun, S. I. (2010). Myco-crystallization of silver ions to nanosized particles by live and dead cell filtrates of *Aspergillus oryzae var. viridis* and its bactericidal activity toward Staphylococcus aureus KCCM 12256. *Industrial Engineering Chemistry Research, 49*, 852–858.

Buseck, P. R., & Pósfai, M. (1999). Process Notational Academic Science U.S.A., 96, 3372. In ur Muneeb, R. (2021). *Nanomaterials: Synthesis, characterization, hazards and safety*. Chapter 2 (pp. 15–29). Elsevier Inc. https://doi.org/10.1016/B978-0-12-823823-3.00007-0

Buzea, C., Pacheco, I. I., & Robbie, K. (2007). Nanomaterials and nanoparticles: Sources and toxicity. *Biointerphases, 2*, MR17–MR71. https://doi.org/10.1116/1.2815690

Cao, S., Zhao, C., Han, T., & Peng, L. (2016). Hydrothermal synthesis and gas sensing properties of the WO3 nanofibers. *Materials Letter, 169*, 17–20.

Chandra, S., Das, P., Bag, S., Laha, D., & Pramanik, P. (2011). Synthesis, functionalization and bioimaging applications of highly fluorescent carbon nanoparticles. *Nanoscale, 3*(4), 1533–1540.

Chen, A., & Holt-Hindle, P. (2010). Platinum-based nanostructured materials: Synthesis, properties, and application. *Chemical Reviews, 110*, 3767–3804.

Couvreur, P., Dubernet, C., & Puisieux, F. (1995). Controlled drug delivery with nano particles: Current possibilities and future trends. *European Journal of Pharmaceutical Biopharmaceutics, 41*, 2–13.

Danks, A. E., Hall, S. R., & Schnepp, Z. (2016). The evolution of sol-gel chemistry as a technique for materials synthesis. *Materials Horizons, 3*(2), 91–112.

Dong, Y., Du, X., Liang, P., & Man, X. (2020). One-pot solvothermal method to fabricate 1D-VS4 nanowires as anode materials for lithium ion batteries. *Inorganic Chemistry Communication, 115*, 107883.

Dreaden, E. C., Alkilany, A. M., Huang, X., Murphy, C. J., & El-Sayed, M. A. (2012). The golden age: Gold nanoparticles for biomedicine. *Chemical Society Reviews, 41*(7), 2740–2779.

Dubas, S. T., & Pimpan, V. (2008). Humic acid assisted synthesis of silver nanoparticles and its application to herbicide detection. *Materials Letters, 62*, 2661–2663.

Ealia, S. A. M., & Saravanakumar, M. P. (2017). A review on the classification, characterization, synthesis of nanoparticles and their application. In *IOP conference series: Materials science and engineering* (p. 32019). IOP Publishing.

EPA. (2002). *Process nanotechnology and the environment: Applications and implications*. Published Feb 2003, p. 83. http://www.nsf.gov/home/crssprgm/nano/GC ENV EPA2002

Proc 03-0204.pdf URL: https://www.epa.gov/pesticide-science-and-assessing-pesticide-risks/epa-2002-guidelines-ensuring-and-maximizing-quality

Fu, Z. D., Cui, Y. S., Zhang, J., Chen, D. P., Yu, D. P., Zhang, S. L., Niu, L., & Jiang, J. Z. (2007). Study on the quantum confinement effect on ultraviolet photoluminescence of crystalline ZnO nanoparticles with nearly uniform size. *Applied Physic Letter, 90*(26), 263113.

Garcıa-Munoz, J., Vazquez, L., Cuerno, R., Garcia-Sa'nchez, A., Castro, M., & Gago, R. (2009). Self-organized surface nanopatterning by ion beam sputtering. In *Toward functional nanomaterials* (pp. 323–398). Springer US.

Garg, V., Mote, R. G., & Fu, J. (2020). Facile fabrication of functional 3D micro-nano architectures with focused ion beam implantation and selective chemical etching. *Applied Surface Science, 526*, 146644.

Goldberg, M., Langer, R., & Jia, X. (2007). Nanostructured materials for applications in drug delivery and tissue engineering. *Journal of Biomaterials Science Polymer, 218*, 241–268.

Griffin, S., Masood, M. I., Nasim, M. J., Sarfraz, M., Ebokaiwe, A. P., Schafer, K., Keck, C. M., & Jacob, C. (2018). Natural nanoparticles: A particular matter inspired by nature. *Antioxidants, 7*(1), 29. https://doi.org/10.3390/antiox7010003

Gujrati, M., Malamas, A., Shin, T., Jin, E., Sun, Y., & Lu, Z. R. (2014). Multifunctional cationic lipid-based nanoparticles facilitate endosomal escape and reduction-triggered cytosolic siRNA release. *Molecular Pharmaceutics, 11*(8), 2734–2744.

Gupta, S. M., & Tripathi, M. (2012). An overview of commonly used semiconductor nanoparticles in photocatalysis. *High Energy Chemistry, 46*(1), 1–9.

Hahens, W. I., Oomen, A. G., DeJong, W. H., & Cassee, F. R. (2007). What do we need to know about the kinetic properties of nanoparticles in the body? *Regulatory Toxicology and Pharmacology, 49*, 217–229.

Hartland, A., Lead, J. R., Slaveykova, V. I., O'Carroll, D., & Valsami-Jones, E. (2013). The environmental significance of natural nanoparticles. *National Education Knowledge, 4*(8), 7.

Hasan, S. (2015). A review on nanoparticles: Their synthesis and types. *Reserved Journal Recent Science, 2277*, 2502–2504.

Illes, E., & Tombacz, E. (2006). The effect of humic acid adsorption on pH-dependent surface charging and aggregation of magnetite nanoparticles. *Journal of Colloid Interface Science, 295*, 115–123.

Jones, A. C., & Hitchman, M. L. (2008). In A. C. Jones & M. L. Hitchman (Eds.), *Chemical vapor deposition* (pp. 1–36). Royal Society of Chemistry. https://doi.org/10.1039/9781847558794-00001

Khan, I., Saeed, K., & Khan, I. (2019). Nanoparticles: Properties, applications and toxicities. *Arab Journal Chemistry, 12*(7), 908–931.

Kohler, N., Fryxell, G. E., & Zhang, M. (2004). A bifunctional poly (ethylene glycol) silane immobilized on metallic oxide-based nanoparticles for conjugation with cell targeting agents. *Journal of the American Chemical Society, 126*(23), 7206–7211.

Kolahalam, L. A., Viswanath, I. V. K., Diwakar, B. S., Govindh, B., Reddy, V., & Murthy, Y. L. N. (2019). Review on nanomaterials: Synthesis and applications. *Materials Today Proceedings Journal, 18*, 2182–2190.

Kumar, P. S., Sundaramurthy, J., Sundarrajan, S., Babu, V. J., Singh, G., Allakhverdiev, S. I., & Ramakrishna, S. (2014). Hierarchical electrospun nanofibers for energy harvesting, production and environmental remediation. *Energy Environmental Science*, *7*, 3192–3222. In ur Muneeb, R. (2021). *Nanomaterials: Synthesis, characterization, hazards and safety*. Chapter two (pp. 15–29). Elsevier Inc. https://doi.org/10.1016/B978-0-12-823823-3.00007-0.

Larson, D. R., Zipfel, W. R., Williams, R. M., Clark, S. W., Bruchez, M. P., & Wise, F. W. (2003). Water-soluble quantum dots for multiphoton fluorescence imaging in vivo. *Science, 300*, 1434–1436.

Li, Y., Cheng, Y., & Xu, T. (2007). Design, synthesis and potent pharmaceutical applications of glycodendrimers: A mini review. *Current Drug Discovery Technology, 4*, 246–2454.

Linak, W. P., Miller, C. A., & Wendt, J. O. J. (2000). Comparison of particle size distribution and elemental portioning from the combustion of pulverized coal and residual fuel oil. *Journal of the Air Waste Management Association, 50*, 1532.

Liu, M., Zhao, F., Zhu, D., Duan, H., Lv, Y., Li, L., & Hue, H. (2018). Ultra-microporous carbon nanoparticles derived from metal–organic framework nanoparticles for high-performance supercapacitors. *Materials Chemistry and Physic, 211*, 234–241.

Long, C. M., Nascarella, M. A., & Valberg, P. A. (2013). Carbon black vs black carbon and other airborne materials containing elemental carbon: Physical and chemical distinctions. *Environment Pollutants, 181*, 271–286.

Lu, K. Q., Quan, Q., Zhang, N., & Xu, Y. J. (2016). Multifarious roles of carbon quantum dots in heterogeneous photocatalysis. *Journal of Energy Chemical, 25*(6), 927–935.

Lyu, H., Gao, B., He, F., Ding, C., Tang, J., & Crittenden, J. C. (2017). Ball-milled carbon nanometer for energy and environmental applications. *ACS Sustainable Chemistry and Engineering, 5*(11), 9568–9585.

Machado, S., Pacheco, J. G., Nouws, H. P. A., Albergaria, J. T., & Delerue-Matos, C. (2015). Characterization of green zero-valent iron nanoparticles produced with tree leaf extracts. *Science Total Environment, 533*, 76–81.

Malik, M. A., Wani, M. Y., & Hashim, M. A. (2012). Microemulsion method: A novel route to synthesize organic and inorganic nanomaterials: 1st nano Update. *Arabian Journal of Chemistry, 5*(4), 397–417.

Mauter, M. S., & Elimelech, M. (2008). Environmental applications of carbon-based nanomaterials. *Environment of Science Technology, 42*(16), 5843–5859.

Mecke, A., Uppuluri, S., Sassanella, T. M., Lee, D. K. N., Ramamoorthy, A., & Baker, J. R., Jr. (2004). Direct observation of lipid bilayer disruption by poly (amidoamine) dendrimers. *Chemistry Physics of Lipids, 132*, 3–14.

Mody, V. V., Siwale, R., Singh, A., & Mody, H. R. (2010). Introduction to metallic nanoparticles. *Journal of Pharmacy Bioallied Sciences, 2*(4), 282.

Moreno-Vega, A. I., Gomez-Quintero, T., Nunez-Anita, R. E., Acosta-Torres, L. S., & Castaño, V. (2012). Polymeric and ceramic nanoparticles in biomedical applications. *Journal Nanotechnology*. https://doi.org/10.1155/2012/936041

Mourato, A., Gadanho, M., Lino, A. R., & Tenreiro, R. (2011). Biosynthesis of crystalline silver and gold nanoparticles by extremophilic yeasts. *Bioinorganic Chemistry and Applications, 546074.*

Mukherjee, P., Ahmad, A., Mandal, D., Senapati, S., Sainkar, S. R., Khan, M. I., Parishcha, R., Ajaykumar, P. V., Alam, M., Kumar, R., & Sastry, M. (2001). Fungus-mediated synthesis of silver nanoparticles and their immobilization in the mycelial matrix: A novel biological approach to nanoparticle synthesis. *Nano Letters, 1*(10), 515–519.

Mulvaney, P. (2015). Nanoscience vs nanotechnology-defining the field. *ACS Nanotechnology, 14*, 18–23.

Nascimento, M. A., Cruz, J. C., Rodrigues, G. D., de Oliveira, A. F., & Lopes, R. P. (2018). Synthesis of polymetallic nanoparticles from spent lithium-ion batteries and application in the removal of reactive blue 4 dye. *Journal of Cleaner Production, 202*, 264–272.

Ng, K. K., & Zheng, G. (2015). Molecular interactions in organic nanoparticles for photothrombotic applications. *Journal of Chemical Reviews, 115*(19), 11012–11042.

Nguyen, T. D. (2013). From formation mechanism to synthetic methods toward shape-controlled oxide nanoparticles. *Nanoscale, 5*(20), 9455–9482.

Nie, M., Sun, K., & Meng, D. D. (2009). Formation of metal nanoparticles by short-distance sputter deposition in a reactive ion etching chamber. *Journal of Applied Physics, 106*, 054314.

Oh, W. K., Yoon, H., & Jang, J. (2010). Size control of magnetic carbon nanoparticles for drug delivery. *Biomaterials, 31*(6), 1342–1348.

Ostermann, R., Cravillon, J., Weidmann, C., Wiebcke, M., & Smarsly, B. M. (2011). Metal-organic framework nanofibers via electrospinning. *Chemical Communication Journal, 47*, 442–444.

Pan, K., & Zhong, Q. (2016). Organic nanoparticles in foods: Fabrication, characterization, and utilization. *Annual Reverse Food Science Technology, 7*, 245–266.

Parashar, M., Shukla, V. K., & Singh, R. (2020). Metal oxide nanoparticles via sol-gel method: A review on synthesis, characterization and application. *Journal of Materials Science: Materials in Electronics, 31*, 3729–3749.

Park, J., Fong, P. M., Lu, J., Russell, K. S., Booth, C. J., Saltzman, W. M., & Fahmy, T. M. (2017). PEGylated PLGA nanoparticles for the improved delivery of doxorubicin. In *Nanomedicine in cancer* (pp. 575–596). Jenny Stanford Publishing.

Patil, C. D., Patil, S. V., Borase, H. P., Salunke, B. K., & Salunkhe, R. B. (2012). Larvicidal activity of silver nanoparticles synthesized using Plumeria rubra plant latex against Aedes aegypti and Anopheles stephensi. *Parasitology Research, 110*, 18151822.

Pimpin, A., & Srituravanich, W. (2012). Review on micro and nanolithography techniques and their applications. *Engineering Journal, 16*, 37–56.

Prasad, K., Jha, A. K., & Kulkarni, A. R. (2007). Lactobacillus assisted synthesis of titanium nanoparticles. *Nanoscale Research Letters, 2*, 248–250.

Rogers, F., Arnott, P., Zielinska, B., Sagebiel, J., Kelly, K. E., Wagner, D., Lighty, J. S., & Sarofim, A. J. (2005). Real-time measurement of jet aircraft engine exhaust. *Journal of the Air and Waste Management Association, 55*, 583–593.

Seshadri, S., Saranya, K., & Kowshik, M. (2011). Green synthesis of lead sulfide nanoparticles by the lead resistant marine yeast, Rhodosporidium diobovatum. *Biotechnology Progress, 27*, 1464–1469.

Shah, K. A., & Tali, B. A. (2016). Synthesis of carbon nanotubes by catalytic chemical vapour deposition: A review on carbon sources, catalysts and substrates. *Materials Science Semiconductor Processing, 41*, 67–82.

Shankar, S. S., Rai, A., Ahmad, A., & Sastry, M. (2004). Biosynthesis of silver and gold nanoparticles from extracts of different parts of the Geranium plant. *Applications in Nanotechnology, 1*, 69–77.

Sharma, N., Pinnaka, A. K., Raje, M., Fnu, A., Bhattacharyya, M. S., & Choudhury, A. R. (2012). Exploitation of marine bacteria for production of gold nanoparticles. *Microbial Cell Factories, 11*, 86.

Shi, Z., Shao, L., Jones, T. P., & Lu, S. (2005). Microscopic and mineralogy of airborne particles collected during severe dust storm episode in Beijing China. *Journal of Geophysical Research: Atmosphere, 110*, D1.

Shiri, M. S. Z., Henderson, W., & Mucalo, M. R. (2019). A review of the laser-studied microemulsion-based synthesis methodologies used for preparing nanoparticle systems of the noble metals, Os, Re, Ir and Rh. *Materials, 12*(12), 1896.

Siddiqi, K. S., Husen, A., & Rao, R. A. K. (2018). A review on biosynthesis of silver nanoparticles and their biocidal properties. *Journal of Nanobiotechnology, 16*, 14.

Sigg, L. (1994). Regulation of trace elements in lakes. In J. Buffle & R. de Vitre (Eds.), *Chemical and biological regulation of aquatic processes* (pp. 177–197). Lewis Pub.

Sigmund, W., Yuh, J., Park, H., Maneeratana, V., Pyrgiotakis, G., & Daga, A. (2006). Processing and structure relationships in electrospinning of ceramic fiber systems. *Journal of the American Ceramic Society, 89*, 395–407.

Sioutas, C., Delfino, R. J., & Singh, M. (2005). Exposure assessment for atmospheric ultrafine particles (UFPs) and implementation in epidemiology research. *Environment Health Perspective, 113*(8), 947.

Sunkar, S., & Nachiyar, C. V. (2012). Microbial synthesis and characterization of silver nanoparticles using the endophytic bacterium Bacillus cereus: A novel source in the benign synthesis. *Global Journal of Medical Research, 12*, 43–49.

Thakkar, K. N., Mhatre, S. S., & Parikh, R. Y. (2010). Biological synthesis of metallic nanoparticles. *Nanomedicine, 6*, 257–262.

Thomas, S., Kumar, M. P., & Talegaonkar, S. (2015). Ceramic nanoparticles: Fabrication methods and applications in drug delivery. *Current Pharmaceutics Discovery, 21*(42), 6165–6188.

Tomalia, D. A. (2004). Birth of a new macromolecular architecture: Dendrimers as quantized building blocks for nanoscale synthetic organic chemistry. *Aldrichimica Acta, 37*(2), 39–57.

Toshima, N., & Yonezawa, T. (1998). Bimetallic nanoparticles—novel materials for chemical and physical applications. *New Journal of Chemistry, 22*(11), 1179–1201.

Tseng, T. K., Lin, Y. S., Chen, Y. J., & Chu, H. (2010). A review of photocatalysts prepared by sol-gel method for VOCs removal. *International Journal Molecular Science, 11*, 2336–2361.

ur Muneeb, R. (2021). *Nanomaterials: Synthesis, characterization, hazards and safety* (Miro and nano technologies series. Chapter 2), (pp. 15–29). Elsevier Inc. https://doi.org/10.1016/B978-0-12-823823-3.00007-0.

Vankar, P. S., & Bajpai, D. (2010). Preparation of gold nanoparticles from Mirabilis jalapa flowers. *Indian Journal of Biochemistry and Biophysics, 47*, 157–160.

Wender, H., Migowski, P., Feil, A. F., Teixeira, R. S., & Dupont, J. (2013). Sputtering deposition of nanoparticles onto liquid substances: Recent advances and future trends. *Coordination Chemistry Review, 257*, 2468–2483.

Wigginton, N. S., Haus, K. L., & Hochella, M. F. (2007). Aquatic environmental nanoparticles. *Journal of Environmental Monitoring, 9*, 1306–1316.

Wu, Z.-S., Ren, W., Gao, L., Zhao, J., Chen, Z., Liu, B., Tang, D., Yu, B., Jiang, C., & Cheng, H.-M. (2009). Synthesis of graphene sheets with high electrical conductivity and good thermal stability hydrogen arc discharge exfoliation. *ACS Nano, 3*(2), 411–417.

Xu, J., Wu, J. & He, Y. (2013). *Functions of natural organic matter in changing environment*. Jointly Published by Zhejiang University Press and Springer. https://doi.org/10.1007/978-94-007-56342_133

Yin, Y., Gates, B., & Xia, Y. (2000). A soft lithography approach to the fabrication of nanostructures of single crystalline silicon with well-defined dimensions and shapes. *Advanced Materials, 12*, 1426–1430.

Yuan, X., Zhang, X., Sun, L., Wei, Y., & Wei, X. (2019). Cellular toxicity and immunological effects of carbon-based nanomaterials. *Particle and Fibre Toxicology, 16*. https://doi.org/10.1186/s12989-019-0299-z

Znaidi, L. (2010). Sol-gel deposited ZnO thin films: A review. *Materials Science and Engineering: B, 174*(1–3), 18–30.

Chapter 4
Nanoparticle Fate and Transport in the Environment

Alhassan Mohammed Alhassan, Shehu Usman Abdulmalik, Abdulmajid Musa Maku, Japhet Gaius Yakubu, Patrick Omoregie Isibor, and Oluwafemi Adebayo Oyewole

4.1 Introduction

A growing concentration of nanoplastic in the water bodies and land has projected its widespread distribution and potential negative health impacts making it an emergent environmental hazard. This is because they are used in so many facets of modern life, such as packaging, electronics, food, cosmetics, pharmaceuticals, and wastewater treatment (Mishra & Sundaram, 2023). One of the reasons why modern technologies have turned toward the use of nanoparticle is because they differ from their bulk or dissolved counterparts in several ways, including particle size, surface area, surface reactivity, charge, and shape, which may have helped its formation. Nanoparticle integration is gaining popularity across a wide range of sectors (Bundschuh et al., 2018). There will surely be an increase in the variety of commercial products using created engineered nanoparticles, which will lead to the emergence of a new category of trash called nano-wastes that includes nano-sized pollutants (Mazayen et al., 2022). The amount of information that is now accessible on the uses and production rate of nanoparticles is insufficient. Nanoparticle production estimations indicate that there were approximately 2000 tons produced in 2004 and that there would be approximately 58,000 tons produced by 2020. Exposures to the environment and people have increased as a result of the exponential rise in nanoparticle manufacturing and use. As a result, several experts are concerned about their potential to have harmful impacts (Keshav, 2022). According to Bundschuh et al. (2018), coatings, paints and pigments, catalytic additives, and

A. M. Alhassan (✉) · S. U. Abdulmalik · A. M. Maku · J. G. Yakubu · O. A. Oyewole
Department of Microbiology, Federal University of Technology, Minna, Nigeria
e-mail: alhassanmuhammadalhassan@gmail.com

P. O. Isibor
Department of Biological Sciences, Covenant University, Ota, Ogun State, Nigeria
e-mail: patrick.isibor@covenantuniversity.edu.ng

P. O. Isibor et al. (eds.), *Environmental Nanotoxicology*,
https://doi.org/10.1007/978-3-031-54154-4_4

cosmetics are notable items containing nanoparticles. In the area of medicine and pharmaceuticals, nanotechnology is a relatively new and rapidly growing sector. Nanoparticles are useful drug delivery systems because they boost treatment effectiveness and reduce adverse drug responses. Notably, the hitherto concept of lessening the toxicity and adverse effects of medications has been cautioned by the knowledge of unaddressed adverse effects of nanoparticles (Mazayen et al., 2022). This chapter thus focuses on the aspect of toxicodynamics of nanoparticles and the health and environmental consequences. Furthermore, the chapter explores the roles of nanoparticles as toxicant vectors and their interaction with biological systems.

4.1.1 Fate of Environmental Nanoparticles

The aggregation, disaggregation, and chemical aging of nanoparticles in the environment are all related. The fate and environmental impact of nanoparticles, particularly in relation to their movement and interactions in the environment and their potential to cause harm to ecosystems and organisms, are determined by the interaction between these processes and the transit of nanoparticles (Mishra & Sundaram, 2023). Because these aging and transit processes are regulated by particle quality and ambient conditions, a straight translation of data to substantially diverse settings is problematic and may potentially be deceptive (Sun et al., 2022).

According to Bundschuh et al. (2018), important chemical transformation mechanisms of nanoparticles include changes in chemical speciation, dissolution, degradation, and changes in surface characteristics caused by precipitation and adsorption or desorption (Fig. 4.1). In both aquatic and soil environments, these processes have been thoroughly explored. Dibyanshu et al. (2023) mentioned in this regard that a general observation is that functionalization of surfaces of nanoparticles, which enhance the particle qualities for usage in industry, will strongly control environmental transformation processes (Lihui et al., 2023). In order to comprehend nanoparticle fate, Sun et al. (2022) suggested that it is required to fully characterize the surface chemistry of nanoparticles (the formation or loss of coating) throughout time.

4.1.2 Toxicodynamics of Nanoparticles and Environmental Consequences

Humans have been using nanoparticles for thousands of years, and they have existed for millions of years. Much attention has been focused on this sort of material in recent years due to enhanced human capacity to produce synthetic nanoparticles. Applications for nanoparticles are numerous and include the electronic sector, health and beauty products, pharmaceuticals, and environmental initiatives. Investments in nanotechnology applications are growing in popularity globally due to the immense potential of this technology (Sayed et al., 2013; Keshav, 2022).

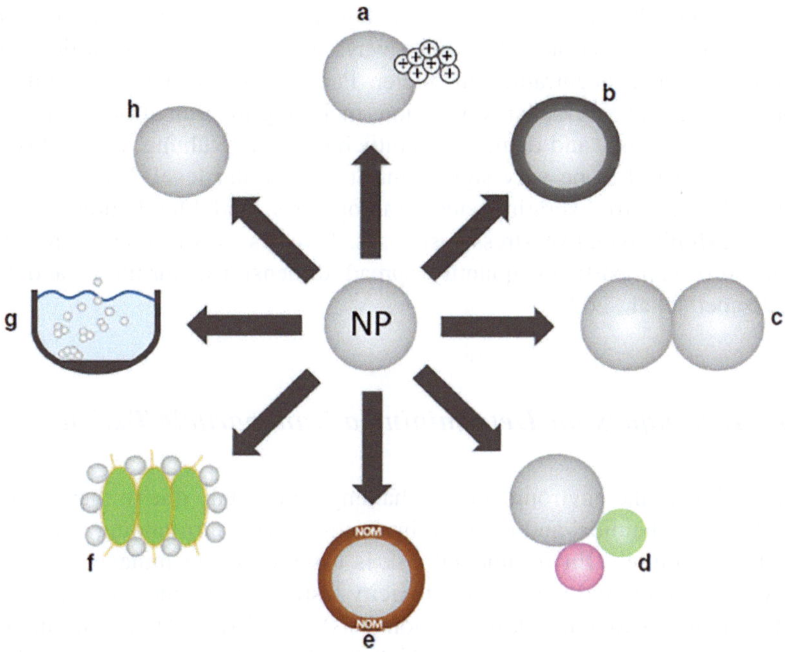

Fig. 4.1 Environmental interplay and nanoparticle fate considering (**a**) dissolution, (**b**) sulfidation, (**c**) homo-aggregation, (**d**) hetero-aggregation, (**e**) coating with NOM, (**f**) NP adsorption on biological surfaces, (**g**) sedimentation/deposition, and (**h**) persistence. (Source: Bundschuh et al., 2018)

All nano-based products have certain toxicological effects, which vary depending on the criteria that induce nanoparticle toxicity. Nanotoxicology is concerned with the negative environmental and health impacts induced by nanoparticles, as well as the effects of nanomaterial exposure. Nanomaterials are found in many significant natural sources, such as pollen grains, dust particles, aerosols, smoke, and ultrafine particles in air pollution. Nanomaterials are also carried in various amounts by a wide range of different natural exhaust systems. Nanoparticle generation is not only restricted to laboratory tests and extremely complex target-oriented investigations (Lihui et al., 2023).

4.1.3 Toxicological Concerns

A key component of understanding a drug's toxicity understands the amount that it can affect both people and the environment. It can be dangerous for humans to be exposed to compounds in the wrong dosage or unintentionally, yet it can be challenging to interpret the findings of studies using materials at the nanoscale. Although various biological applications have been proposed for nanomaterials, little is understood about their toxicity, possible mutagenesis consequences, or total harm to human health. The environment is not always threatened by nanoparticles; exposure, dosage, and concentration all matter. Particle toxicity has been linked to a number of potential mechanisms of action, including inflammation, allergies, oxidative stress response, and damage to epithelial tissue (Ruhung, 2012). Oxidative stress is considered to be very significant at the cellular level. When exposed to nanoparticles in vitro, keratinocytes, macrophages, and blood monocytes were shown to exhibit oxidative stress responses. Five Ds mostly dictate the toxicity mechanism of nanoparticles: quantity, spread, dimensions, sturdiness, and protection (Elizabeth et al., 2023).

4.1.4 Techniques for Determining a Nanoparticle Toxicity

Nanomaterials in the environment are challenging to study due to their minuscule size and erratic chemical and physical interactions with surrounding materials and the biological milieu. Environmental risk is assessed using mutagenicity testing, eco-toxicity testing, and mammalian toxicity testing (acute and chronic tests, oral toxicity, cutaneous toxicity, skin irritation tests) (Jin, 2012). Analyzing the materials' physicochemical characteristics, which enable them to interact with and potentially harm biological systems, is the first step in a logical approach to nanomaterial toxicity (Laura et al., 2023).

Currently, there are numerous spectroscopic and microscopic methods for finding nanomaterials in the environment. The maximum dimension of nanoparticle toxicity is between 65 nm and 200 nm (Keshav, 2022). Finally, any test approach should incorporate the proper use of animal models, which could be the most challenging test to take into account.

4.2 Health and Environmental Problems Associated with Nanoparticles

Despite their wide range of applications, very little is known about nanoparticles' toxicity, possible health risks, and environmental hazards. Indeed, it is common to ignore the toxicity of nanoparticles used in environmental remediation methods and nanomedicine (Chidambaram & Krishnasamy, 2012). The usage and disposal of nanomaterials or commodities derived from nanotechnology carry a number of health and environmental dangers that are responsible for the toxicological concerns related to nanoparticles.

4.2.1 Health Risk

Nanomaterials are absorbed by the human body far more quickly due to their low surface-to-volume ratio, which increases the particles' biological activity and chemical reactivity (Ruhung, 2012). There are two types of harmful effects that nanoparticle interactions with biological systems can produce: (i) the quicker delivery of chemicals that are proven to be toxic and (ii) the hazardous consequences of comparatively safe substances resulting from biological structures being disrupted due to size.

There are several ways that nanoparticles might enter the body, which makes determining the risks connected to any one product much more difficult. Inhalation is the most frequent method of introducing nanoparticles into the body, followed by eating them or applying them to the skin, either on purpose or accidentally (Bakand et al., 2012). Workers may be exposed to nanomaterials or mixtures of nanomaterials during production and use, the general public may be exposed through workplace releases of nanomaterials during production, and the general public may be directly exposed when using commercially available products containing nanomaterials. The main pathways for occupational and ambient nanoparticle exposure include the respiratory tract, gastrointestinal tract, and epidermis (Lihui et al., 2023). It is significant that both patients and employees get exposed to nanomedical items. The human body absorbs nanoparticles more quickly through a variety of methods, including injection, cutaneous, oral, and inhalation (Elizabeth et al., 2023).

4.2.2 Inhalation

One of the main concerns is the abundance of nanoparticles in the atmosphere. Since the respiratory system is the main route of exposure for airborne nanoparticles, it can serve as a tool to emphasize fundamental ideas in nanotoxicology, such as the importance of dosage, rate, metrics, and bio-kinetics (Laura et al., 2023). In the more distant respiratory region of the lung, particles with a diameter of 0.5–5 m can pass through the airways near the small terminal bronchioles, filtering out larger particles. A portion of these particles might potentially settle in the major conducting airways or higher up in the nose. Following their deposition in the respiratory system, nanoparticles may then go to the brain, liver, spleen, lung interstitium, and, in the case of pregnant women, the fetus. One human investigation showed that inhaled nanoparticles could enter the circulation (Bakand et al., 2012).

In the alveoli, gases from breathed air and blood are exchanged (McKinney et al., 2012). A mucus layer keeps the airways from the nasal cavity to the bronchioles moist. The destiny of inhaled nanoparticles is determined by their physical and chemical properties. Depending on whether they are soluble in fat or water, gases and vapors are either directly absorbed into the bloodstream or dissolve in the mucous in the airways, while particles of nanoparticulate matter are either trapped in the mucous layer or the alveoli, depending on their size. Additionally, breathing patterns and depth, as well as the irritating properties of material inhaled, may affect the deposition of nanoparticles (Lihui et al., 2023; Fatima et al., 2023).

4.2.3 Dermal

Many naturally occurring chemicals are blocked by the skin. Certain liquids and dissolved compounds, on the other hand, can penetrate the skin. The most essential aspects are lipid solubility and molecular size, which indicates that high lipid solubility and tiny molecular size improve skin penetration (Fatima et al., 2023). Dermal exposure to nanoparticles has attracted a lot of attention, possibly because of occupational concerns and the addition of nanoparticles to cosmetic and pharmaceutical products, such as titanium dioxide at the nanoscale.

Because of their ability of penetration into the epidermal layer, compared to macro-sized particles, it is more likely for nanomaterials to enter the body through the skin. Consumer items, such as TiO_2 and ZnO nanoparticles in sunscreens, also expose the skin to nanoparticles. According to some research, nanoparticles can enter the skin through skin appendages, such as the dermis, stratum corneum, and epidermis. They can also penetrate the skin by going between or through epithelial cells. Experiments conducted in vitro and in vivo demonstrated that nanoparticles are dermatoxic. The toxicity of titanium dioxide (TiO_2), which is frequently used as a physical sunscreen due to its ability to reflect and scatter UVB (290–320 nm) and UVA (320–400 nm) light rays – which are the most detrimental to skin – has been called into question. Reactive oxygen species produced by titanium dioxide particles exposed to UV light has been demonstrated to inhibit the formation of tumors in human bladder cancer cells that have been grown (Ritu & Huan, 2018).

4.2.4 Injection

Studies on drug delivery and nanomedicine have examined the injection of nanoparticles, that is, nanomedicine. Nanoparticles administered intravenously remained in the body longer than those consumed. After 1 week of exposure, 90% of functionalized fullerenes that are injected, for instance, remain intact. The negative health effects of nanoparticle injection are determined by particle chemistry and charge (Sayed et al., 2013). Hypersensitivity is a frequent adverse reaction to intravenous injection of nanoparticles, which is likely due to complement activation in many recipients. After intravenous injection, it was shown that the nanoparticles had an impact on the lymphatics, liver, spleen, bone marrow, colon, and lungs. When injected intramuscularly, the lymphatic system and neurons are absorbed; when injected intradermally, lymph nodes are absorbed; and when delivered intravenously, nanoparticles swiftly circulate throughout the circulatory system and eventually arrive at the organs (Ritu & Huan, 2018).

4.2.5 Oral

Nanoparticles in the workplace, for example, and oral nanomedicine, can also enter the neurological system through the blood-brain barrier (BBB) via oral absorption. After being consumed orally, nanoparticles enter the lungs, brain, liver, spleen, kidneys, and gastrointestinal (GI) system. Certain nanoparticles can cross the GI barrier and enter the systemic circulation because they can pass through the GI tract and are quickly eliminated in urine and feces. Nonetheless, some of the system's nanoparticles may concentrate in the liver during first-pass metabolism (Ritu & Huan, 2018).

4.3 Environmental Risk Assessment of Nanoparticles

The use and manufacture of nano-based materials are growing and thus expose the environment to traditional exposure pathways for assessing conventional chemicals, such as waste cycle, release from products over the product life cycle, and industrial wastes (liquid, solid, and airborne) (Lihui et al., 2023). Environmental health and safety research has placed a strong emphasis on the sources, fate, transportation, and toxicity of artificial nanomaterials throughout the past 10 years (Fatima et al., 2023). However, respiratory health is the main area of concern; epidemiological studies unequivocally demonstrate the effect of ultrafine particles on respiratory conditions such as inflammation, coughing, wheezing, and so on.

Discharges from consumer products like detergents, lotions, and cosmetics as well as nanoparticles like titanium dioxide, zinc oxide, and silver are a key source and are predicted to rise as nanotechnology develops (Danail et al., 2012). Direct interactions occur between intentionally produced nanoparticles and chemically complex combinations found in a range of environmental media, including soil, water, ambient air, and combustion emissions, as a consequence of nanotechnology-based environmental technologies (Helinor et al., 2012).

4.3.1 Toxicity of Nanoparticles Through Environmental Exposure

The focus of environmental nanotoxicology is on ecological interactions within ecosystems, communities, and population levels. The chemical characteristics of engineered nanomaterials (ENMs), their biochemical pathways, and the physical and ecological processes that control their effects at the ecosystem level and the services they provide are all intricately linked by eco-nanotoxicology (Samreen et al., 2015). Concerns concerning possible hazards from exposure to engineered nanomaterials are raised by uncertainties on the impact on health, ecology, and the environment (Fatima et al., 2023).

The main health problems, however, are respiratory implications. Epidemiological studies have linked ultrafine particles to respiratory conditions like inflammation, coughing, wheezing, and other symptoms (Lihui et al., 2023; Fatima et al., 2023). Consumer goods like cosmetics, creams, and detergents include nanoparticles of silver, zinc oxide, and titanium dioxide, which are major sources of discharges that are expected to rise as nanotechnology advances (Danail et al., 2012). To investigate the interactions and exposure of nanoparticles in the environment, life cycle assessment is required.

4.3.2 Exposure Through the Soil

Terrestrial ecosystems are what give soil its distinctive character and are the main source of biodiversity. The great diversity of soil particles in their different chemical and physical forms makes it challenging to examine the behavior of nanomaterials as toxicants through their interactions with soil.

Different soil textures interact with nanoparticles in different and unanticipated ways, depending on whether they contain soil media, soil organic matter, soil solutions, or pre-existing wastes (Lihui et al., 2023). The chemical and physical properties of the nanomaterial will determine how much is released into the soil. Because of their high surface area to mass ratio, the nanoparticles bind to the silt and soil particles, immobilizing them. Numerous soil parameters, including pH, ionic strength, zeta potential, and soil texture, determine what happens to nanoparticles produced as waste in a given type of environment (Fatima et al., 2023).

4.3.3 Exposure Through the Air

Depending on the circumstances, the amount of nanoparticles in the air in urban and rural locations can vary greatly, with concentrations in the latter area reaching as high as 106–108 np/L (nanoparticles per liter). The fate of airborne nanomaterials is influenced by multiple mechanisms, in addition to their basic geometrical and chemical properties. These mechanisms affect how long the particles stay in the air, how they interact with other molecules or airborne particles, and how far they can fly before being destroyed. Airborne particles are classified into three groups or modes based on their size and behavior: the coarse mode (diameters >2 μm), the accumulation mode (diameters >80 nm and <2 μm), and the coagulation mode (diameters <80 nm). Because they are respirable, they can settle in the alveolar areas of the lung and travel large distances through the atmosphere via Brownian diffusion and then disseminate to other parts of the body (Keshav, 2022).

4.3.4 *Exposure Through the Water*

The key to determining a nanomaterial's aquatic nanotoxicity is to examine how its nanoparticles interact with water. Wastewater spills or overflows from wastewater treatment facilities where nanoparticles are not sufficiently contained or broken down could further expose people to the environment.

Spills resulting from the manufacture, shipping, and disposal of nanomaterials or their byproducts are additional sources of environmental exposure. Though it may also be the one causing the greatest concerns, spontaneous nanoparticle remediation represents one of the most potential environmental nanotechnologies (Keshav, 2022). The Royal Society and the Royal Academy of Engineering advise against using free nanoparticles in environmental cleanup or other uses unless it is proven that the advantages exceed the hazards. As a byproduct of industrial manufacturing, nanoparticles are most commonly released into the environment through the air or fluid waste streams as industrial waste. Consumer goods, biological excretion, and the degradation of infrastructure that contains nanoparticles are other sources of nanopollution (Keshav, 2022).

4.4 Nanoparticles as a Carrier Vector Ecotoxicity of Co-occurring Contaminants

Fears regarding the possible for nanoparticles to serve as transporters of chemical stimuli, both inorganic and organic with anthropogenic origin, were highlighted since aquatic and terrestrial ecosystems are frequently subjected to combinations of chemical stressors. For instance, Bundschuh et al. (2018) reported a higher herbicide (diuron)-induced toxicity when it was administered at environmentally relevant quantities while carbon-based nanoparticles were present. Similar to this, the addition of fullerene nanoparticles increased the acutely poisonous effect of the pesticide "bifenthrin" while having no influence on its chronic effects (Yang et al., 2017; Bundschuh et al., 2018).

In agreement with these findings, a recent review proposed that carbon-based nanoparticles might potentially serve as a trap for metal ions. The properties of the surrounding environment have a significant impact on the ability of carbon-based nanoparticles to serve as a "Trojan horse" for other pollutants (Lochab et al., 2021). Raju et al. (2023) dispute this conclusion, reporting antagonistic effects on bacteria and daphnids for the majority of carbon-based nanoparticles and organic co-contaminant combinations. Additionally, conflicting findings were seen when metallic nanoparticles and other chemical stressors were combined.

For instance, Ramirez et al. (2019) reported that phenols and polycyclic aromatic hydrocarbon absorption was decreased in the presence of TiO_2 nanoparticles and that this does not constantly lessen those chemical substances' cumulative harmful effects and TiO_2 nanoparticles. More so, it indicates a major role of the nanoparticles to the biological responses. On the other hand, Xin et al. (2021) suggested that the presence of TiO_2 nanoparticles promoted the buildup of perfluorooctanesulfonate in fish. Due to perfluorooctanesulfonate's adsorption on TiO_2 surfaces of nanoparticles and their ensuing agglomeration and deposition of perfluorooctanesulfonate-loaded nanoparticles, this pattern was most obvious near the bottom of the experimental systems. Similar to this, TiO_2 nanoparticles boost biological responses by increasing the uptake of metal ions (Hu et al., 2017) from sediments as well as from the watery phase and at high OM concentrations (Xin et al., 2021) in biota.

Other studies show the reverse pattern, showing a decrease in metal ion-induced effects in algae or mussels when $nTiO_2$ or Al_2O_3 nanoparticles are present, with an even greater decrease when DOM (dissolved organic matter) is present (Hu et al., 2017). Subsequent studies have demonstrated that the direction and strength of effects caused by co-exposure to metal ions and TiO_2 nanoparticles are determined by the charge of the most dangerous metal ion (Ramirez et al., 2019).

Field settings are likely to witness interactions between nanoparticles and co-contaminants, even though the underlying mechanisms remain incompletely understood. Furthermore, the concentrations of nanoparticles utilized in those tests are frequently at least a factor of 10 higher than those that are environmentally relevant. It is currently unclear if the direction and amount of these interactions matter in relation to how co-contaminants affect animals (Bundschuh et al., 2018).

4.5 Impact of Nanoparticles on Biotic Interactions with Communities

The ways in which metal-based nanoparticles interact in soils were reviewed by McKee and Filser; they also emphasized the significance of abiotic variables as well as species interactions for the destiny and bioavailability of these nanoparticles. Their results suggest that the detrimental impacts of nanoparticles on ecosystem processes like carbon dioxide emission and nitrogen or phosphorus fluxes may be attributed, in particular, to biotic interactions (Wang et al., 2017), which were not discernible from short-term, single species testing (Bundschuh et al., 2018). Similarly, Cao et al. (2022) showed that Ag nanoparticles at 50 and 100 mg/L significantly reduced TN and nanoparticle effectiveness of removal and raised reactive oxygen species levels in a flocculated activated sewage system and ROS (reactive oxygen species). According to what he claimed, AgNPs at 1 and 5 mg per liter for a 285-day incubation also changed the makeup of the microbial communities and reduced the effectiveness of phosphorus removal in membrane bioreactors that are anaerobic, anoxic, and aerobic (A2O-MBR). Additionally, in tidal flow-created wetland systems, AgNPs at 50 milligrams per liter for 5- and a 60-day considerably reduced nitrogen removal rates from TN and ammonia, modifications to bacterial community architectures, and adjustments to nitrogen metabolic pathways.

In contrast to these results, Button et al. (2016) discovered that 28 days of exposure to ionic Ag and citrate- and PVP-coated Ag nanoparticles (at 100 g/L) had no negative effects on the composition or functionality of the microbial community in wetland settings. The consequences of this adaptation are still unknown. On the other hand, these microbial communities became tolerant to Ag, demonstrating that they could respond to these new stresses by increasing the synthesis of extracellular polymers (Bundschuh et al., 2018). But below a particular threshold concentration – which varies based on the AgNP's original form – ex situ toxicity is noted; nonetheless, Button et al. (2016) suggested that in a comparatively short amount of time, wetland microbial communities can withstand and become resistant to lower levels of in situ AgNP exposure.

In addition, the rate of nitrite formation was decreased by roughly 30% when 200 mg of magnetite were present (Fe_3O_4) NP/L (Michels et al., 2017), a level considered to be relatively low in soils when organic pollution has been remedied (Bundschuh et al., 2018). Numerous other research findings reveal sporadically unexpectedly strong adverse impacts on microorganisms subjected to iron oxide nanoparticles, supporting this fairly low response threshold (iron oxides are typically regarded safe to beneficial) (Zeng et al., 2021; Gabrielyan et al., 2019).

AgNPs were applied foliarly at concentrations of 20, 40, and 60 mg/l. This resulted in an increase in the content of indole acetic acid (IAA) and photosynthetic pigment (chlorophyll a and b and carotenoids). This in turn increased the quantity and quality of yield (number of pods/plant, number of seeds/pod, the amount of seeds/plant, and seed indcx) as well as photosynthetic pigment in agricultural soil (Cao et al., 2017). Chavan and Nadanathangam (2019) have observed impacts of nanoparticles on microorganisms that promote plant development. They reported that the presence of silver nanoparticles in the soil affected the relative abundances of the main and agriculturally important taxa, Proteobacteria, Actinobacteria, and Firmicutes. The three phyla's abundance was altered by silver nanoparticles by 25–45%. At the phylum level, zinc oxide nanoparticles had very little of an impact.

In a study using *Zea mays* and Fe_3O_4 nanoparticles, the variety of arbuscular mycorrhizal fungi was shown to be reduced, along with plant biomass, phosphorus content, carbon transfer to the fungus, and root colonization as well as phosphatase activity (Cao et al., 2017). These examples show how intricate the relationships are between soil bacteria, plants, and nanoparticles, highlighting the need for greater research. Decisions on potential environmental issues with nanoparticles when utilized in agricultural fields will be supported in the long term by addressing the underlying causes. Recent studies have demonstrated that nanoparticles may affect invertebrate species indirectly through microorganisms (Bundschuh et al., 2018).

According to Wang and Liu (2022), nanoparticles like Ag-, ZnO-, and CuO-NPs are frequently used as bactericides. However, unintended consequences that are dispersed into the environment and waste may potentially harm nontarget animals. A review of the literature showed that protozoa, nematodes, crustaceans, fish, or cell cultures were the subject of these impacts. For both terrestrial and aquatic invertebrates, TiO_2 nanoparticle uptake through contaminated food appears to be highly relevant (Lüderwald et al., 2019).

This direct absorption pathway enhances nanoparticle exposure and may have an impact on food quality by affecting the microbes that are associated with food (Wu et al., 2021). Although frequently at fairly high concentrations, structural (Wang & Liu, 2022) and functional (Bundschuh et al., 2018) alterations, such as photosynthetic efficiency and leaf breakdown, have been seen in aquatic environments. These illustrations demonstrate that nanoparticles can impact species at different trophic levels as well as species interactions (Wu et al., 2021). However, neither the mechanisms causing these changes nor their effects on the larger food web or entire ecosystems have even been touched upon (Lüderwald et al., 2019).

4.5.1 Roles of Nanoparticles as Toxicant Vectors and the Interactions with Biological Systems

According to Bundschuh et al. (2018), since the beginning of the study of "nanoecotoxicology" about 10 years ago, nanoparticles have the potential to have negative effects on the biota by producing reactive oxygen species (ROS) that could destroy biological structures. In addition, Sharma et al. (2021) mentioned the possibility that nanoparticles could serve as a carrier for other contaminants. The last 10 years of research have demonstrated that nanoparticles have the capacity to operate via numerous pathways, among which the induction of oxidative stress is one. It is clear from literatures that oxidative stress can certainly be a driver for many effects generated by nanoparticles (Kumah et al., 2023). The key elements allegedly responsible for the ecotoxicity of NP are briefly highlighted in the sections that follow.

There is no single nanoparticles' toxicity mechanism that may be regarded as universal (Bundschuh et al., 2018). However, oxidative stress in animals is a condition that is commonly seen. Physiological effects that can result in reproductive failure by altering hormones or hatching enzymes were documented as a few other pertinent processes (Xuan et al., 2023). These results point to population development implications and the possibility of transgenerational consequences. In addition, algae and aquatic plants (Jiang et al., 2017) have different photosynthetic pigment compositions and shown effects on photosystem II (Bundschuh et al., 2018). Similarly, particle size accumulation in plants that live on land may bring about physiological and metabolic alteration (Gao et al., 2023). For instance, how exposure to CeO_2 nanoparticles affects photosynthesis efficiency in fixing carbon and utilizing water has been reported by Cao et al. (2017). The latter may alter soil moisture levels, which could have an indirect impact on soil organisms.

4.5.2 Roles of Nanoparticles as Toxicant Vectors in Human

One of the crucial topics that must be determined is how nanoparticles interact with one another within a human being. The growing trend of slowly degrading or non-degradable nanoparticles in the organs is a serious reason to be concerned. Phagocytes, which are the cells where foreign objects are ingested and eliminated, may get overloaded by nanoparticles, triggering stress responses. These responses eventually cause inflammation and a weakened immune system that leaves the body vulnerable to various diseases (Jassby et al., 2019). Additionally, when exposed to tissue and other biological fluids, nanoparticles can absorb a wide range of macromolecules onto their surface, which may have an effect on the regulatory mechanisms of other proteins and enzymes. Functional groups, surface charges, surface topologies, solubility and aggregation, and chemical compositions of nanoparticles are other features that can be harmful (Rajput et al., 2020).

Because the toxicity of nanoparticles relies on a wide range of parameters, it is exceedingly difficult to generalize about the health issues connected with nanoparticle exposure. The human body can tolerate most naturally occurring chemicals and compounds. Since there is no natural defense against the novel chemicals, nanoparticles are probably more hazardous than previously believed.

Human exposure to nanoparticles can be divided into four different categories: (a) cutaneous (e.g., by applying cosmetics containing nanoparticles), (b) inhalation, (c) ingestion, and (d) injection (e.g., using medicine based on nanotechnology). The lungs, digestive tract, and skin are the three entry points for nanoparticles into the body, which may make it easier for "free radicals" to form that might harm cells (Kumah et al., 2023). Another worry is that once in the bloodstream, nanoparticles might be able to pass across the blood-brain barrier (Najahi-Missaoui et al., 2020). Rajput et al. (2020) reported an investigation on the impact of nanoparticles on the environment and human health. He explained that every aspect of the risk associated with exposure to nanoparticles through a variety of methods was examined, including cutaneous absorption, ingestion, inhalation, and drug delivery by designed nanoparticles. He further went on to say that it is necessary to study the toxicological consequences of nanoparticles at every stage of a products' lifecycle, from production to use to disposal, and concluded that through the food chain, negative impacts on human health can be avoided either directly or indirectly (Sairam et al., 2023).

Najahi-Missaoui et al. (2020) identified three key factors that contribute to the toxicity of nanoparticle's interactions with living cells. The chemical toxicity of the raw materials is used to make the nanoparticles. For instance, cadmium selenide nanoparticles (NPs) emit Cd2+. Due to their minute sizes in nanoscale, they are tiny in nature. Nanoparticles can adhere to biological membranes and penetrate cells. Even chemically inert nanoparticles, which do not break down or interact with other matrix components, can hinder cellular function when they are attached to membranes and stored inside of cells' form. The authors talked about many ways that nanoparticles might enter the human body, how they can be used in daily life, and how nanoparticles manifest their toxicity. Additionally, they investigate how nanoparticles might enter ecosystems in the air, soil, and water (Xuan et al., 2023).

It may be risky for NPs to come into touch with people or the environment. By inhaling Fe nanoparticles and TiO_2 NPs (20 nm) in several studies, it has been demonstrated that nanoparticles can harm rats' lungs (Sharma et al., 2021). Furthermore, it has been demonstrated that TiO_2 nanoparticles and fullerene cause brain injury in fish and dogs (Sutomo et al., 2022). They revealed that some studies done on rats and mice revealed that NPs have real negative effects on the brain. Both silver and titanium dioxide nanoparticles (TiO_2) have cytotoxic and genotoxic effects on mouse and even human cells. The effects are brought on by the production of reactive oxygen species (ROS), which damages DNA and promotes cell proliferation (Sadak, 2019).

According to Rajput et al. (2020), ZnO and CuO nanoparticles have toxicological effects on human health when they interact with plants and soil. These nanoparticles were subjected to a variety of chemical and metabolic interactions, which may have had an impact on biological nitrogen fixing in addition to harming plant cells. Consequently, there can also be a serious risk to human well-being. Other research (Assadian et al., 2018; Azizi et al., 2017) has demonstrated the deadly effects of CuO-based nanoparticles on human blood lymphocyte mitochondria and lysosomes, as well as their ability to produce reactive oxygen species (ROS) that cause oxidative stress in MDA-MB 231 cell lines of breast cancer cells. To address these difficulties, the authors proposed investigations on safety assessments and toxicological risk evaluation, involving exposure route and safe exposure doses of ZnO and CuO nanoparticles.

Through the integration of substance scientists, toxicologists, medical professionals, and ecological engineers in a basic multidisciplinary research endeavor, we can enhance our comprehension of the detrimental impacts on health and safety issues associated with nanoparticles. Chemical toxicity was taken into consideration throughout the creation and standardization of almost all toxicity evaluation techniques. However, a variety of unique physicochemical properties of nanoparticles may hinder or complicate conventional toxicity assessments. In other words, uses of nanoparticles that entail their direct introduction into the natural environment will probably be contentious until the unanswered problems regarding toxicity, destiny, and transport are resolved (Zhu et al., 2013; Zhu et al., 2022; James et al., 2022).

4.5.3 Roles of Nanoparticles as Toxicant Vectors in Plants

Concerningly, the buildup of nanoparticles in edible plant tissues poses a risk to human health via the food chain. The nanoparticles move to aerial portions of the plant after interacting with the roots, where they aggregate in cellular or subcellular organelles. The first phase in the bioaccumulation of nanoparticles can be thought of as their adsorption by plant roots from the soil (Rajput et al., 2017). CuO NPs, for instance, may be heavily adsorbed on the surface of the root, at least in part during mechanical attachment. The competition for ions prevented the adsorbed CuO nanoparticles in this situation from being desorbed (Rajput et al., 2020). In their evaluations of various nanoparticles, several studies proposed that adsorption in the roots caused buildup in plants. After which there is diffusion across plant tissues as a result of a number of changes, including biotransformation, bioaccumulation, and crystal phase disintegration. The nanoparticles can be taken in and stored in plant cells (Bundschuh et al., 2018; Rajput et al., 2020). These findings imply that plant roots and shoot tissues serve as hosts for the uptake of nanoparticles.

The buildup of nanoparticles in plants generally has the potential to change physiological processes by slowing down photosynthesis and transpiration, upsetting the consistency of organelles within cells and subcells, and impairing plant growth and productivity. Decreases in photosystem quantum yield and transpiration rate have occasionally been seen as well (Rajput et al., 2017). Numerous studies have suggested that nanoparticles may have an impact on crops by lowering germination rates, shortening roots and shoots, changing photosynthesis, causing oxidative stress and antioxidant imbalance, and lowering the nutritional value of crops that are edible and quality of production (Gao et al., 2018; Ochoa et al., 2018).

It appears that nanoparticles reach plant cells at the cellular level, either by securing themselves in cell walls or building up in subcellular organelles which includes vacuoles and chloroplasts. These buildups could happen as metal ions or in their original state as nanoparticles. Physicochemical characteristics, however, could vary (Minkina et al., 2019). Because plants are physiologically and morphologically diverse, they have different capacities for nanoparticles' absorption (Zhu et al., 2013; Zhu et al., 2022; Reza et al., 2023). For instance, the various root and vascular morphologies through which nanoparticles enter plant tissue may lead to varying absorption methods (Cao et al., 2017). By producing hydroxyl radicals, the buildup of NPs in plant tissues also has the potential to negatively alter the composition of proteins, lipids, and nucleic acids.

4.6 Conclusion

It is frightening and evident that nanoparticles are present everywhere, including in air, water, and soil used to grow food. Our knowledge of the causes, course, and impacts of NP on the environment has significantly improved during the last 10 years. Understanding the destiny of nanoparticles in the environment is significantly hampered by their nanoscale size. Due to recent environmental nanoparticle releases, known adverse effects of nanoparticles, and a lack of information that makes risk assessment and management challenging, the destiny and effects of nanoparticles in nature are becoming more significant. In addition to the request to take into account NP concentrations that are relevant to the environment and to keep track of what happens to NP during biological testing, there are a number of unanswered issues that require more thought. Finding out how soil characteristics (including saturated and unsaturated systems) affect NP destiny and, consequently, the risk of groundwater pollution by NP requires a more methodical approach immediately.

References

Assadian, E., Zarei, M. H., Gilani, A. G., Farshin, M., Degampanah, H., & Pourahmad, J. (2018). Toxicity of copper oxide (CuO) nanoparticles on human blood lymphocytes. *Biological Trace Element Research, 184*, 350–357.

Azizi, M., Ghourchian, H., Yazdian, F., Dashtestani, F., & AlizadehZeinabad, H. (2017). Cytotoxic effect of albumin coated copper nanoparticle on human breast cancer cells of MDA-MB 231. *PLoS One, 12*(11), e0188639.

Bakand, S., Hayes, A., & Dechsakulthorn, F. (2012). Nanoparticles: A review of particle toxicology following inhalation exposure. *Inhalation Toxicology, 24*(2), 125–135. https://doi.org/10.3109/08958378.2010.642021

Bundschuh, M., Filser, J., Lüderwald, S., McKee, M. S., Metreveli, G., Schaumann, G. E., Schulz, R., & Wagner, S. (2018). Nanoparticles in the environment: Where do we come from, where do we go to? *Environmental Sciences Europe, 30*(1), 1–17.

Button, M., Auvinen, H., Van Koetsem, F., Hosseinkhani, B., Rousseau, D., Weber, K. P., & Du Laing, G. (2016). Susceptibility of constructed wetland microbial communities to silver nanoparticles: A microcosm study. *Ecological Engineering, 97*, 476–485.

Cao, Z., Stowers, C., Rossi, L., Zhang, W., Lombardini, L., & Ma, X. (2017). Physiological effects of cerium oxide nanoparticles on the photosynthesis and water use efficiency of soybean (Glycine max (L.) Merr.). *Environmental Science Nano, 4*(5), 1086–1094.

Cao, C., Huang, J., & Yan, C. (2022). Unveiling changes of microbial community involved in N and P removal in constructed wetlands with exposing to silver nanoparticles. *Journal of Hazardous Materials, 432*, 128642.

Chavan, S., & Nadanathangam, V. (2019). Effects of nanoparticles on plant growth-promoting bacteria in Indian agricultural soil. *Agronomy, 9*(3), 140. MDPI AG.

Chidambaram, M., & Krishnasamy, K. (2012). Nanotoxicology: Toxicity of engineered nanoparticles and approaches to produce safer nanotherapeutics, International. *Journal of Pharmaceutical Sciences, 2*(4), 117–122.

Danail, R. H., Stefania, G., Andrea, C., & Antonio, M. (2012). Risk assessment of engineered-nanomaterial's: A review of available data and approaches from a regulatory perspective. *Nanotoxicology, 6*(8), 880–898. https://doi.org/10.3109/17435390.2011.626534

Dibyanshu, K., Chhaya, T., & Raychoudhury, T. (2023). A review on the fate and transport behavior of engineered nanoparticles: Possibility of becoming an emerging contaminant in the groundwater. *International Journal of Environmental Science and Technology, 20*, 4649–4672.

Elizabeth, A. K., Raoul, D. F., Saeed, H., Paul, B., Fatemeh, V. Z., & Tannaz, P. (2023). Human and environmental impacts of nanoparticles: A scoping review of the current literature. *BMC Public Health, 23*, 1–28.

Fatima, C. F. S., Rudo, A. V., Cornelis, A. M. V., & Monica, J. B. A. (2023). Toxicokinetics and toxicodynamics of Ag nanomaterials (NM300K) in the soil environment–impact on *Enchytraeus crypticus* (Oligochaeta). *Ecotoxicology and Environmental Safety, 252*, 1–9.

Gabrielyan, L., Hovhannisyan, A., Gevorgyan, V., Ananyan, M., & Trchounian, A. (2019). Antibacterial effects of iron oxide (Fe_3O_4) nanoparticles: Distinguishing concentration-dependent effects with different bacterial cells growth and membrane-associated mechanisms. *Applied Microbiology and Biotechnology, 103*(6), 2773–2782.

Gao, X., Avellan, A., Laughton, S., Vaidya, R., Rodrigues, S. M., Casman, E. A., & Lowry, G. V. (2018). CuO nanoparticle dissolution and toxicity to wheat (Triticum aestivum) in rhizosphere soil. *Environmental Science & Technology, 52*(5), 2888–2897.

Gao, M., Chang, J., Wang, Z., Zhang, H., & Wang, T. (2023). Advances in transport and toxicity of nanoparticles in plants. *Journal of Nanobiotechnology, 21*(1), 75.

Helinor, J., Giulio, P., & Stefano, Z. (2012). Engineered nanomaterial risk, lessons learnt from completed nanotoxicology studies: Potential solutions tocurrent and future challenges. *Critical Reviews in Toxicology, 43*(1), 1–20. https://doi.org/10.3109/10408444.2012.738187

Hu, M., Lin, D., Shang, Y., Hu, Y., Lu, W., Huang, X., Ning, K., Chen, Y., & Wang, Y. (2017). CO2-induced pH reduction increases physiological toxicity of nano-TiO2 in the mussel *Mytilus coruscus*. *Scientific Reports, 7*(1), 40015.

James, Y., Liu, M. S., & Christie, M. S. (2022). A toxicological profile of silica nanoparticles. *Toxicology Research, 11*, 565–582.

Jassby, D., Su, Y., Kim, C., Ashworth, V., Adeleye, A. S., Rolshausen, P., Roper, C., & White, J. (2019). Delivery, uptake, fate, and transport of engineered nanoparticles in plants: A critical review and data analysis. *Environmental Science: Nano, 8*(6), 2311–2331.

Jiang, H. S., Yin, L. Y., Ren, N. N., Zhao, S. T., Li, Z., Zhi, Y. W., Shao, H., Li, W., & Gontero, B. (2017). Silver nanoparticles induced reactive oxygen species via photosynthetic energy transport imbalance in an aquatic plant. *Nanotoxicology, 11*(2), 157–167.

Jin, S. K. (2012). Genotoxicity, acute oral and dermal toxicity, eye and dermal irritation and corrosion and skin sensitisation evaluation of silver nanoparticles. *Nanotoxicology*. https://doi.org/10.3109/17435390.2012.676099

Keshav, S. (2022). Role of nanotechnology and nanomaterials for water treatment and environmental remediation. *International Journal of New Chemistry, 9*(3), 373–398.

Kumah, E. A., Fopa, R. D., Harati, S., Boadu, P., Zohoori, F. V., & Pak, T. (2023). Human and environmental impacts of nanoparticles: A scoping review of the current literature. *BMC Public Health, 23*, 1059.

Laura, N. V., Swati, D. G. R., Daniel, A. A., Deborah, H. B., Phil, B., Courtney, C. C., Nicholas, C., Miriam, L. D., Rashmi, J., Bhavna, S., Kristin, S., Wilma, A. S., Ken, Z., & Tracey, J. W. (2023). Addressing systemic problems with exposure assessments to protect the public's health. *Environmental Health, 21*, 121.

Lihui, X., Zhao, J., Magdalena, S., Ping-Kun, Z., & Ruixue, H. (2023). Nanoparticles-induced potential toxicity on human health: Applications, toxicity mechanisms, and evaluation models. *MedComm, 4*(40), e327.

Lochab, A., Sharma, R., Kumar, S., & Saxena, R. (2021). Recent advances in carbon based nanomaterials as electrochemical sensor for toxic metal ions in environmental applications. *Materials Today: Proceedings, 45*, 3741–3753.

Lüderwald, S., Schell, T., Newton, K., Salau, R., Seitz, F., Rosenfeldt, R. R., Dackermann, V., Metreveli, G., Schulz, R., & Bundschuh, M. (2019). Exposure pathway dependent effects of titanium dioxide and silver nanoparticles on the benthic amphipod Gammarus fossarum. *Aquatic Toxicology (Amsterdam, Netherlands), 212*, 47–53.

Mazayen, Z. M., Ghoneim, A. M., Elbatanony, R. S., Basalious, E. B., & Bendas, E. R. (2022). Pharmaceutical nanotechnology: From the bench to the market. *Future Journal of Pharmaceutical Science, 8*, 12.

McKinney, W., Jackson, M., & Sager, T. M. (2012). Pulmonary and cardiovascular responses of rats to inhalation of a commercial antimicrobial spray containing titanium dioxide nanoparticles. *Inhalation Toxicology, 24*(7), 447–457. https://doi.org/10.3109/08958378.2012.685111

Michels, C., Perazzoli, S., & Soares, H. M. (2017). Inhibition of an enriched culture of ammonia oxidizing bacteria by two different nanoparticles: Silver and magnetite. *The Science of the Total Environment, 586*, 995–1002.

Minkina, T., Rajput, V., Fedorenko, G., Fedorenko, A., Mandzhieva, S., Sushkova, S., Morin, T., & Yao, J. (2019). Anatomical and ultrastructural responses of Hordeum sativum to the soil spiked by copper. *Environmental Geochemistry and Health, 42*, 45–58.

Mishra, S., & Sundaram, B. (2023). Fate, transport, and toxicity of nanoparticles: An emerging pollutant on biotic factors, process safety and environmental protection. *Official Journal of the European Federation of Chemical Engineering, 174*, 595–607.

Najahi-Missaoui, W., Arnold, R. D., & Cummings, B. S. (2020). Safe nanoparticles: Are we there yet? *International Journal of Molecular Sciences, 22*(1), 385.

Ochoa, L., Zuverza-Mena, N., Medina-Velo, I. A., Flores-Margez, J. P., Peralta-Videa, J. R., & Gardea-Torresdey, J. L. (2018). Copper oxide nanoparticles and bulk copper oxide, combined with indole-3-acetic acid, alter aluminum, boron, and iron in Pisum sativum seeds. *Science of the Total Environment, 634*, 1238–1245.

Rajput, V. D., Minkina, T., Sushkova, S., Tsitsuashvili, V., Mandzhieva, S., Gorovtsov, A., et al. (2017). Effect of nanoparticles on crops and soil microbial communities. *Journal of Soils and Sediments, 18*(6), 2179–2187.

Rajput, V., Minkina, T., Mazarji, M., Shende, S., Sushkova, S., Mandzhieva, S., Burachevskaya, M., Chaplygin, V., Singh, A., & Jatav, H. (2020). Accumulation of nanoparticles in the soil-plant systems and their effects on human health. *Annals of Agricultural Sciences, 65*(2), 137–143.

Raju, C. V., Cho, C. H., Rani, G. R., Manju, V., Umapathi, R., Huh, Y. S., & Park, J. P. (2023). Emerging insights into the use of carbon-based nanomaterials for the electrochemical detection of heavy metal ions. *Coordination Chemistry Reviews, 45*(3), 3741–3753.

Ramirez, L., Ramseier Gentile, S., Zimmermann, S., & Stoll, S. (2019). Behavior of TiO2 and CeO2 nanoparticles and polystyrene nanoplastics in bottled mineral, drinking and Lake Geneva waters. Impact of water hardness and natural organic matter on nanoparticle surface properties and aggregation. *Water, 11*(4), 721.

Reza, A., Ghazal, S., Mohammadmahdi, M., Sarah, D., & Lobat, T. (2023). Structural parameters of nanoparticles affecting their toxicity for biomedical applications: A review. *Journal of Nanoparticles Research, 25*(3), 43.

Ritu, G., & Huan, X. (2018). Nanoparticles in daily life: Applications, toxicity and regulations. *Journal of Environmental Pathology, Toxicology and Oncology, 37*(3), 209–230.

Ruhung, W. (2012). Generation of toxic degradation products by sonication of Pluronic dispersants: Implications for nanotoxicity testing. *Nanotoxicology, 7*, 1272–1281.

Sadak, M. S. (2019). Impact of silver nanoparticles on plant growth, some biochemical aspects, and yield of fenugreek plant (*Trigonella foenum-graecum*). *Bulletin of the National Research Centre, 43*, 38.

Sairam, A. B., Sanmugam, A., Pushparaj, A., Kumar, G. M., Sundarapandian, N., Balaji, S., Nallal, M., & Park, K. H. (2023). Toxicity of polymeric nanodrugs as drug carriers. *ACS Chemical Health & Safety, 30*(5), 236–250.

Samreen, H. K., Fulekar, M. H., & Bhawana, P. (2015). Nanotoxicology- health and environmental impacts: A review. *Journal of Environmental Nanotechnology, 4*(3), 55–73.

Sayed, M. T., Mahdiye, M., Maryam, A., Mohammad, A., Faramarz, S., Sayed, A. T., Marzieh, S., & Mohsen, K. (2013). Effects of nanoparticles on the environment and outdoor workplaces. *Electronic Physician, 5*(4), 706–712.

Sharma, S., Parveen, R., & Chatterji, B. P. (2021). Toxicology of nanoparticles in drug delivery. *Current Pathobiology Reports, 9*(4), 133–144.

Sun, H., Zhou, S., Jiang, Y., Xi, X., Tan, Y., Zhang, G., Jiang, N., Zhou, T., Yin, X., Wang, M., & Gao, B. (2022). Chapter 7 - Fate and transport of engineered nanoparticles in soils and groundwater. In *Emerging contaminants in soil and groundwater systems* (pp. 205–251). Elsevier.

Sutomo, D., Ahsani, D., & Fidianingsih, I. (2022). Toxicity of nanoparticles on the spleen in animal studies: A scoping review. *Jurnal Ilmu Kefarmasian Indonesia, 20*(2), 158–168.

Wang, T., & Liu, W. (2022). Emerging investigator series: Metal nanoparticles in freshwater: transformation, bioavailability and effects on invertebrates. *Environmental Science. Nano, 9*(7), 2237–2263.

Wang, X., Li, J., Liu, R., Hai, R., Zou, D., Zhu, X., & Luo, N. (2017). Responses of bacterial communities to CuO nanoparticles in activated sludge system. *Environmental Science & Technology, 51*(10), 5368–5376.

Wu, J., Bosker, T., Vijver, M. G., & Peijnenburg, W. J. G. M. (2021). Trophic transfer and toxicity of (Mixtures of) Ag and TiO_2 nanoparticles in the lettuce–terrestrial snail food chain. *Environmental Science & Technology, 55*(24), 16563–16572.

Xin, X., Huang, G., Zhang, B., & Zhou, Y. (2021). Trophic transfer potential of nTiO2, nZnO, and triclosan in an algae-algae eating fish food chain. *Aquatic Toxicology, 235*, 105824.

Xuan, L., Ju, Z., Skonieczna, M., Zhou, P. K., & Huang, R. (2023). Nanoparticles-induced potential toxicity on human health: Applications, toxicity mechanisms, and evaluation models. *MedComm, 4*(4), e327.

Yang, Y., Lin, Y., & Liao, C. (2017). Toxicity-based toxicokinetic/toxicodynamic assessment of bioaccumulation and nanotoxicity of zerovalent iron nanoparticles in *Caenorhabditis elegans*. *International Journal of Nanomedicine, 12*, 4607–4621.

Zeng, Q., Xu, J., Hou, Y., Li, H., Du, C., Jiang, B., & Shi, S. (2021). Effect of Fe3O4 nanoparticles exposure on the treatment efficiency of phenol wastewater and community shifts in SBR system. *Journal of Hazardous Materials, 407*, 124828.

Zhu, M., Guangjun, N., Huan, M., Xia, T., Andre, N., & Yuliang, Z. (2013). Physicochemical properties determine nanomaterial cellular uptake, transport and fate. *Accounts of Chemical Research*. https://doi.org/10.1021/ar300031y

Zhu, X., Teng, J., Xu, E. G., Zhao, J., Shan, E., Sun, C., & Wang, Q. (2022). Toxicokinetics and toxicodynamics of plastic and metallic nanoparticles: A comparative study in shrimp. *Environmental Pollution, 312*, 120069.

Chapter 5
Nanoparticle-Organism Interactions: Cellular Uptake and Biodistribution

Japhet Gaius Yakubu, Patrick Omoregie Isibor, Ameh Simon Sunday, Frances Iseghohi, Olugbenga Samson Taiwo, and Oluwafemi Adebayo Oyewole

5.1 Introduction

One of the most rapidly expanding fields of cutting-edge technology is the development and production of nanomaterials, which offers a variety of novel applications in various fields including the electronic, healthcare, cosmetic, agronomy, engineering, and food industries (Maurizi et al., 2018; Yang et al., 2021). The advent of nanotechnology has increased the number of nanoparticles in the environment, exposing people to them continuously through uncontrolled skin or inhalation contact as well as oral administration or medication injection (Maurizi et al., 2018). Nanomaterials are used for a variety of medical purposes in the healthcare and life sciences, including drug delivery systems, antibacterial agents, MRI agents, hyperthermia treatment media, and biotechnological applications (such as biosensors) (Singh & Nalwa, 2011).

Engineered nanoparticles (also known as nanoparticles) are materials that have been commercially generated and have at least one dimension smaller than 100 nm (Bhatt & Tripathi, 2011). The industrial sector has undergone a major upheaval thanks to nanotechnology. A lot of attention is recently channeled to application of

J. G. Yakubu (✉) · O. A. Oyewole
Department of Microbiology, Federal University of Technology, Minna, Nigeria
e-mail: japhetgaiusyakubu12@gmail.com

P. O. Isibor
Department of Biological Sciences, Covenant University, Ota, Ogun State, Nigeria
e-mail: patrick.isibor@covenantuniversity.edu.ng

A. S. Sunday · O. S. Taiwo
Department of Biochemistry, Federal University of Technology, Minna, Nigeria

F. Iseghohi
Medical Biotechnology Department, National Biotechnology Development Agency, Abuja, Nigeria

P. O. Isibor et al. (eds.), *Environmental Nanotoxicology*,
https://doi.org/10.1007/978-3-031-54154-4_5

nano-size in the fields of electronics, biology, and aerospace engineering due to their unique physicochemical and electrical properties (Bahadar et al., 2016). The expanding usage of nanomaterials has sparked worries that human exposure to them could pose hazards to both the environment and the human body. Micelles, polymers, liposomes, protein-based, liposomes, nanomaterials, and inorganic nanomaterials are just a few of the many types of nanomaterials currently being used in a variety of industries. Because they are not bio-persistent and are more likely to be metabolized and eliminated from the body, organic nanomaterials are typically regarded as less harmful than inorganic nanomaterials (Bahadar et al., 2016; Sun et al., 2021).

Nanoparticles have been widely used in biomedical applications because of their unique features. It has the potential to cure challenging diseases with poor prognoses, including malignancies, intricate infections, autoimmune diseases, and other clinical issues (Zhang et al., 2022). As a result, even while exposure routes vary for various organisms, the widespread use and rising manufacturing of nanoparticles has inevitably increased human and environmental exposure to these substances. Concerns about possible safety issues have been raised as a result of the anticipated increase in exposure in both human and environmental organisms (Swartzwelter et al., 2021).

Nanoparticles can quickly enter a biological system, evade multiple obstacles in the body, and reach the most delicate parts of the organism. Nanoparticles being smaller than 10 nanometer expressly penetrate membranes and may interfere with the regular biochemical processes of cells. Studies on both humans and animals have shown that after inhalation and oral exposure, nanoparticles are transmitted to the brain, spleen, heart, and liver via the gastrointestinal tract and the lungs. The invasion of nanoparticles to the various systems triggers the immune system's action against the antibody (Bahadar et al., 2016). The resultant effects of the interaction between the physiological processes and the invasive nanoparticles include immunogenic reactions, nephrotoxicity, and neurotoxicity. The most notable impacts of nanoparticle toxicity are characterized by high reactivity, prolonged circulation time, off-target nanomaterial accumulation, and inadvertent exposure. Although there have been many toxicological studies on nanoparticles, the potential acute and chronic risks to biota are poorly understood. The safe application of nanoparticles requires a close analysis of the mechanisms underlying the toxicity (Zhang et al., 2022).

5.2 Biophysical Interaction of Nanoparticles

The biophysical interaction of nanoparticles in general is a current topical aspect of profound importance in nanotoxicology studies. A highly heterogeneous NP-cell interface is created when an NP docks on a cell membrane, starting a series of kinetic events as well as dynamic physiochemical interactions. Various biophysical interactions have various effects on interactions, associated energy landscapes, and nanoparticle endocytosis. As depicted in Fig. 5.1, the cell membrane plays a key role in wrapping up the target nanoparticle which in turn moves the membrane toward the wrapping site despite the membrane tension (Zhang et al., 2015). The key components of the adhesion force that propels endocytosis are other forces, such as electrostatic, van der Waals (vdW), hydrophobic, and ligand-receptor interaction, among others (Egbuna et al., 2021). The adhesion force may result from interactions that are specific, general, or both. The recognition and binding of the ligands coated on the nanoparticle surface to the complementary receptors on the cell membrane constitute specific interactions.

Nanomaterials are used in dendritic drug, prodrug transporters, graphene vapor sensors, quantum dot tags, tear-resistant carbon nanotube fabrics, carbon nanotube transistors, radios, displays, and a contrast-enhancing agent for MRIs (Enrico, 2021).

5.2.1 Interaction of Nanomaterial with Biological System (Nano-bio-interaction)

Numerous types of nanomaterials have been developed and researched as potential diagnostic tools or pharmaceutical alternatives to enhance the quality of human life and to open up further application opportunities in highly specific fields such as agriculture, medicine, biotechnology, and pharmaceuticals (Enrico, 2021). Nanomaterials' antibacterial, anticancer, anti-inflammatory, immune system

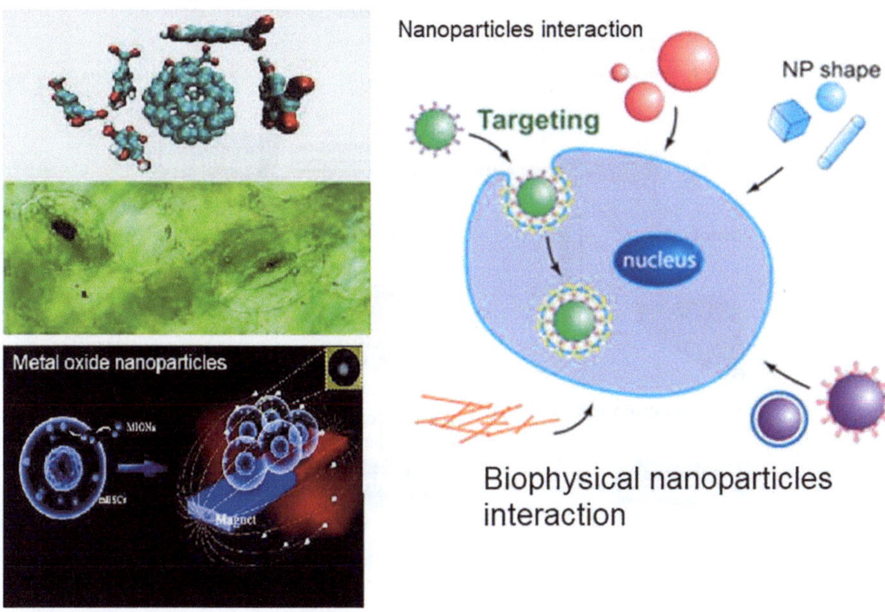

Fig. 5.1 Biophysical interaction of nanoparticles (Enrico, 2021)

modulation, antidiabetic, antioxidant, membrane permeability, and biocompatibility all underscore its novelty (de Souza, 2023). However, despite the devoted research explorations on the biosafety of nanoparticles, the processes underlying the interactions with biological systems (nano-bio-interaction) have been unsatisfactorily characterized as the interactions with biological systems and are not entirely predictable at the nanoscale. The paucity of knowledge limits the wide applicability of nanomaterials (de Souza, 2023). Furthermore, interactions can be altered by engineering techniques, which can alter the properties of nanomaterials (de Souza, 2023).

5.2.2 Toxicokinetics of Nanoparticles

The toxicokinetics of nanomaterials has recently taken center stage in nanotoxicological research. Nanoparticles undergo the conventional toxicokinetic processes such as absorption, distribution, metabolism, and excretion (Bahadar et al., 2016; Maurizi et al., 2018; Sun et al., 2021; Yang et al., 2021; de Souza, 2023). Figure 5.2 illustrates the organismal incorporation of nanoparticles through medical procedures, cutaneous absorption, oral ingestion, and nasal inhalation which is followed by homeostatic toxicokinetic processes that dictate the resultant impacts on the

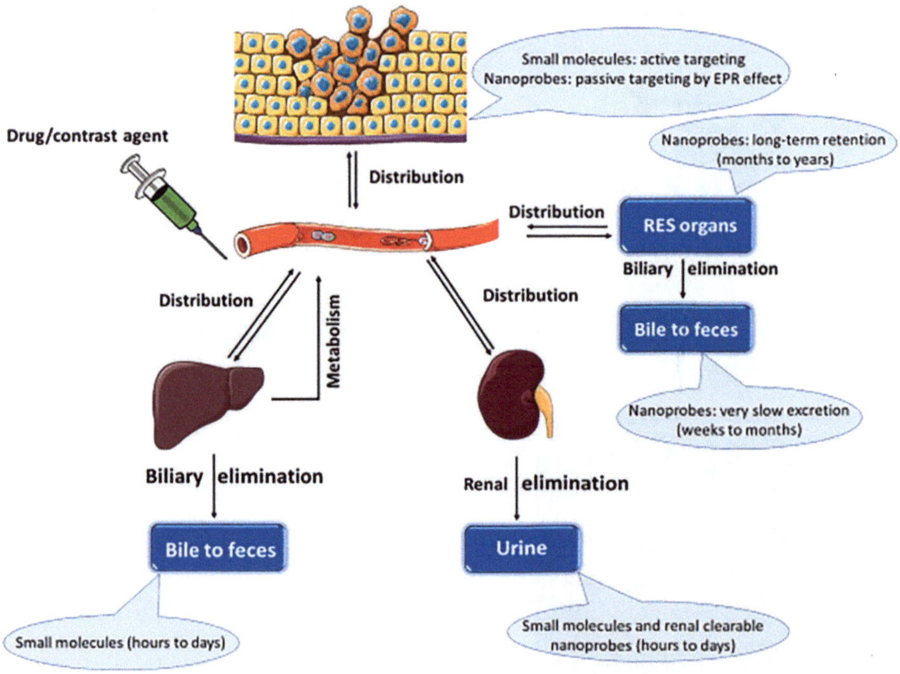

Fig. 5.2 Schematic presentation of the toxicokinetics of nanoparticles (Enrico, 2021)

organism. Major research attention has been focused on metals and metal oxide nanoparticles, while the indispensable polymer nanomaterials have conversely received much less attention (Badrigilan et al., 2020; Enrico, 2021).

5.2.3 *Absorption of Nanoparticles*

The boundaries and crucial intracellular environment of the cell are maintained by the selectively permeable plasma membrane. Oxygen and carbon dioxide are examples of small, nonpolar molecules that permeate the lipid bilayer. Conversely, bigger, polar molecules such as ions and nanomaterials cannot independently penetrate the plasma membrane (Patil & Shete, 2019; Badrigilan et al., 2020). In nature, significant ions and proteins with a size of a few nanometers are transported across the lipid bilayer by specialized membrane-transport protein channels, whereas most nanomaterials are absorbed by active processes like pinocytosis or phagocytosis depending on a dynamic series of physicochemical properties (Fig. 5.3). The picture below provides a schematic illustration of the cellular uptake of (a) big and (b) tiny NPs.

The uptake of nanoparticles by cells is significantly influenced by the hybrid nanostructures' surface functioning, surface charge, shape, and size (Zhang et al., 2022). Gold nanoparticles of 14, 50, and 74 nm are readily taken up by HeLa cell lines. It was shown that the cellular uptake varied with different-sized NPs, with 50-nm NPs being the most effective in their uptake, suggesting there may be a preferred size for efficient cellular absorption (Enrico, 2021). Spherical NPs absorbed 500% more than rod-shaped NPs in terms of cellular uptake, which was also influenced by the NP structure (Zhang et al., 2015; Ferdous & Nemmar, 2020). This is consistent with the longer membrane wrapping times needed for the elongated particles. Various cellular absorption paths for NPs are schematically depicted together

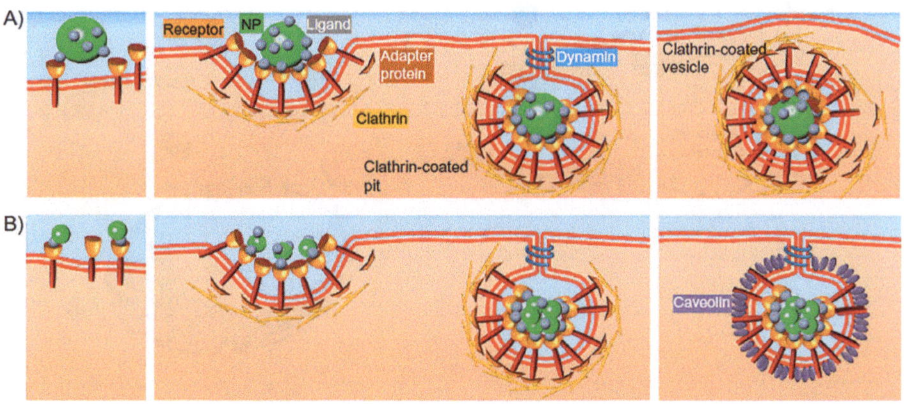

Fig. 5.3 Cellular absorption of small and large nanoparticles (Patil & Shete, 2019)

with intracellular trafficking mechanisms and potential endocytosis pathways (Fig. 5.4).

5.2.4 Distribution of Nanoparticles

A nanoparticle can be systemically dispersed throughout organismal body by the vascular and lymphatic systems upon entry into the circulatory system where it interacts with the white and red blood cells, coagulating factors, plasma proteins, and platelets in a variety of ways (Brandelli, 2020; Ferdous & Nemmar, 2020; Enrico, 2021). The level of cardiac output that travels through a tissue is characterized by the distribution of nanoparticles in the tissue. Unhindered access of nanoparticle into the vascular tissue may result in high concentrations in the tissues and organs with enhanced blood flow in the spleen, kidneys, lungs, gut, heart, liver, and brain (Hoshyar et al., 2016; Ferdous & Nemmar, 2020). Consequently, a physiological factor (cardiac output) may function as a filter for the distribution of nanomaterials. Long circulatory residence periods and delayed extravasation of the vasculature to the interstitial regions are characteristics of minute nanomaterials that range 1–20 nm in size. Intravenous administration may mitigate optimum volume of distribution or result in changed volume to bigger particles. It was discovered that gold nanoclusters and quantum dots with a diameter of less than 10 nm assemble in the

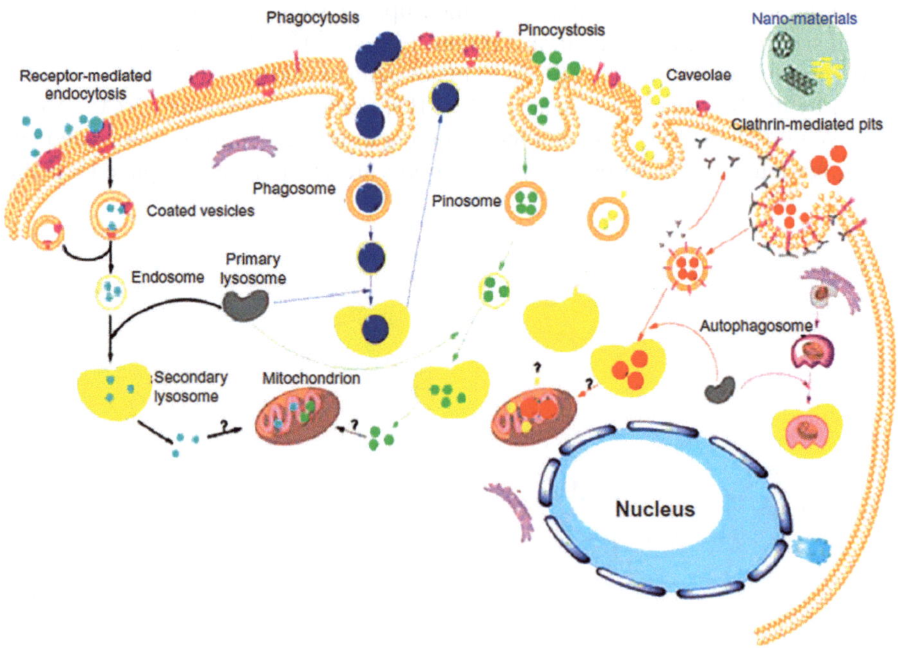

Fig. 5.4 Mechanisms for NP intracellular uptake (Patil & Shete, 2019)

plasma membrane before gradually moving within the cell. Contrastingly, the larger polystyrene nanoparticles (100 nm) were directly ingested without any observable buildup in the plasma membrane beforehand (Ferdous & Nemmar, 2020; Swartzwelter et al., 2021).

Smaller nanoparticles interact with receptors less frequently than larger ones. In order to activate the membrane's wrapping, numerous tiny nanoparticles must engage simultaneously with nearby receptors. Large individual nanoparticles, on the other hand, can link to groups of receptors and encourage absorption (Ray et al., 2021). There is likely an ideal size for active uptake because different cell lines have shown size-dependent uptake of silica mesoporous nanoparticles, polystyrene, iron oxide, and Au. The highest cellular uptake was reported at a nanoparticle core size in the range of 30–50 nm. In addition to their sizes, nanoparticles can target particular cells or organs based on their surface properties (Ray et al., 2021). Several examples of active targeting occur, but passive targeting depends on a physiological parameter acting as a distributive filter. It is possible to attach biological markers to the surface of the nanomaterials, such as antibodies, aptamers, or internalization peptides. The bioconjugation of nanoparticles with peptides that penetrate the cells may facilitate the direct transport of particles across cell membranes (Enrico, 2021; Liu & Sayes, 2022).

5.2.5 Nanoparticle Adsorption

In biologically relevant circumstances, nanomaterials not only display dynamic colloidal and physicochemical behavior but also dynamic changes in surface chemistry. In actual exposure circumstances, proteins, lipoproteins, and dissolved organic debris may adsorb onto metallic nanomaterial surfaces and form a molecular "corona" (Fig. 5.5) (Podkolodnaya et al., 2012; Fleischer & Payne, 2014). Protein coronas, which have received the most attention, have been discovered to significantly modify the colloidal behavior of nanomaterials and their interactions with biological systems in comparison to the pristine nanomaterial by changing factors such as surface chemistry, surface charge, and nanomaterial size (Enrico, 2021).

The amount of internalization of nanomaterials by cells can be influenced by the structure of the protein corona, such as the glycosylation state of proteins that have been adsorbed. A recent study discovered that adsorbate proteins can also affect how cells respond to immunological stimuli (Brandelli, 2020). It has also been demonstrated that adsorption of other molecules changes the colloidal behavior and toxicity of nanomaterials. The dynamic composition of the protein corona as well as coronas formed from less-researched molecules like dissolved organic matter must be thoroughly characterized to understand the interface between biological systems and nanomaterials under environmental and biologically relevant exposure scenarios (Fleischer & Payne, 2014; Sabourian et al., 2020). The molecule's orientation, thickness, surface coverage, and composition of the nanomaterial surface, as well as the binding constants controlling molecular adsorption to such surfaces, have all

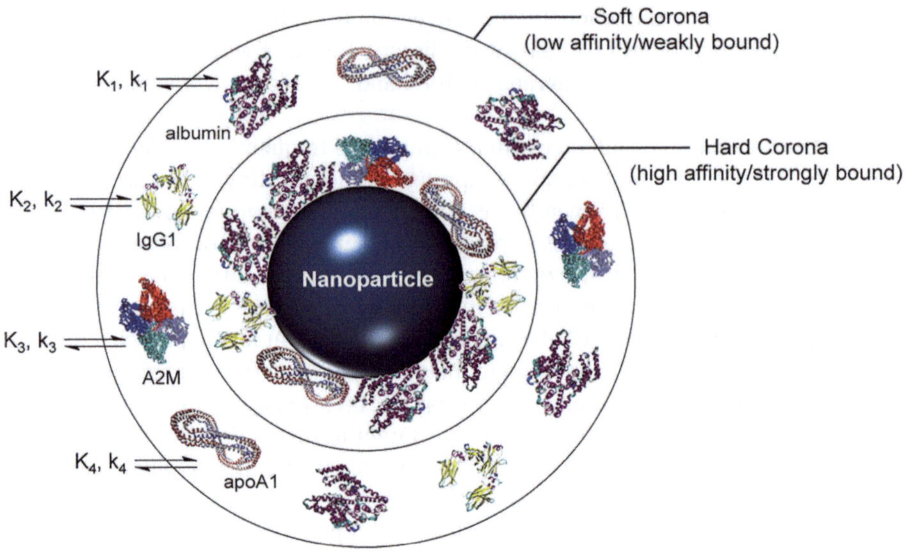

Fig. 5.5 The creation of a protein corona on a nanoparticle surface (Fleischer & Payne, 2014)

been investigated. Signal interference is a substantial barrier to the in situ assessment of molecular adsorption on nanomaterial surfaces due to the high chemical complexity of biologically relevant media, which may comprise proteins, lipids, and dissolved organic matter (Brandelli, 2020; Sabourian et al., 2020; Ray et al., 2021).

Furthermore, comparable chemical fingerprints might pose challenges to distinguishing between species that are adsorbed to the nanoparticle surface and those that are in solution. Before analysis, these issues can be resolved by isolating metallic nanomaterial subpopulations or by separating nanomaterials from interfering colloidal species in the sample matrix; nevertheless, separation techniques have not been used very frequently in published investigations. Column chromatography and its variations, as well as field flow fractionation, are currently used separation techniques. The development of more practical and generalizable methods to remove nanomaterials from complex biological matrixes would facilitate accurate nanomaterial characterization (Brandelli, 2020; Sun et al., 2021).

5.2.6 Metabolism of Nanoparticles

The MPS (mononuclear phagocyte systems), as well as the renal and hepatobiliary systems, eliminate nanomaterials from the body. The movement of nanomaterials within cells can be summed up as follows: after uptake, the nanomaterials are often confined in vesicular structures like endosomes. Most endosomal components are transferred to lysosomes after endosome maturation, where they are destroyed.

Nanomaterials occasionally manage to escape from endosomes and go after intracellular organelles (Patil & Shete, 2019; Poon et al., 2019; Enrico, 2021). In lysosomes, nanomaterials can be broken down either enzymatically or nonenzymatically (Patil & Shete, 2019; Sun et al., 2021). The lysosomes of liver cells contain numerous xenobiotic metabolizing enzymes with overlapping substrates, which are typically categorized as either phase I enzymes (such as CYP450 and flavin-containing monooxygenases) or phase II enzymes (such as glutathione S-transferases, glucuronyl transferases, and sulfotransferases) with the primary function of conjugating xenobiotics (Poon et al., 2019; Enrico, 2021).

Proteases, hydrolases, dehydrogenases, prostaglandin synthetase, and epoxide hydrolases are additional minor phase I enzymes. When metal ions bond to acid radicals and dissolve nanomaterials and their oxides, metal salt products are created. For instance, in L-929 fibroblast cells, Ag nanoparticles dissolve and transform into insoluble silver species like AgCl or AgS. Acids may be produced during the degradation of nonmetallic nanoparticles and their oxides (Enrico, 2021). For instance, SiO_2 and Si nanoparticles undergo sequential hydration, hydrolysis, and ion exchange processes before being converted into silicic acid-Si $(OH)_4$ and Si (OH_4), respectively. Phosphoric acids like H_3PO_4 can be produced when BP nanoparticles are broken down (Awashra & Młynarz, 2023). Targeting intracellular organelles is possible with the release of lysosomal nanomaterial breakdown products. Finally, exocytosis, which is often mediated by lysosomes, is used to excrete degradation products of nanomaterials from cells into the extracellular space in the form of membrane-wrapped vesicles (Sun et al., 2021). The hepatic filtration mechanism for nanoparticles and the intracellular movement of nanomaterials are schematically shown in Fig. 5.6.

5.2.7 Excretion of Nanoparticle

Due to worries regarding chronic buildup and patient safety, the destiny of nanoparticles is crucial for regulatory purposes (Poon et al., 2019). Excretion or clearance, a crucial function in biological processes that may aid in preventing damage and toxicity, is how unwanted substances are removed from the body. The two primary excretion paths for agents are the hepatic and renal pathways (Patil & Shete, 2019; Poon et al., 2019; Yang et al., 2021). Better options include renal clearable substances since they can leave the body more quickly and cause only minor cellular metabolism and internalization. Therefore, it is a frequent practice to lessen the body's exposure to such agents. Based on glomerular filtration in the kidneys, size, surface charge, and shape of materials are important elements for the excretion of these substances through the urine system. The glomerulus capillary walls typically have a filtration-size threshold (FST) of around 6 nm, meaning that substances with hydrodynamic diameters (HDs) larger than the FST cannot be filtered by the kidneys (Brandelli, 2020; Enrico, 2021). When a material that cannot be cleared by the kidneys is ingested, it remains in the bloodstream and is sucked up by phagocytosis

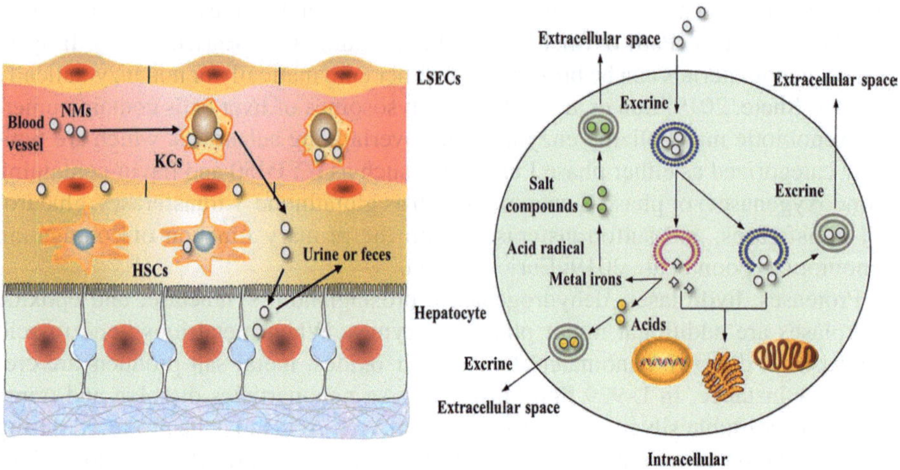

Fig. 5.6 Hepatic and intracellular transport of nanoparticles shown schematically (Sun et al., 2021)

cells such as Kupffer cells (Ray et al., 2021; Liu & Sayes, 2022). The reticuloendothelial system (RES) contains these cells, also known as resident macrophages, which are responsible for recognizing and phagocytosing foreign antigens and substances (>10 nm) in the blood to remove them from the body. As a part of the immune system, RES are dispersed throughout the body, primarily in the liver, spleen, and other reticular connective tissues, where foreign substances are broken down and removed by the hepatic pathway. The liver controls metabolism, biliary excretion in hepatic excretion, and finally the expulsion of foreign chemicals in feces (Badrigilan et al., 2020). The hepatic route, which has a more convoluted process and a longer retention length than the renal route, increases the risk to health; hence renally excreted probes are preferred for clinical purposes (Badrigilan et al., 2020).

5.2.8 Nanomaterials' Translocation Through Tissues

Numerous entry points, including the skin, respiratory system, gastrointestinal tract, or parenteral injection, are available for nanomaterials to enter the body. The body is shielded from hazardous environmental elements by these organs. In other words, they play á crucial role in the transportation of oxygen, water, and nutrients. Except for specific elements like oxygen for the retina and UV radiation for vitamin D synthesis, the skin serves as a barrier against chemicals (Enrico, 2021). Some organs, including the lungs, gut, and skin, can be penetrated and entered by nano-sized particles. Some can reach the dermis' lowest levels. Size and the characteristics of the nanoparticle surface affect how deeply they penetrate. It should be mentioned that *in vitro* testing on nanoparticle toxicity must be conducted before *in vivo* tests are carried out (Enrico, 2021).

All nutrition is exchanged in the mouth, except from gas, where nanomaterials enter the body. These three organs differ from other locations in terms of their histology. Keratin cells make up the body's skin surface area, which has a surface area of approximately 2 m^2 and a thickness of approximately 10 m (Adewale et al., 2019). These cells create a wall that prevents ions from moving through. Organ, age, and other factors, as well as other agents, all influence penetration. Nanomaterials, plasma, and blood components can interact thanks to the translocation of nanoparticles into the bloodstream. When metallic nanoparticles interact with plasma proteins, a "protein corona" is created that could change the biological and pharmacological characteristics of the nanoparticles (Ray et al., 2021).

5.3 Toxicity of Nanoparticles

The body of knowledge regarding the harmful effects of these tiny molecules is still developing quickly. Numerous studies have demonstrated the toxicity of certain nanoparticles in biological systems. As a result, both internal and external environmental researches are required; external studies can guide internal environment research (Badrigilan et al., 2020). According to several studies (Brandelli, 2020; Badrigilan et al., 2020; Enrico, 2021), most nanoparticles can emit active oxygen, which causes the reticuloendothelial system to experience oxidative stress and inflammation. The *in vivo* effects of nanoparticle-induced acute 17 toxicity have been studied. The findings suggest that the toxicity of metallic nanoparticles relies on their size, coating, and chemical makeup (Sabourian et al., 2020; Enrico, 2021). Additionally, various organs and tissues have demonstrated the systemic impacts of nanoparticles. Lung, liver, heart, and brain oxidative stress or pre-inflammatory cytotoxin activity are possible consequences on the inflammatory and immunological systems (Enrico, 2021; de Souza, 2023). Prethrombosis effects and paradox effects on heart function are just two examples of the impacts on the circulatory system. The effects of nanoparticles may lead to genotoxicity, carcinogenicity, and teratogenicity. More research is obviously needed; however, some nanoparticles may be able to cross the blood-brain barrier and cause brain damage. For macromolecules to pass through the digestive tract, there will be a rise in macromolecule absorption due to the increased nanoparticle loading (de Souza, 2023).

Researchers have examined the use of nanoparticles in diagnosis and treatment, including carbon nanotubes, iron oxides, silver, zirconia, and titania. Many of these nanoparticles might harm cells. Most nanoparticles may have hazardous consequences, yet biocompatible coatings have not been found to have any toxic impacts (Badrigilan et al., 2020). The effective design of medical materials also makes use of nanoparticles with biodegradable coatings. Even when constructed of inert substances like gold, nanomaterials exhibit significant levels of activity. Regarding the investigation and appropriate use of metallic nanoparticles, nanotoxicology is essential (Badrigilan et al., 2020). Because cells are unable to withstand the stress and repair the damage, nanotoxicity can cause cell apoptosis or the suppression of growth (through cell cycle arrest). Toxic effects brought on by exposure to metallic nanoparticles and nanomaterials may be mediated by several cell systems (Badrigilan et al., 2020; Liu & Sayes, 2022). Cellular injuries are caused by the disruption of cell signaling pathways brought on by nanoparticles, which disturb cellular homeostasis. The reduction of proliferation that results reduces the number of cells from one generation of cells to the next when the processes that cause which cell cycle phase could become stalled are numerous (Liu & Sayes, 2022). Expanding our understanding of nanoparticle toxicity will make it easier to build safer metallic nanocomposites and use them in advantageous ways. The capacity of nanomaterials to interact with the body on the cellular, subcellular, and molecular levels presents new problems for biocompatibility (Ray et al., 2021).

5.4 Nanoparticle-Mediated Immunotoxicity

The usage of the various nanomaterials discussed in the previous sections for a wide range of therapeutic applications has increased significantly since the development of nanomedicine. Although nanoparticles have a few characteristics, such as a high payload, low dosage, and targeting capability, which make them desirable candidates for use in the treatment of cancer, an important factor for clinical evaluation is their immunotoxicity (Brandelli, 2020). The interaction between biological agents and nanoparticles during *in vivo* application, particularly for systemic therapeutic drug administration, is crucial to the efficient transport, clearance, and accumulation of nanomaterials. The body is primarily protected by the immune system from external substances that are seen as dangers (Brandelli, 2020; Ray et al., 2021).

The immune system can be divided into two primary groups: innate and adaptive. When foreign bioactive agents, such as bacterial and viral strains, encounter innate immunity, it results in an all-around inflammatory response. Antigen-presenting cells (APCs), which have many receptors that serve as pathogen recognition agents in this system, activate the acquired immune system. As a result of the antigens, the adaptive immune system oversees manufacturing antibodies, which has a complicated route. Based on their chemical composition and surface features, immune systems can recognize some nanoparticles. These qualities can therefore be used as controls to regulate how nanoparticles interact with the immune system (Ray et al., 2021).

The protein corona particle, also known as the crown effect, is a phenomenon that occurs when nanoparticles first encounter biofluids. Although other biomolecules, such as DNA, RNA, and ribose sugars, also interact with nanoparticles, their study in literature and the resulting scientific research has been underrepresented. The interactions of the nanoparticles with plasma proteins and other biomolecules change how they behave, much as different biochemical alterations influence the bioactivity of the particles (Behzadi et al., 2017). The first cells in the immune system to respond to any foreign material, including nanoparticles, are called phagocytes. As a result, one of the two types of interactions – immunosuppression or immunostimulation – occurs. While immunosuppression renders the host organism vulnerable to many diseases since the immune system is unable to defend against any invasion, immunostimulation produces a wide range of disorders, including inflammation (Ray et al., 2021).

Immunostimulation or immunosuppression of the immune system by nanoparticles has frequently been reported to have undesirable effects (Fig. 5.7). These effects may result in inflammatory or autoimmune disorders, increasing the likelihood that the receptor's body will contract an infection. There are numerous ways to avoid immune recognition, but the most well-known one is utilizing a polymeric substance to produce a hydrophilic environment. This, however, cannot prevent the formation of antibodies. On the other hand, antigen-presenting cells are directly activated, or antigen is supplied to the targeted cellular compartment via targeting approaches to elicit the desired immune response (Ray et al., 2021).

Inhaling carbon nanotubes (CNTs) to inhibit B cell function is one of various methods that nanoparticles have been used to reduce immunity. An important component of the reported immunosuppressive mechanism is the cytokine transforming growth factor- (TGF-), which is generated by alveolar macrophages (Enrico, 2021). On the other hand, activation of the complement by the nanoparticles in subcutaneous or intradermal administration of nanomedicine can increase the effectiveness of

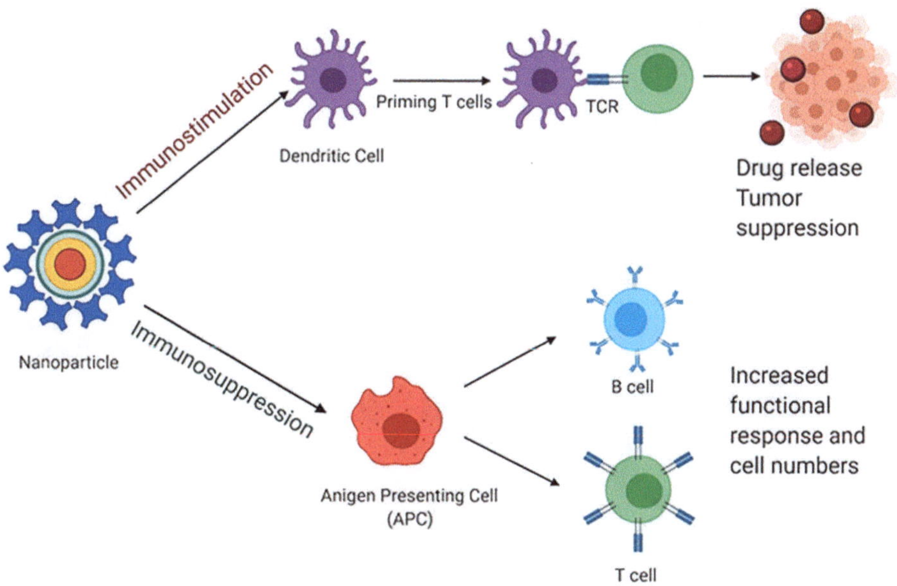

Fig. 5.7 The two key ways that nanoparticles and the immune system interact are depicted schematically (Ray et al., 2021)

the treatment (Enrico, 2021). While manufactured nanosystems have also demonstrated inherent antigenic characteristics, several systems have been reported to increase the antigenicity of conjugated weak antigens and thereby serve as adjuvants (Badrigilan et al., 2020).

By altering their physicochemical characteristics, nanoparticles can control the many paths through which they activate the complement system. Mast cells can contribute to inflammation and the negative consequences of specific nanoparticles, and they play a critical role in the biological processes that take place following exposure to nanoparticles. The size, crystallinity, aggregation properties, composition, shape, and surface charge of nanoparticles all have an impact on how they interact with the immune system (Fig. 5.8) (Ray et al., 2021).

5.5 Factors That Influence the Toxicity of Nanoparticles

The toxicity of nanoparticles has been demonstrated to be affected by varying factors, which are discussed below.

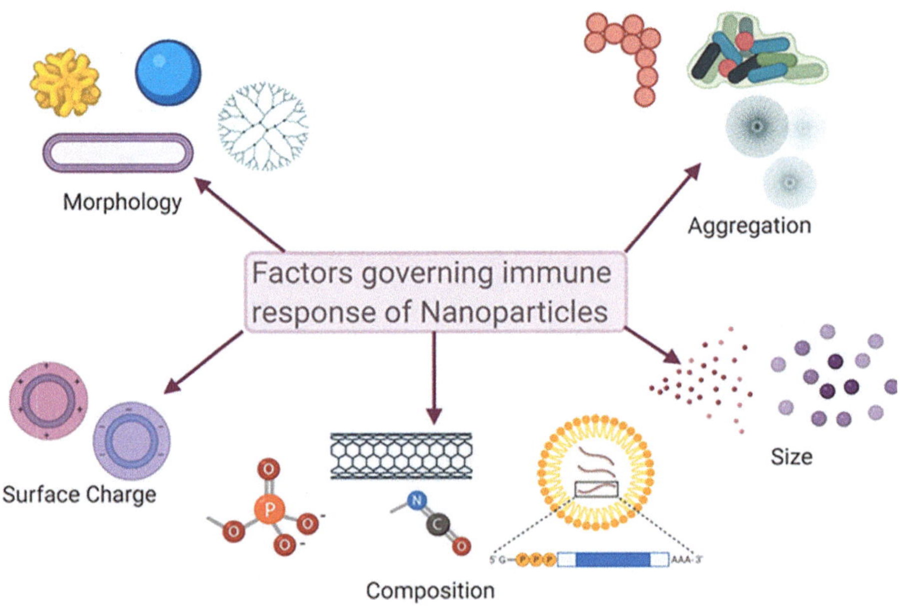

Fig. 5.8 Several elements controlling the immunological response brought on by nanoparticles are shown schematically (Ray et al., 2021)

5.5.1 *Size and Surface Area*

As nanoparticle size decreases, the ratio of surface area to volume climbs exponentially, enhancing biological and chemical reactivities. For instance, when the NP was shrunk from 30 nm to 3 nm in size, the proportion of expressed surface molecules rose from 10% to 50% (Egbuna et al., 2021). Nanomaterials become harmful when their surface interacts with certain components of cells. Therefore, nanoparticles with the same chemical composition might have significantly different degrees of cytotoxicity depending on surface area and particle size. NPs are more dangerous than larger particles with similar compositions, to put it another way (Egbuna et al., 2021).

5.5.2 *Shape*

Shape is a crucial component of nanoparticles that determines both their biological reactivity and toxicity. According to Yang et al. (2021), nanoparticles typically have the following shapes: rod, sheet, cube, cylinder, and sphere (Fig. 5.9).

A nanoparticle's cellular uptake is significantly influenced by its form (Brandelli, 2020; Egbuna et al., 2021; Sun et al., 2021). Spherical carbon nanomaterial and tubes constructed of multi-graphitic sheets were absorbed by cells in the epithelial tissues of the gut and gill, while cube-shaped carbon nanoparticles were not. Silver nanoplates were discovered to be more harmful than silver nanospheres in zebrafish (*Danio rerio*) embryos. In comparison to other shapes, spherical nanoparticles are

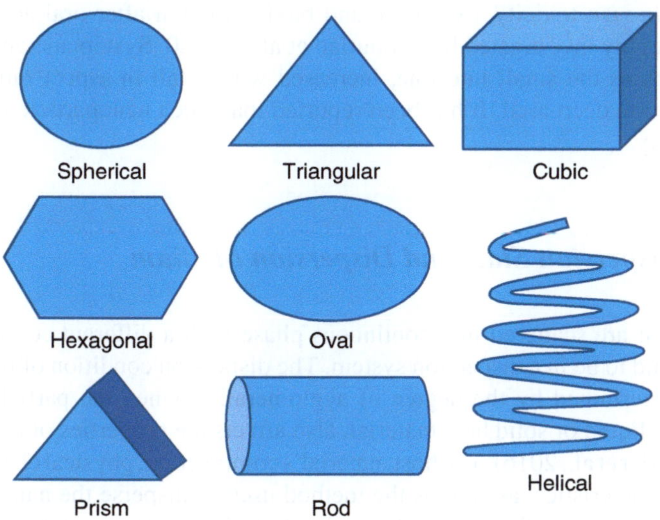

Fig. 5.9 Typical shapes of nanomaterials (Egbuna et al., 2021)

absorbed more frequently by cells (Egbuna et al., 2021). Compared to gold nanospheres, gold nanorods do not lead to as much autophagosome accumulation. In a recent study, the cytotoxic effects of gold nanoparticles on human fetal osteoblast, osteosarcoma, and pancreatic duct cell lines were evaluated using the MTT assay. Star-shaped gold nanoparticles are the most lethal to human cells (Egbuna et al., 2021).

Gold nanoparticles' cytotoxicity and anticancer properties are shaped-dependent. Mesoporous silica nanoparticles have shown promise as an oral pharmaceutical delivery mechanism. Needle-shaped nanoparticles are more dangerous than those with spherical forms due to greater multiple endocytic processes, internalization rates, and more effective adhesiveness to the surface of the target cell (Egbuna et al., 2021).

5.5.3 Aspect Ratio

The ratio of width to height is known as a nanoparticle aspect ratio. A nanotube has an aspect ratio that is very close to zero, whereas a spherical particle has an aspect ratio of 1. The toxicity of the nanoparticles increases with their aspect ratio. In the lungs, aspect ratio-dependent toxicity is typically seen. The nanofibers with dimensions of 150 nm in thickness and 2 m, 5 m, and 10 m in length, respectively, exhibit lung cancer, mesothelioma, and asbestosis. Zhang et al. (2022) investigated the pulmonary toxicity of high aspect ratio carbon nanotubes administered directly into the trachea in Sprague-Dawley rats. The carbon nanotube samples induced significant protein exudation and granulomas on the peritoneal side of the diaphragm. The effects of mesoporous silica nanoparticles with different aspect ratios of 5, 1.75, and 1 on their *in vivo* toxicity, excretion, and biodistribution after oral administration were thoroughly investigated by Sabourian et al. in 2020. Systematic absorption by organs, such as the small intestine, increased with a fall in aspect ratio, whereas urine excretion decreased. It has been reported that silica nanoparticle form affects renal toxicity.

5.5.4 Dispersion State and Dispersion Medium

Particles that are scattered in a continuous phase with a different composition (or state) are said to be in a dispersion system. The dispersion condition of the particles, which is determined by the degree of agglomeration when the particles are submerged in a liquid or solid host material, also affects the properties of nanocomposites (Schneider et al., 2016). The host material's composition, physical characteristics, and NP characteristics, as well as the method used to disperse the nanoparticles in the host material, all influence the dispersion state. To ensure constant product quality during fabrication, the dispersion state must be monitored often (Sabourian et al., 2020). The biodistribution of hybrid nanostructures in tissues and cells is significantly influenced by their surface charge, which is primarily determined by the hybrid nanostructures' dispersion state and dispersion medium. In comparison to their qualities in powder form, hybrid nanostructures suspended in aqueous and organic media have differing surface (chemical and physical) properties. The characteristics of hybrid nanostructure suspensions made with surfactants, multivalent ions, polyelectrolytes, and salts may differ, which in turn affects how well the identical hybrid nanostructures are absorbed by cells and distributed throughout the body. Surface characteristics can vary depending on the dispersion state (colloid or gel, suspension, or solution), much like a dispersion medium can. According to Patil and Shete (2019), the dispersion state mostly influences the optical characteristics of nanoparticles.

5.5.5 Crystallinity

The toxicity of nanomaterials may vary depending on the type of crystalline structure. According to Egbuna et al. (2021), polymorphs are various crystalline forms with the same chemical makeup that exhibit variable chemical and physical properties. In both *in vitro* and *in vivo* experiments, 10-hydroxycamptothecin (HCPT) nanoparticle dispersions were found to be cytotoxic, depending on the polymorph, according to Egbuna et al. (2021). They created and characterized three 10-hydroxycamptothecin polymorphic nanoparticle dispersions, including pancake, prismatic, and needle morphologies. The cytotoxicity data demonstrated that the cellular toxicity of all the various HCPT nanoparticles relied on size and shape. Although they are taken up by cells similarly to prismatic nanoparticles, needle-shaped HCPT nanoparticles are more effective at inducing an apoptotic response in cancer cells. The preference for polymorphs with various thermodynamic properties, such as lattice energy, may be explained by this effect. Studies on the polymorph-dependent absorption and toxicity of titanium dioxide NPs in A549 lung epithelial cells emphasize the importance of accurately characterizing nanoparticles' polymorphic shape (crystalline structure) for appropriate toxicity evaluation (Awashra & Młynarz, 2023).

5.5.6 Surface Coating or Surface Functionalization

Nanoparticle surface coatings are used to change a material's characteristics. A particle's surface (the "core") is encased in several layers (the "shell"). The surface coating's goal could be to modify the material's functioning, wettability, stability, or disintegration. According to Patil and Shete (2019), the surface coating can make dangerous particles innocuous while making less toxic particles more toxic due to bioavailability.

5.5.7 Dissolution

One essential property of nanoparticles that influences their safety, absorption, and associated detrimental mechanisms is their capacity to disintegrate. Depending on the type of surface modification, two identical NPs with comparable content and size may behave very differently throughout the dissolving process. The preferred cellular entrance route for nanoparticles that dissolve in the media before being taken up by organisms is typically through clear ion channels and ion transporters (Awashra & Młynarz 2023).

5.5.8 Agglomeration

Nanomaterials' high free surface energy makes them prone to aggregating in solutions. Nanomaterials are protected with protective agents to prevent agglomeration. Whether or not aggregation took place also affects how hazardous nanoparticles are. The accumulation of nanoparticles may be a factor in human inflammatory lung diseases. Carbon nanotubes and oxide nanoparticles are the nanomaterials where the toxicity is most frequently seen to be agglomeration dependent. Awashra and Młynarz (2023) state that well-dispersed carbon nanotubes are less dangerous than agglomerated carbon nanotubes. By showing that large silver nanoparticles aggregate substantially less hemolytic toxicity than small agglomerates in 2023, Awashra and Mynarz illustrated the importance of agglomeration management.

5.6 Mechanisms of Nanoparticle Toxicity

The physicochemical characteristics control the toxicity of metal-based NP. To evaluate the toxicity of NPs, numerous *in vitro* and *in vivo* models have already been used, including cell lines, three-dimensional cell culture, zebrafish, rats, and others (Sabourian et al., 2020). Uncertainty surrounds the precise mechanism of NP toxicity. The oxidative stress caused by the production of ROS, mitochondrial dysfunction, DNA damage, and other processes all contribute to the toxicity of metal-based NPs (Sabourian et al., 2020; Zhang et al., 2022). The results of ROS formation are correlated with NP toxicity in recent investigations. Most of the research has demonstrated that the underlying cause of NP toxicity is ROS production and the subsequent oxidative stress. Numerous nanoparticles have already been found to cause toxicity in living organisms by producing ROS.

The production and elimination of ROS are balanced in cells that are in their normal state. According to Poljak and Dahmane (2012), the primary components of ROS are hydrogen peroxide, hydroxyl radicals, singlet oxygen, hypochlorous acids, and superoxide anion. These compounds are important oxidative stress triggers.

Under physiological circumstances, ROS is essential for controlling several cellular functions, such as cell division, proliferation, and death. The oxidation of macromolecules like proteins, lipids, and nucleic acids caused by an elevated ROS level can cause DNA damage, disrupted signal transduction, cytotoxicity, and cell death. However, the disruption of redox balance may be caused by increased ROS generation and/or lower antioxidant levels (Awashra & Młynarz, 2023). According to the underlying mechanisms, there are basically two forms of oxidative stress caused by NPs. Direct oxidative stress, also known as primary oxidative stress, is one type. The more reactive sites that are present on the surface of NPs with a higher surface area are shown to have higher chemical reactivity, which increases ROS formation. The production of ROS can also be considerably accelerated by the surface's reactivity with oxidants and free radicals (Poljšak & Dahmane, 2012). For instance, crystalline silica particles with surface-bound free radicals might increase oxidative stress by producing ROS in a surface reaction (Awashra & Młynarz, 2023). Ray et al. (2021), and Behzadi et al. (2017) have demonstrated that some metal-based nanoparticles, such as TiO2 nanoparticles, can produce ROS through a photocatalytic process when exposed to light.

Additionally, through the Fenton and Haber-Weiss reactions, transition metals in nanoparticles like copper (Cu) and chromium (Cr) can contribute to the production of ROS, ultimately escalating oxidative stress. Indirect oxidative stress, often known as secondary oxidative stress, is an additional form (Fig. 5.10) (Ray et al., 2021). In this case, oxidative stress is not directly brought on by nanoparticles. Important organelles implicated in oxidative stress caused by NP include mitochondria. Internalized metal-based nanoparticles can release intracellular metal ions that depolarize the mitochondrial membrane and disrupt the electron transport chain, which ultimately results in mitochondrial malfunction. In the physiological process, a series of linked proton and electron transport events in mitochondria use molecular oxygen to produce adenosine triphosphate. The mitochondrial electron-transport chain (METC) would be damaged because of exposure to metal-based nanoparticles, which would raise the level of intracellular ROS (Awashra & Młynarz, 2023).

5.7 Size of Nanoparticles and Their Function in Nanotoxicology

Size and chemical composition are the two key factors that can cause nanotoxicity in nanoparticles (Brandelli, 2020; Egbuna et al., 2021; Sun et al., 2021). The particle surface area of nanoparticles grows as their size decreases. More chemical molecules could thus adhere to this surface, increasing its reactivity and amplifying its harmful effects (Egbuna et al., 2021). When cells are stopped in one or more cell cycle phases, metallic nanoparticles can inhibit cell proliferation. These effects have been investigated in numerous research on the absorption of metallic nanoparticles

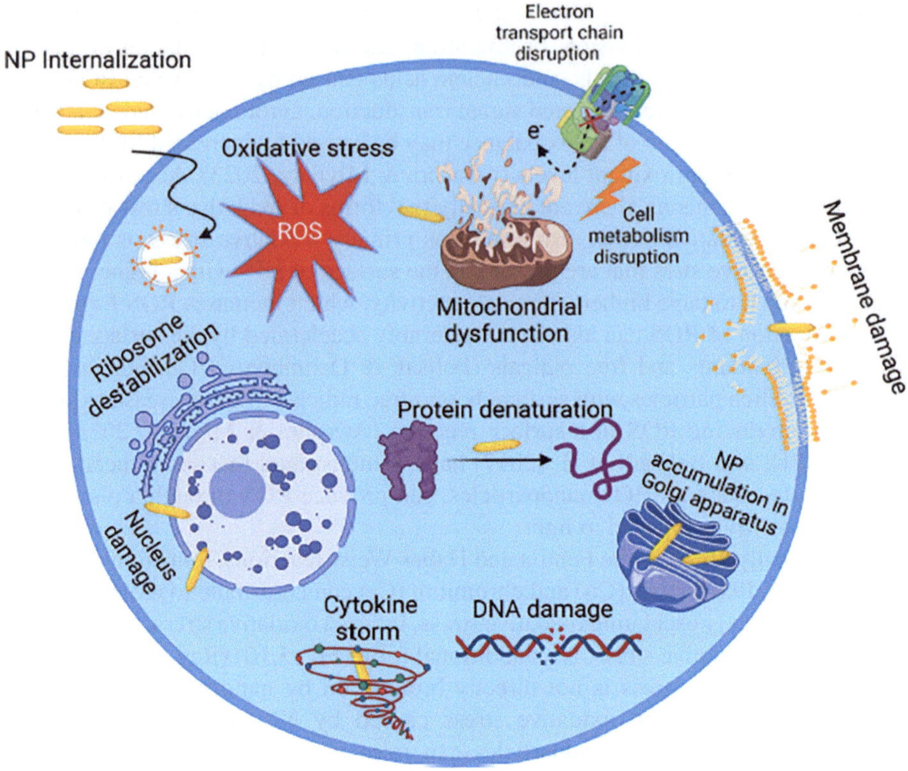

Fig. 5.10 Various modes of action of NPs on cells (Awashra & Młynarz, 2023)

through various biological barriers. Nanoparticles are absorbed, travel to the bloodstream, and then spread throughout the tissue (Awashra & Młynarz, 2023).

According to one study, the colon's mucosal and lymphatic tissues contain 33% 50 nm, 26% 100 nm, and 10% 500 nm particles. In lymphatic tissues, nanoparticles larger than 1 m were sporadically found, as were nanoparticles larger than 3 m. Researchers have found that intestinal cells can only absorb metallic nanoparticles smaller than 100 nm; while intestinal cells can absorb larger metallic nanoparticles (300 nm), only smaller nanoparticles (100 nm) can enter the circulatory system, and lymphatic tissue can absorb smaller metallic nanoparticles (100 nm) more efficiently than intestinal cells (Sun et al., 2021; Awashra & Młynarz, 2023).

There is disagreement among scientists regarding the relationship between particle size, metallic content, and their penetration into the mesenteric lymphatic glands. Nanoparticles are smaller than larger particles, which allows them to penetrate cell membranes, enter the bloodstream, and reach other organs. In comparison to larger particles, they also have a higher surface-to-volume ratio. Because more metal molecules reside on the surface, nanoparticles are typically more hazardous than larger particles with the same composition (Brandelli, 2020; Enrico, 2021). This may be one of the reasons why. Different types of platelet activation altered rate variability, vasomotor dysfunction, and pulmonary and systemic inflammation are some of the mechanisms of nanotoxicity. In cytotoxicity assays, changes in cell numbers show the effects of nanotoxicity and include both cell killing and cell cycle arrest, which inhibits cell reproduction. Studies on the cell cycle arrest brought on by nanoparticles help explain why there are fewer viable cells (Awashra & Młynarz, 2023).

5.8 Particle Surface

The cellular absorption of nanoparticles and the way they interact with the intracellular environment of organelles and biomolecules are both significantly influenced by the particle surface charge. Consequently, cytotoxicity is influenced by particle surface charge. *In vitro* studies have shown that very small particles have more pathogenic and damaging power on the lungs than the same particles of smaller size due to their larger surface area, greater tendency to conjugate, and longer energy sustainability (Patil & Shete, 2019). Two elements that are connected to cytotoxicity are the rising atomic number of the element and the alteration in cell viability because of particle surface charge, accessible binding site on a particle surface, and particle metal dissolution, but not because of bandgap energy (Patil & Shete, 2019).

5.9 Conclusion and Future Perspectives

The field of nanoparticles is one that is expanding quickly and has numerous potential uses. However, a significant issue that needs to be addressed is the toxicity of nanoparticles. Numerous molecular and physicochemical factors, such as particle size, shape, and crystallinity, affect how hazardous nanoparticles behave. Before final application, freshly created nanoparticles are tested for potential toxicity *in vitro* to see whether they pose a risk to the health of both producers and consumers. Though the findings acquired do not clearly depict the impact of nano-carriers, the interaction of nanoparticles with skin, and particularly skin models, is an exciting subject. According to the literature, various outcomes, primarily on pig or murine skin, have either shown substantial penetration or the contrary.

Understanding the mechanisms of toxicity, as well as the mechanisms governing their dispersion and accumulation in cells and tissues and their removal from the body, is a difficult topic involving the toxicity of nanoparticles. When compared to larger particles of the same chemical compound, nanoparticles are more harmful to human health, according to a review of recent toxicological research. The pace at which a material enters the body through various exposure pathways and its fate once within the body are both described by the toxicokinetics of nanoparticles. The rate of absorption, distribution, metabolism, and excretion of nanoparticles determines the level of concentration in the human system. The task of predicting the toxicological effects of manufactured nanoparticles is difficult and necessitates the creation of fresh, impartial toxicological models that are universally accepted. As a result, studying the mechanisms underlying nanoparticle toxicity and creating new toxicological models to evaluate the danger they pose to both human health and the environment are important.

References

Adewale, O. B., Davids, H., Cairncross, L., & Roux, S. (2019). Toxicological behavior of gold nanoparticles on various models: Influence of physicochemical properties and other factors. *International Journal of Toxicology, 38*(5), 357–384.

Awashra, M., & Młynarz, P. (2023). The toxicity of nanoparticles and their interaction with cells: An in vitro metabolomic perspective. *Nanoscale Advances, 5*(10), 2674–2723.

Badrigilan, S., Heydarpanahi, F., Choupani, J., Jaymand, M., Samadian, H., Hoseini-Ghahfarokhi, M., et al. (2020). A review on the biodistribution, pharmacokinetics and toxicity of bismuth-based nanomaterials. *International Journal of Nanomedicine, 15*, 7079–7096.

Bahadar, H., Maqbool, F., Niaz, K., & Abdollahi, M. (2016). Toxicity of nanoparticles and an overview of current experimental models. *Iranian Biomedical Journal, 20*(1), 1.

Behzadi, S., Serpooshan, V., Tao, W., Hamaly, M. A., Alkawareek, M. Y., Dreaden, E. C., et al. (2017). Cellular uptake of nanoparticles: Journey inside the cell. *Chemical Society Reviews, 46*(14), 4218–4244.

Bhatt, I., & Tripathi, B. N. (2011). Interaction of engineered nanoparticles with various components of the environment and possible strategies for their risk assessment. *Chemosphere, 82*(3), 308–317.

Brandelli, A. (2020). The interaction of nanostructured antimicrobials with biological systems: Cellular uptake, trafficking and potential toxicity. *Food Science and Human Wellness, 9*(1), 8–20.

Egbuna, C., Parmar, V. K., Jeevanandam, J., Ezzat, S. M., Patrick-Iwuanyanwu, K. C., Adetunji, C. O., et al. (2021). Toxicity of nanoparticles in biomedical application: Nanotoxicology. *Journal of Toxicology, 2021*, 1–21.

Enrico, E. (2021). Biophysical interaction, nanotoxicology evaluation, and biocompatibility and biosafety of metal nanoparticles. *arXiv preprint arXiv:2108.05964*.

Ferdous, Z., & Nemmar, A. (2020). Health impact of silver nanoparticles: A review of the biodistribution and toxicity following various routes of exposure. *International Journal of Molecular Sciences, 21*(7), 2375.

Fleischer, C. C., & Payne, C. K. (2014). Nanoparticle–cell interactions: Molecular structure of the protein corona and cellular outcomes. *Accounts of Chemical Research, 47*(8), 2651–2659.

Hoshyar, N., Gray, S., Han, H., & Bao, G. (2016). The effect of nanoparticle size on in vivo pharmacokinetics and cellular interaction. *Nanomedicine, 11*(6), 673–692.

Liu, J. Y., & Sayes, C. M. (2022). A toxicological profile of silica nanoparticles. *Toxicology Research, 11*(4), 565–582.

Maurizi, L., Papa, A. L., Boudon, J., Sudhakaran, S., Pruvot, B., Vandroux, D., et al. (2018). Toxicological risk assessment of emerging nanomaterials: cytotoxicity, cellular uptake, effects on biogenesis and cell organelle activity, acute toxicity and biodistribution of oxide nanoparticles. In *Unraveling the safety profile of nanoscale particles and materials-from biomedical to environmental applications* (pp. 17–36). Intech.

Patil, R. M., & Shete, P. B. (2019). Biodistribution and cellular interaction of hybrid nanostructures. In *Hybrid nanostructures for cancer theranostics* (pp. 63–86). Elsevier.

Podkolodnaya, O. A., Ignatieva, E. V., Podkolodnyy, N. L., & Kolchanov, N. A. (2012). Routes of nanoparticle uptake into mammalian organisms, their biocompatibility and cellular effects. *Biology Bulletin Reviews, 2*, 279–289.

Poljšak, B., & Dahmane, R. (2012). Free radicals and extrinsic skin aging. *Dermatology Research and Practice*, 135206. https://doi.org/10.1155/2012/135206

Poon, W., Zhang, Y. N., Ouyang, B., Kingston, B. R., Wu, J. L., Wilhelm, S., & Chan, W. C. (2019). Elimination pathways of nanoparticles. *ACS Nano, 13*(5), 5785–5798.

Ray, P., Haideri, N., Haque, I., Mohammed, O., Chakraborty, S., Banerjee, S., et al. (2021). The impact of nanoparticles on the immune system: A gray zone of nanomedicine. *Journal of Immunological Sciences, 5*(1).

Sabourian, P., Yazdani, G., Ashraf, S. S., Frounchi, M., Mashayekhan, S., Kiani, S., & Kakkar, A. (2020). Effect of physico-chemical properties of nanoparticles on their intracellular uptake. *International Journal of Molecular Sciences, 21*(21), 8019.

Schneider, S., Eppler, F., Weber, M., Olowojoba, G., Weiss, P., Hübner, C., et al. (2016). Multiscale dispersion-state characterization of nanocomposites using optical coherence tomography. *Scientific Reports, 6*(1), 31733.

Singh, R., & Nalwa, H. S. (2011). Medical applications of nanoparticles in biological imaging, cell labeling, antimicrobial agents, and anticancer nanodrugs. *Journal of Biomedical Nanotechnology, 7*(4), 489–503.

Souza, A. O. D. (2023). Overview of nanomaterials and cellular interactions. *Biointerface Research in Applied Chemistry. Biointerface Research in Applied Chemistry, 13*(4).

Sun, T., Kang, Y., Liu, J., Zhang, Y., Ou, L., Liu, X., et al. (2021). Nanomaterials and hepatic disease: Toxicokinetics, disease types, intrinsic mechanisms, liver susceptibility, and influencing factors. *Journal of Nanobiotechnology, 19*, 1–23.

Swartzwelter, B. J., Mayall, C., Alijagic, A., Barbero, F., Ferrari, E., Hernadi, S., et al. (2021). Cross-species comparisons of nanoparticle interactions with innate immune systems: A methodological review. *Nanomaterials, 11*(6), 1528.

Yang, W., Wang, L., Mettenbrink, E. M., DeAngelis, P. L., & Wilhelm, S. (2021). Nanoparticle toxicology. *Annual Review of Pharmacology and Toxicology, 61*, 269–289.

Zhang, S., Gao, H., & Bao, G. (2015). Physical principles of nanoparticle cellular endocytosis. *ACS Nano, 9*(9), 8655–8671.

Zhang, N., Xiong, G., & Liu, Z. (2022). Toxicity of metal-based nanoparticles: Challenges in the nano era. *Frontiers in Bioengineering and Biotechnology, 10*, 1001572.

Chapter 6
Mechanism of Nanoparticle Toxicity

Patrick Omoregie Isibor, Ameh Simon Sunday, Adamu Binta Buba, and Oluwafemi Adebayo Oyewole

6.1 Introduction

Because of their small size and special qualities, nanoparticles which are defined as particles with at least one dimension between 1 nm and 100 nm have enormous potential in biomedicine (Hano & Abbasi, 2021). Nevertheless, the toxicity of nanoparticles continues to be a worry, which has hampered their clinical translation. Unintended side effects and a lack of biocompatibility have surfaced as problems for nanoparticles, despite their considerable potential in medication administration, imaging, diagnostics, and therapy (Dobrovolskaia & McNeil, 2015). It is essential to clarify the processes by which nanoparticles cause toxicity in order to create safer designs and focused mitigation techniques.

NPs might not appear to have any harmful effects. Reactive oxygen species (ROS) are produced more frequently as a result of these particles' enhanced chemical reactivity due to their larger surface area-to-volume ratio (Flores-López et al., 2019). Because of how their surfaces interact with biological systems, nanoparticles' surface area does have a significant role in their intrinsic toxicity (Canta & Cauda, 2020). One of the harmful mechanisms of NPs is the generation of ROS, which can lead to oxidative stress, inflammation, and subsequent damage to DNA,

P. O. Isibor (✉)
Department of Biological Sciences, Covenant University, Ota, Ogun State, Nigeria
e-mail: patrick.isibor@covenantuniversity.edu.ng

A. S. Sunday
Department of Biochemistry, Federal University of Technology, Minna, Niger State, Nigeria

A. B. Buba
National Biotechnology Development Agency, Abuja, Nigeria

O. A. Oyewole
Department of Microbiology, Federal University of Technology, Minna, Nigeria

P. O. Isibor et al. (eds.), *Environmental Nanotoxicology*,
https://doi.org/10.1007/978-3-031-54154-4_6

proteins, and cell membranes. Thus, in biomedical applications such as drug delivery systems, gene delivery, and therapeutic applications, evaluation of nanoparticles' toxicity is required (Ivask et al., 2014).

Prooxidants are substances that cause oxidative stress by blocking antioxidants or generating reactive oxygen species (Ozougwu, 2016). By activating NADPH-dependent enzyme systems and causing intracellular ROS formation through mitochondrial respiration, NPs interact with cells to produce their prooxidant effects (Fard et al., 2015). A thorough understanding of the underlying mechanisms of nanoparticle toxicity will clarify risk factors and structure-function correlations. With less harmful effects, safer nanoparticles can be designed for a variety of biomedical applications thanks to this knowledge. A mechanistic view of toxicity will be essential as nanoparticle technologies develop in order to fully realise their potential and protect human health.

Generally, particles with at least one dimension in the 1–100 nm size range are considered nanoparticles (Joudeh & Linke, 2022). Compared to bulk materials, nanoparticles have special qualities due to their small size, such as an increased surface area-to-volume ratio and quantum effects (Joudeh & Linke, 2022).

Because of their capacity to interact with biomolecules at both the cellular and molecular levels, nanoparticles in biomedicine are promising for uses like as medication administration, medical imaging, biosensing, and cancer therapy (Nejati et al., 2021). But the same qualities that help nanoparticles become therapeutically beneficial can also cause unintentional harm (Souri et al., 2022). Toxicology is an important topic to understand as nanoparticle engineering develops. The intricate relationships between the biological environment and the physicochemical characteristics of the nanomaterial cause nanoparticle toxicity. Size, shape, surface charge, chemistry, solubility, and aggregation propensity are important factors that affect toxicity (Magdolenova et al., 2014).

6.2 Mechanism of Action of Nanoparticles in Cells

6.2.1 Oxidative Stress

According to Abdul Dayem et al. (2017), nanoparticles have the ability to produce free radicals and reactive oxygen species (ROS) that overwhelm the body's natural antioxidant defences and result in oxidative stress. Proteins, DNA, and lipid membranes can all be harmed by ROS. Reactive oxygen species (ROS) overpower the body's natural antioxidant defences, causing oxidative stress and subsequent cellular damage (He et al., 2017). Superoxide, hydrogen peroxide, hydroxyl radicals, and peroxynitrite are examples of common ROS. ROS aid in cell signalling at physiological levels. On the other hand, elevated ROS causes DNA damage, lipid peroxidation, protein oxidation, and inflammatory activation (Fig. 6.1). This leads to further tissue damage and fibrosis. It has been demonstrated repeatedly that reactive

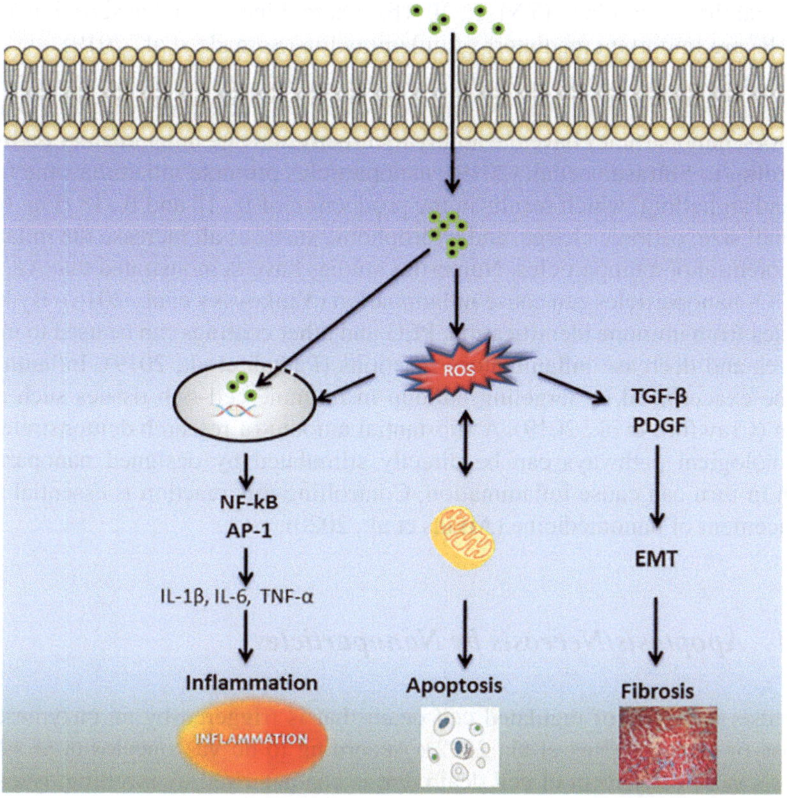

Fig. 6.1 Various toxicity effects of nanoparticles in the cell. (Source: Saifi et al., 2018)

species dissolution and particle-cell interactions cause nanoparticles to produce reactive oxygen species (ROS) (Nel et al., 2006). There are several suggested processes, including mitochondrial disruption, NADPH oxidase activation, inflammatory reactions, and direct redox reactions at the particle surface. The composition of nanoparticles, surface redox activity, type of exposed cell, and intracellular particle interactions are factors that affect oxidative stress (Horie & Tabei, 2021). One of the main causes of nanoparticle toxicity for silica, metal oxides, carbon-based, and polymeric nanoparticles is oxidative stress (Sharma et al., 2019).

6.2.2 *Pro-inflammatory Effects*

According to Chen et al. (2018), inflammation is an immunological response brought on by cytokine signalling that attracts and activates immune cells to areas of tissue injury or infection. While low-level inflammation is a typical physiological reaction, persistent inflammation can lead to fibrosis and detrimental tissue damage.

Prostaglandins, cytokines (TNF-α, IL-1β), chemokines, and transcription factors (NF-kB) are important mediators of inflammation (Somade et al., 2019).

Pro-inflammatory cytokines can be released when pro-inflammatory pathways such as NF-κB are activated by nanoparticles. By interacting with immune cell receptors, nanoparticles have the ability to directly activate inflammatory pathways. According to Shirasuna et al. (2019), nanoparticles promote inflammasome formation and signalling, which results in the production of IL-1β and IL-18 (Fig. 6.1).

Small size, cationic charge, and hydrophobic surfaces all increase the inflammatory potential of nanoparticles. Numerous studies have demonstrated that Ag, SiO_2, and TiO_2 nanoparticles can cause inflammation (Yankovsky et al., 2016). By hiding particles from immune identification, PEG and other coatings can be used to modify surfaces and decrease inflammatory reactions (Pathak et al., 2019). Inflammation may be exacerbated by targeting buildup in immune cell-rich tissues such as the spleen (Crawford et al., 2019). A substantial amount of research demonstrates that immunological pathways can be directly stimulated by designed nanoparticles, which in turn can cause inflammation. Controlling this reaction is essential to the advancement of nanomedicine (Aikins et al., 2020).

6.2.3 Apoptosis/Necrosis by Nanoparticles

Apoptosis is a type of regulated cell death that is triggered by an enzyme called caspase or protease (Rex et al., 2022). According to de Vasconcelos et al. (2019), necrosis is an early form of cell death that is characterised by swelling, lysosomal leakage, and rupture of the plasma membrane. While apoptosis can be physiological, excessive activation causes disease and tissue damage. Usually, necrosis is pathological.

According to Lanone et al. (2009), high quantities of intracellular nanoparticles can cause apoptosis by damaging DNA, mitochondria, the ER, and oxidative stress. Apoptosis could result from nanoparticles physically disturbing organelles like mitochondria and lysosomes within cells. Inflammatory reactions to nanoparticles also trigger apoptotic pathways. By severely damaging membranes and organelles, extreme nanoparticle exposures can result in uncontrollably necrotic cell death (Albanese et al., 2012). Dose-dependent apoptosis/necrosis has been seen in carbon nanotubes, metal oxides, and polymeric nanoparticles (Elsaesser & Howard, 2012). Minimising intracellular accumulation and damage can be achieved by reducing the dosage of nanoparticle delivery and designing biocompatible surfaces (De Jong & Borm, 2008). Anti-inflammatory formulations have the ability to block nanoparticle-stimulated apoptotic signalling (Naseer et al., 2022). There is evidence that nanoparticles cause cell death pathways, which has significant effects on safe dosing. According to Andón and Fadeel (2013), specific toxicity against sick cells may be increased by focusing on apoptotic pathways.

Nanoparticles have demonstrated the capacity to induce apoptosis, a regulated form of cell death, through various mechanisms upon interaction with biological systems. This process involves intricate pathways and molecular signalling cascades leading to cell demise. Nanoparticles can trigger apoptosis through oxidative stress (Fig. 6.1). Some nanoparticles generate reactive oxygen species (ROS) upon interaction with cells, leading to oxidative stress. Elevated ROS levels can disrupt cellular functions and activate apoptotic pathways.

Nanoparticles may interfere with mitochondrial function, causing disruptions in the electron transport chain or inducing mitochondrial membrane permeabilisation, consequently triggering apoptotic pathways. Certain nanoparticles can cause DNA damage within cells, activating signalling pathways that regulate cell cycle progression or initiate apoptotic responses to prevent the propagation of damaged cells. Nanoparticles can activate inflammatory responses and modulate cellular signalling pathways, leading to the activation of apoptotic cascades in affected cells. Upon internalisation by cells, nanoparticles can induce cellular stress responses, disrupt cellular structures, or interfere with specific cellular processes, ultimately leading to apoptosis (Crawford et al., 2019).

The induction of apoptosis by nanoparticles is a complex phenomenon influenced by various nanoparticle properties, including size, shape, surface charge, and chemical composition. Understanding the mechanisms through which nanoparticles trigger apoptosis is crucial for evaluating their potential toxicity and designing nanoparticles with reduced adverse effects in biomedical and environmental applications.

6.2.4 Fibrosis by Nanoparticles

Understanding the mechanisms underlying nanoparticle-induced fibrosis is crucial for assessing the potential risks associated with their use in various applications. It emphasises the importance of designing nanoparticles with reduced fibrotic potential to ensure their safety in biomedical and industrial settings. Nanoparticles have been associated with the potential to induce fibrosis, a pathological condition characterised by the excessive deposition of extracellular matrix components, notably collagen, leading to tissue scarring and dysfunction. Several mechanisms contribute to nanoparticle-induced fibrosis (Saifi et al., 2018). As explained earlier, nanoparticles can trigger inflammatory reactions upon interaction with cells or tissues. Prolonged or unresolved inflammation can stimulate fibroblast activation and collagen production, promoting fibrotic changes (Fig. 6.1).

Nanoparticles can stimulate the activation of fibroblasts, leading to their transformation into myofibroblasts, which are responsible for excessive collagen deposition and tissue remodelling in the fibrotic process. Certain nanoparticles induce oxidative stress within cells or tissues, leading to the activation of profibrotic signalling pathways. Elevated oxidative stress can contribute to the progression of fibrosis. Epithelial-mesenchymal transition may be triggered by nanoparticles in epithelial cells, promoting their transformation into mesenchymal cells. This transition can contribute to fibrosis by enhancing the production of extracellular matrix components. Nanoparticles with specific properties or characteristics may directly injure tissues, triggering a repair response that inadvertently leads to fibrotic changes (Choi et al., 2007).

Prolonged exposure to certain nanoparticles, especially those with poor clearance from the body, may result in their accumulation within tissues, contributing to persistent tissue damage and fibrotic responses.

6.2.5 Membrane Disruption

A lipid bilayer that makes up cell membranes acts as a selective barrier to control the movement of molecules (Sakuragi & Nagata, 2023). Cell death, cytoplasmic leakage, uncontrolled diffusion, and ion inflow can all result from membrane integrity disruption. Surfactants, peptides, and environmental pollutants are known to cause membrane disruption (Bezza et al., 2020). Nanoparticles directly interact with lipid bilayers by insertion, electrostatic and hydrophobic interactions, and pore creation. Because they attract the negatively charged cell surface, cationic nanoparticles exhibit increased toxicity and disruption of membranes. Small nanoparticle-induced high local curvature can also interfere with membrane packing and permeability. Membrane-damaging effects have been reported for dendrimers, metal nanoparticles, and carbon nanotubes (Elsaesser & Howard, 2012). Hydrophilic coatings, such as PEG, when applied to the surface, minimise toxicity and minimise membrane adherence and insertion.

6.2.6 DNA Damage

According to Huang and Zhou (2021), DNA damage can take the form of cross-links, structural distortions of the double helix, single- and double-strand breaks, and chemical changes of bases. Damage to DNA can result in mutations, genomic instability, changed gene expression, and even cell death if it is not correctly repaired. When exposed to and absorbed by DNA strands, nanoparticles have the potential to cause direct physical harm. Metal oxide nanoparticles have the ability to dissolve inside cells, releasing ions that bind with DNA and produce adducts. Nanoparticle-induced ROS is a common source of oxidative damages to DNA, such as 8-OHdG (Lanone et al., 2009). Metal oxides, quantum dots, and carbon-based nanoparticles have all demonstrated genotoxic properties (Magdolenova et al., 2014). By reducing nanoparticle interactions and disintegration, surface coatings such as aminosilanes can lessen damage to DNA.

6.2.7 Protein Denaturation

Protein denaturation is the process of changing a folded protein's structure, which results in function loss. According to Ahkarume et al. (2021), it is brought on by variations in temperature, pH, or chemical exposure. According to Wijayanti et al. (2014), denaturation results from the loss of stabilising intramolecular connections, which unfolds and aggregates proteins. According to Frank et al. (2022), the consequences of protein denaturation include thrombus development, inhibited signalling pathways, protein fibrillation, and enzyme inhibition. Through hydrophobic and electrostatic interactions that destabilise native conformations, the surfaces of nanoparticles can denature proteins. Protein denaturing and fibrillating effects have been reported for carbon nanotubes and metal nanoparticles. PEG surface modification reduces denaturing surface interactions with proteins by encircling nanoparticles in a hydration layer. Reduced dosages reduce the effects of protein binding and aggregation. Research shows that proteins can be easily unfolded and misfolded by nanoparticles via surface interaction, which makes this mechanism essential for biocompatibility.

6.2.8 Mitochondrial Toxicity

Important organelles called mitochondria use oxidative phosphorylation to produce ATP (Iwasaki et al., 2020). Energy balance and cell respiration can be negatively impacted by disruptions in mitochondrial activity (Seungyoon & Pekkurnaz, 2018). Apoptosis activation, increased ROS production, and modified membrane potential are some of the main consequences of mitochondrial malfunction (Sun et al., 2022). Through direct contacts and membrane disruption, nanoparticles can cause damage to mitochondria. Nanoparticle-induced oxidative stress can also harm mitochondrial DNA, lipids, and proteins. Cationic charge, redox activity, and lipophilicity are factors that increase mitochondrial toxicity. Nanoparticles' direct interactions with mitochondria can be reduced by surface modification and coatings. Another important tactic is to minimise particle interactions with mitochondria by optimising size and charge (Stern et al., 2012). According to research, one of the main toxicity mechanisms of manufactured nanoparticles is its direct impairment of mitochondrial structure and function through oxidative damage and membrane rupture (Bathi et al., 2022).

6.2.9 Immunotoxicity

Through a network of cells, signalling proteins, and chemical mediators, the immune system defends against infection and illnesses (Zhao et al., 2021). Exogenous substances that unintentionally stimulate or depress the immune system can cause immunotoxicity, which can lead to inflammation, hypersensitivity, immunosuppression, and autoimmune. By interacting with proteins, receptors, and cells, nanoparticles have the ability to trigger immunological responses (Ilinskaya & Dobrovolskaia, 2013a, b). According to Italiani and Boraschi (2016), immunogenic pollutants, cationic nanoparticles, and the application of immunostimulatory coatings all increase immunotoxicity. According to Dobrovolskaia and McNeil (2015), Ag, Au, SiO_2, carbon nanotubes, and polymeric nanoparticles have demonstrated pro-inflammatory, immunosuppressive, and adjuvant properties. Unwanted immune detection, activation, and sequestration can be reduced with the use of biocompatible surface coatings. An increasing amount of evidence suggests that nanoparticles interact with immunological components, leading to immunotoxicity that is not intended and needs to be carefully designed around.

6.2.10 Reticuloendothelial System Toxicity

A crucial factor for the effectiveness of a drug involves its sustained presence in the bloodstream to execute its intended action or reach specific targeted sites. Similarly, the success of nanoparticles (NPs) depends on their adequate distribution and persistence in the body to facilitate imaging or drug delivery to intended locations (Choi et al., 2007).

However, the reticuloendothelial system (RES) swiftly removes nanoparticles from the bloodstream, directing them to organs like the liver and spleen known for their heightened RES activity. When administered intravenously, the predominant route for nanoparticle delivery, most nanoparticles fail to reach their targeted action sites. Instead, they are sequestered by the liver and spleen or eliminated via kidney filtration, resulting in less than 5% of nanoparticles effectively reaching diseased areas (Murdock et al., 2009; Zhang et al., 2016; Sharma et al., 2018).

Furthermore, kidneys significantly influence nanoparticle blood levels, particularly for nanoparticles smaller than 6 nm, which are swiftly filtered and eliminated. This rapid clearance by the kidneys nullifies the intended actions of these nanoparticles (Navya & Daima, 2016; Simonin et al., 2017; De Matteis et al., 2020). The reticuloendothelial system (RES) comprises an integral component of the body's immune defence, involving diverse phagocytic cells such as macrophages, Kupffer cells, and monocytes. These cells predominantly reside in organs like the liver, spleen, and bone marrow, equipped to engulf and degrade nanoparticles (NPs). Within the liver's sinusoids, resident tissue macrophages known as Kupffer cells play a significant role in eliminating synthetic particles from circulation (Yu & Zheng, 2015).

Pristine NPs cannot be directly eliminated from the bloodstream. However, upon encountering opsonins in the blood, these proteins bind to the NP surface, rendering them susceptible to recognition by macrophages. Consequently, NPs undergo rapid clearance from the bloodstream, accumulating inadvertently in RES organs, thereby potentially inducing toxicity in these organs. Studies involving gold NPs demonstrated an increase in Kupffer cells within the liver of mice, indicating liver toxicity, alongside evident splenic toxicity attributed to these gold NPs (Astashkina et al., 2014). Besides the liver, the spleen has also been identified as contributing to the removal of nanoparticles from the bloodstream. Nanoparticles larger than 200 nm are notably preferentially eliminated by the spleen due to specific physiological mechanism and the presence of a tight reticular mesh system of 200 nm width (Shannahan et al., 2015). Studies revealed that approximately 50% of polystyrene nanoparticles, measuring 250 nm in diameter and coated with poloxamine 908, exhibit an inclination to accumulate within the spleen within a 24-h timeframe subsequent to injection, owing to the spleen's physical filtration capability. Despite the deterrence of nanoparticle interactions with the reticuloendothelial system (RES), thereby prolonging their circulation duration and diminishing toxicity, a notable proportion of nanoparticles still tend to accumulate in the liver and spleen, signifying potential for long-term adverse effects. Furthermore, the introduction of a neutral charge on nanoparticles has demonstrated potential in evading RES clearance and mitigating associated toxicities (Crosera et al., 2009; Rajani et al., 2022).

6.3 Factors Influencing Nanoparticle Toxicity

6.3.1 Nanoparticle Size

The surface area-to-volume ratio increases exponentially with decreasing nanoparticle size. This increases reactivity on the surface. According to Sharma (2019), smaller nanoparticles have a higher capacity to interact with biomolecules, enter cells and tissues, and concentrate inside cells. Nonetheless, the mononuclear phagocyte system and renal excretion may be able to quickly remove very tiny nanoparticles.

Size-dependent toxicity of metal, metal oxide, and polymeric nanoparticles has been demonstrated *in vitro*. Additionally, studies conducted *in vivo* show that smaller nanoparticles are more harmful to animal models. Increased surface contacts, catalytic reactivity, and particle number at equivalent mass dosages are some of the suggested causes (Nel et al., 2006). However, the kidneys often remove nanoparticles smaller than 10 nm quickly, which lessens long-term accumulation and consequences (Choi et al., 2007).

Numerous studies attest to the fact that size plays a crucial role in determining how harmful nanoparticles are. In order to create safe nanomedicines, smaller sizes raise the possibility of negative effects but also need to be balanced with pharmacokinetic considerations (Domingues et al., 2022).

6.3.2 Shape of Nanoparticle

Many interactions in biological settings are governed by the shape of nanoparticles in addition to their size (Kinnear et al., 2017). The interface between nanoparticles and cells and biomolecules is influenced by aspect ratios, angles, surface curvature, edges/vertices, and symmetry (Gwo et al., 2016). Circulation time, biodistribution, and clearance patterns are all impacted by shape.

Nonspherical particles such as nanorods, nanowires, and nanotubes are more hazardous *in vitro* than spherical nanoparticles. According to Gratton et al. (2008), shape affects the kinetics of cell membrane attachment, cellular uptake, and the generation of inflammatory cytokines and oxidative stress. *In vivo* investigations have also reported increased toxicity of high aspect ratio particles in animal models. Increased membrane adherence, thwarted phagocytosis, and reduced clearance for nonspherical shapes are some of the hypothesised mechanisms.

A rising corpus of research indicates that the form of nanoparticles profoundly affects behaviour, interactions, and toxicity in biological systems (Malhotra et al., 2020). Optimal nanomedicines could be achieved through rational form design.

6.3.3 Surface Chemistry

Numerous interactions between nanoparticles and proteins, lipid membranes, immune cells, and other biomolecules are determined by their surface chemistry (Auría-Soro et al., 2019). Biological responses are mediated by the chemical composition, which includes charge, functional groups, and hydrophobicity/hydrophilicity. Changing the surface chemistry of a particle offers a way to reduce its toxicity and increase its biocompatibility. Because of their favourable electrostatic attachment to negatively charged membranes, cationic nanoparticles are more hazardous than their anionic or neutral counterparts. Hydrophobic surfaces exhibit enhanced toxicity via attaching to and integrating into protein hydrophobic pockets and lipid bilayers in cells. Protein adsorption, water solubility, and catalytic activity patterns are determined by surface functional groups such as amines, carboxyls, and hydroxyls (Nel et al., 2006). To increase biocompatibility, surfaces can be modified using PEG, albumin, or chitosan to lessen protein binding and membrane interactions (Yankovsky et al., 2016). There is compelling evidence that the surface chemistry of nanoparticles has a significant impact on toxicity, biological interactions, and functional performance. For safer design, surface property control is essential.

6.3.4 Aggregation State

Because of their hydrophobic contacts, high ionic strength, and protein binding, nanoparticles have a tendency to cluster in biological fluids. Different processes based on diffusion, sedimentation, and cellular absorption are used by agglomerates and individual nanoparticles to mediate toxicity. Surface chemistry, dosage, dispersion effects, and ambient factors all affect aggregation.

Because of their lower cellular absorption, aggregates exhibit less toxicity *in vitro* than well-dispersed nanoparticles (Murdock et al., 2009). Aggregates, however, might increase the concentration of nanoparticles locally, which can worsen certain consequences such as oxidative stress. While single particles can traverse membranes, aggregates have the potential to physically restrict blood arteries (Albanese et al., 2012).

By boosting solubility and steric repulsion, functionalisation with PEG, albumin, and surfactants reduces aggregation. The processes of nanotoxicity are mediated by both single and aggregated nanoparticles. To reduce toxicity, it is critical to rationally design dispersion and aggregation behaviour.

6.3.5 Dose/Exposure

When it comes to the toxicity of xenobiotics, including nanoparticles, exposure route and dose are crucial factors. It is anticipated that toxicity would follow the traditional dose-response model, in which effects grow with dosage. Exposure routes like intravenous, oral, cutaneous, and inhalation affect tissue distribution. Higher nanoparticle doses cause more toxicity both *in vivo* and *in vitro* by increasing cellular absorption and cumulative damage. While inhalation leads to significant nanoparticle accumulation in the lungs, oral treatment produces limited absorption. Although intravenous injection causes quick distribution in the bloodstream, it can also cause immediate blood-mediated responses. Particularly for nonbiodegradable particles, repeated exposures as opposed to one-time exposures can worsen toxicity (Zhang et al., 2016). The intricate interaction of dosage concentration, exposure route/duration, and nanoparticle characteristics mediates toxicity. A wealth of data suggests that the method of exposure and the dose of given nanoparticles have a significant impact on the toxicity, accumulation, and distribution of the particles. It is essential to carefully regulate these factors.

6.3.6 Persistence

The toxicity of nanoparticles is determined by their rate of degradation and their profile of clearance following exposure. When given repeatedly, biopersistent particles build up and resist destruction. Biodegradable nanoparticles, on the other hand, dissolve and leave the body more easily. According to research conducted *in vivo*, when administered frequently, biopersistent nanoparticles such as SiO_2 and TiO_2 exhibit greater toxicity in comparison to biodegradable nanoparticles (Zhang et al., 2016). Over time, accumulation of persistent particles aggravates mechanisms such as membrane rupture, inflammation, and oxidative stress. Increased retention of nanoparticles in organs is also caused by impaired phagocyte and renal clearance. Soft nanoparticles and biodegradable polymerics reduce the toxicity and steadily increasing tissue concentrations over time. Considerable data suggests that long-term accumulation and toxicity are mediated by the biodegradability and durability of nanoparticles.

6.4 Toxicity Assessment Methods

6.4.1 In Vitro *Models*

The main benefits of using *in vitro* models are their affordability, ease of use, and capacity to examine toxicity pathways at the cellular level. To evaluate nanoparticle cyto-compatibility, a variety of cell types, culture formats, and test endpoints have been used. Immortalised cell lines and primary cells are used in the majority of prevalent *in vitro* toxicity models because they offer repeatable, scalable platforms. The main cell types, which span a variety of organs, are epithelial cells, fibroblasts, macrophages, lymphocytes, hepatocytes, neurons, and stem cells (Lanone et al., 2009). Measurements are frequently made of viability, apoptosis, oxidative stress, inflammation, genotoxicity, and mitochondrial damage. Prior to animal research, *in vitro* methods provide invaluable platforms for quick, economical screening and mechanistic insights into the cellular toxicity of nanoparticles.

6.4.2 In Vivo *Models*

It is still necessary to evaluate the effects of nanomaterials in a physiological environment using *in vivo* toxicity models in complete organisms. Analysing biodistribution, clearance, metabolism, and toxicity in a living system is made possible by the use of animal models. The most popular animals used to assess nanotoxicity in the beginning are mice and rats exposed by intravenous, oral, cutaneous, and inhalation routes (De Jong & Borm, 2008). Organ accumulation, biochemical and haematological indicators, histopathology, oxidative stress, inflammation, genotoxicity, and mortality are examples of toxicity end goals.

6.5 Surface Modification

Protein adsorption, immunological interactions, hydrophobicity, and enzymatic activity that drives toxicity are all determined by the surface of the nanoparticle. Optimising surface chemistry via functionalisation seeks to maintain function while enhancing biocompatibility. PEGylation reduces protein interaction and obscures nanoparticles from immune detection by attaching polyethylene glycol chains. The toxicity of cationic particles can be reduced by adding charged functional groups. Citric acid-based ionic coatings are frequently used. By scavenging reactive oxygen species, antioxidant functionalisation with substances such as catechols reduces oxidative damage. Coatings containing chitosan and albumin reduce interactions with hydrophobic regions on biomolecules and cell membranes. For a wide variety of nanoparticle types, surface modification reliably reduces adhesion, uptake, inflammation, and membrane disruption (Shi et al., 2017). Multipurpose "stealth" coatings preserve functionality such as imaging or targeting while optimising biocompatibility. Numerous studies show that surface engineering greatly improves translational potential by reducing unwanted interactions that cause nanotoxicity.

6.6 Controlled Release

Cytotoxic buildup and overload may result from burst release and inadequate control over cellular nanoparticle uptake. The goal of stimuli-responsive formulations and sustained release systems is to reduce excessive intracellular levels. According to De Jong and Borm (2008), nanoparticles enclosed in porous matrices or shells permit gradual leakage over time, lowering toxicity in comparison to bolus distribution. To regulate release kinetics, stimuli-responsive systems take advantage of external (heat, light, ultrasound, etc.) or internal (pH, enzymes, redox, etc.) stimuli. Ligand-targeted prodrug nanoparticles reduce systemic exposure by releasing medicines only after binding cancer biomarkers. *In vitro* and *in vivo*, sustained release nanoparticles outperform burst release formulations in terms of biocompatibility and survivability (Xia et al., 2014). By restricting medication levels outside of target areas, stimulus-triggered systems increase safety. Managed release techniques reduce the buildup of hazardous nanoparticles and get around toxicity barriers, providing strong design choices to enhance translation (Weng et al., 2008).

6.7 Combination Therapies

Combination therapies and multicomponent systems have the potential to reduce the toxicity of nanoparticles through mutually reinforcing effects. Antioxidants, immune suppressants, co-administration of free medicines, and endpoint inhibitors can reduce the risk of side effects. Oxidative stress and inflammation are reduced when free radical scavengers like vitamin E and N-acetylcysteine are co-delivered with nanoparticles. To avoid immunotoxicity, immunosuppressive medications such as cyclosporine A, rapamycin, and corticosteroids are given in combination with nanodrugs (Sharma, 2019). To reduce toxicity endpoints, apoptosis inhibitors (Z-VAD-FMK), necrosis inhibitors (necrostatins), and anti-inflammatory medications (DFMO) are combined. When compared to nanoparticles alone, combinations show reduced levels of oxidative damage, immune cell infiltration, and cell death indicators both *in vitro* and *in vivo*. Reduced nanoparticle doses can produce therapeutic results with fewer negative consequences when co-treatments are used (Orr et al., 2011). Research suggests that by guarding against important toxicity mechanisms, combination techniques have a great deal of promise to enhance the safety and translational viability of nanotherapeutics.

6.8 Conclusion

Although there are now many promising opportunities for sophisticated nanomedicines and diagnostics thanks to nanotechnology, toxicity concerns are still a major obstacle to clinical translation. To develop safer designs and reduce dangers, a thorough understanding of the mechanisms by which nanoparticles cause harmful consequences is essential. The existing body of research on the main molecular and cellular mechanisms of nanoparticle toxicity was summarised in this study. Oxidative stress, inflammation, disruption of membranes, DNA damage, protein misfolding, mitochondrial malfunction, immunotoxicity, and pathways leading to cell death are important processes. Understanding the connections between the physicochemical characteristics of nanoparticles and these biological interactions offers structure-function information that can direct efforts to increase biocompatibility. Size, shape, surface chemistry, and aggregation are some of the factors that substantially influence a product's propensity towards toxicity and point the way towards rational design. As nanoparticle technologies progress, a mechanistic viewpoint on risk factors, mitigation measures, and processes of nanotoxicity will prove to be useful. To fully realise the promise of nanomedicine, the expanding body of knowledge in this area will speed up the development of safer nanotherapies and nanodiagnostics by shedding light on the underlying causes of side effects and solutions to overcome obstacles.

References

Abdal Dayem, A., Hossain, M. K., Lee, S. B., Kim, K., Saha, S. K., Yang, G. M., & Cho, S. G. (2017). The role of reactive oxygen species (ROS) in the biological activities of metallic nanoparticles. *International Journal of Molecular Sciences, 18*(1), 120.

Aikins, M. E., Xu, C., & Moon, J. J. (2020). Engineered nanoparticles for cancer vaccination and immunotherapy. *Accounts of Chemical Research, 53*(10), 2094–2105.

Akharume, F. U., Aluko, R. E., & Adedeji, A. A. (2021). Modification of plant proteins for improved functionality: A review. *Comprehensive Reviews in Food Science and Food Safety, 20*(1), 198–224.

Albanese, A., Tang, P. S., & Chan, W. C. (2012). The effect of nanoparticle size, shape, and surface chemistry on biological systems. *Annual Review of Biomedical Engineering, 14*, 1–16.

Andón, F. T., & Fadeel, B. (2013). Programmed cell death: Molecular mechanisms and implications for safety assessment of nanomaterials. *Accounts of Chemical Research, 46*(3), 733–742.

Astashkina, A. I., Jones, C. F., & Thiagarajan, G. (2014). Nanoparticle toxicity assessment using an in vitro 3-D kidney organoid culture model. *Biomaterials, 35*(24), 6323–6331.

Auría-Soro, C., Nesma, T., Juanes-Velasco, P., Landeira-Viñuela, A., Fidalgo-Gomez, H., Acebes-Fernandez, V., & Fuentes, M. (2019). Interactions of nanoparticles and biosystems: Microenvironment of nanoparticles and biomolecules in nanomedicine. *Nanomaterials, 9*(10), 1365.

Bathi, J. R., Wright, L., & Khan, E. (2022). Critical review of engineered nanoparticles: Environmental concentrations and toxicity. *Current Pollution Reports, 8*(4), 498–518.

Bezza, F. A., Tichapondwa, S. M., & Chirwa, E. M. (2020). Fabrication of monodispersed copper oxide nanoparticles with potential application as antimicrobial agents. *Scientific Reports, 10*(1), 16680.

Canta, M., & Cauda, V. (2020). The investigation of the parameters affecting the ZnO nanoparticle cytotoxicity behaviour: A tutorial review. *Biomaterials Science, 8*(22), 6157–6174.

Chen, L., Deng, H., Cui, H., Fang, J., Zuo, Z., Deng, J., et al. (2018). Inflammatory responses and inflammation-associated diseases in organs. *Oncotarget, 9*(6), 7204.

Choi, K. S., Cai, Q. Y., Kim, S. H., Kim, S. Y., Byun, S. J., Kim, K. W., et al. (2007). Colloidal gold nanoparticles as a blood-pool contrast agent for X-ray computed tomography in mice. *Investigative Radiology, 42*(12), 797–806.

Crawford, A., Haber, L., Kelly, M. P., Vazzana, K., Canova, L., Ram, P., et al. (2019). A Mucin 16 bispecific T cell–engaging antibody for the treatment of ovarian cancer. *Science Translational Medicine, 11*(497), eaau7534.

Crosera, M., Bovenzi, M., Maina, G., Adami, G., Zanette, C., Florio, C., & Filon Larese, F. (2009). Nanoparticle dermal absorption and toxicity: A review of the literature. *International Archives of Occupational and Environmental Health, 82*, 1043–1055.

De Jong, W. H., & Borm, P. J. (2008). Drug delivery and nanoparticles: Applications and hazards. *International Journal of Nanomedicine, 3*(2), 133–149.

De Matteis, V., Cascione, M., Rizzello, L., Liatsi-Douvitsa, E., Apriceno, A., & Rinaldi, R. (2020). Green synthesis of nanoparticles and their application in cancer therapy. In *Green synthesis of nanoparticles: Applications and prospects* (pp. 163–197). Springer.

de Vasconcelos, N. M., Van Opdenbosch, N., Van Gorp, H., Parthoens, E., & Lamkanfi, M. (2019). Single-cell analysis of pyroptosis dynamics reveals conserved GSDMD-mediated subcellular events that precede plasma membrane rupture. *Cell Death & Differentiation, 26*(1), 146–161.

Dobrovolskaia, M. A., & McNeil, S. E. (2015). Strategy for selecting nanotechnology carriers to overcome immunological and hematological toxicities challenging clinical translation of nucleic acid-based therapeutics. *Expert Opinion on Drug Delivery, 12*(7), 1163–1175.

Domingues, C., Santos, A., Alvarez-Lorenzo, C., Concheiro, A., Jarak, I., Veiga, F., et al. (2022). Where is nano today and where is it headed? A review of nanomedicine and the dilemma of nanotoxicology. *ACS Nano, 16*(7), 9994–10041.

Elsaesser, A., & Howard, C. V. (2012). Toxicology of nanoparticles. *Advanced Drug Delivery Reviews, 64*(2), 129–137.

Fard, J. K., Jafari, S., & Eghbal, M. A. (2015). A review of molecular mechanisms involved in toxicity of nanoparticles. *Advanced Pharmaceutical Bulletin, 5*(4), 447.

Flores-López, L. Z., Espinoza-Gómez, H., & Somanathan, R. (2019). Silver nanoparticles: Electron transfer, reactive oxygen species, oxidative stress, beneficial and toxicological effects. Mini review. *Journal of Applied Toxicology, 39*(1), 16–26.

Frank, D., Zlotnik, A., Boyko, M., & Gruenbaum, B. F. (2022). The development of novel drug treatments for stroke patients: A review. *International Journal of Molecular Sciences, 23*(10), 5796.

Gratton, S. E., Ropp, P. A., Pohlhaus, P. D., Luft, J. C., Madden, V. J., Napier, M. E., & DeSimone, J. M. (2008). The effect of particle design on cellular internalization pathways. *Proceedings of the National Academy of Sciences, 105*(33), 11613–11618.

Gwo, S., Chen, H. Y., Lin, M. H., Sun, L., & Li, X. (2016). Nanomanipulation and controlled self-assembly of metal nanoparticles and nanocrystals for plasmonics. *Chemical Society Reviews, 45*(20), 5672–5716.

Hano, C., & Abbasi, B. H. (2021). Plant-based green synthesis of nanoparticles: Production, characterization and applications. *Biomolecules, 12*(1), 31.

He, L., He, T., Farrar, S., Ji, L., Liu, T., & Ma, X. (2017). Antioxidants maintain cellular redox homeostasis by elimination of reactive oxygen species. *Cellular Physiology and Biochemistry, 44*(2), 532–553.

Horie, M., & Tabei, Y. (2021). Role of oxidative stress in nanoparticles toxicity. *Free Radical Research, 55*(4), 331–342.

Huang, R., & Zhou, P. K. (2021). DNA damage repair: Historical perspectives, mechanistic pathways and clinical translation for targeted cancer therapy. *Signal Transduction and Targeted Therapy, 6*(1), 254.

Ilinskaya, A. N., & Dobrovolskaia, M. A. (2013a). Nanoparticles and the blood coagulation system. Part II: Safety concerns. *Nanomedicine, 8*(6), 969–981.

Ilinskaya, A. N., & Dobrovolskaia, M. A. (2013b). Nanoparticles and the blood coagulation system. Part I: Benefits of nanotechnology. *Nanomedicine, 8*(5), 773–784.

Italiani, P., & Boraschi, D. (2016). Engineered nanoparticles and the immune system: Interaction and consequences. In *Environmental influences on the immune system* (pp. 205–226). Springer.

Ivask, A., Juganson, K., Bondarenko, O., Mortimer, M., Aruoja, V., Kasemets, K., et al. (2014). Mechanisms of toxic action of Ag, ZnO and CuO nanoparticles to selected ecotoxicological test organisms and mammalian cells in vitro: A comparative review. *Nanotoxicology, 8*(sup1), 57–71.

Iwasaki, Y., Takeshima, Y., & Fujio, K. (2020). Basic mechanism of immune system activation by mitochondria. *Immunological Medicine, 43*(4), 142–147.

Joudeh, N., & Linke, D. (2022). Nanoparticle classification, physicochemical properties, characterization, and applications: A comprehensive review for biologists. *Journal of Nanobiotechnology, 20*(1), 262.

Kinnear, C., Moore, T. L., Rodriguez-Lorenzo, L., Rothen-Rutishauser, B., & Petri-Fink, A. (2017). Form follows function: Nanoparticle shape and its implications for nanomedicine. *Chemical Reviews, 117*(17), 11476–11521.

Lanone, S., Rogerieux, F., Geys, J., Dupont, A., Maillot-Marechal, E., Boczkowski, J., et al. (2009). Comparative toxicity of 24 manufactured nanoparticles in human alveolar epithelial and macrophage cell lines. *Particle and Fibre Toxicology, 6*(1), 1–12.

Magdolenova, Z., Collins, A., Kumar, A., Dhawan, A., Stone, V., & Dusinska, M. (2014). Mechanisms of genotoxicity. A review of in vitro and in vivo studies with engineered nanoparticles. *Nanotoxicology, 8*(3), 233–278.

Malhotra, N., Lee, J. S., Liman, R. A. D., Ruallo, J. M. S., Villaflores, O. B., Ger, T. R., & Hsiao, C. D. (2020). Potential toxicity of iron oxide magnetic nanoparticles: A review. *Molecules, 25*(14), 3159.

Murdock, R. C., Hussain, S. M., Braydich-Stolle, L. K., Schrand, A. M., Yu, K. O., Mattie, D. M., et al. (2009). Toxicity evaluation for safe use of nanomaterials: Recent achievements and technical challenges. *Advanced Materials, 21*(16), 1549–1559.

Naseer, F., Ahmed, M., Majid, A., Kamal, W., & Phull, A. R. (2022). Green nanoparticles as multifunctional nanomedicines: Insights into anti-inflammatory effects, growth signaling and apoptosis mechanism in cancer. In *Seminars in cancer biology*. Academic Press.

Navya, P. N., & Daima, H. K. (2016). Rational engineering of physicochemical properties of nanomaterials for biomedical applications with nanotoxicological perspectives. *Nano Convergence, 3*, 1–14.

Nejati, K., Dadashpour, M., Gharibi, T., Mellatyar, H., & Akbarzadeh, A. (2021). Biomedical applications of functionalized gold nanoparticles: A review. *Journal of Cluster Science, 33*, 1–16.

Nel, A., Xia, T., Madler, L., & Li, N. (2006). Toxic potential of materials at the nanolevel. *Science, 311*(5761), 622–627.

Orr, G. A., Chrisler, W. B., & Cassens, K. J. (2011). Cellular recognition and trafficking of amorphous silica nanoparticles by macrophage scavenger receptor A. *Nanotoxicology, 5*(3), 296–311.

Ozougwu, J. C. (2016). The role of reactive oxygen species and antioxidants in oxidative stress. *International Journal of Research, 1*(8).

Pathak, S., Pham, T. T., Jeong, J. H., & Byun, Y. (2019). Immunoisolation of pancreatic islets via thin-layer surface modification. *Journal of Controlled Release, 305*, 176–193.

Rajani, C., Borisa, P., Bagul, S., Shukla, K., Tambe, V., Desai, N., & Tekade, R. K. (2022). Developmental toxicity of nanomaterials used in drug delivery: Understanding molecular biomechanics and potential remedial measures. In *Pharmacokinetics and toxicokinetic considerations* (pp. 685–725). Academic Press.

Rex, D. A. B., Keshava Prasad, T. S., & Kandasamy, R. K. (2022). Revisiting regulated cell death responses in viral infections. *International Journal of Molecular Sciences, 23*(13), 7023.

Saifi, M. A., Khan, W., & Godugu, C. (2018). Cytotoxicity of nanomaterials: Using nanotoxicology to address the safety concerns of nanoparticles. *Pharmaceutical Nanotechnology, 6*(1), 3–16. https://doi.org/10.2174/2211738505666171023152928

Sakuragi, T., & Nagata, S. (2023). Regulation of phospholipid distribution in the lipid bilayer by flippases and scramblases. *Nature Reviews Molecular Cell Biology*, (8):576–5961.

Seungyoon, B. Y., & Pekkurnaz, G. (2018). Mechanisms orchestrating mitochondrial dynamics for energy homeostasis. *Journal of Molecular Biology, 430*(21), 3922–3941.

Shannahan, J. H., Podila, R., & Aldossari, A. A. (2015). Formation of a protein corona on silver nanoparticles mediates cellular toxicity via scavenger receptors. *Toxicological Sciences, 143*(1), 136–146.

Sharma, J. N. (2019). *Electrochemical determination of surface area-to-volume ratio for metal nanoparticle analysis*. University of Louisville.

Sharma, D., Sharma, N., Pathak, M., Agrawala, P. K., Basu, M., & Ojha, H. (2018). Nanotechnology-based drug delivery systems: Challenges and opportunities. In *Drug targeting and stimuli sensitive drug delivery systems* (pp. 39–79). William Andrew.

Sharma, S., Singh, V. K., Kumar, A., & Mallubhotla, S. (2019). Effect of nanoparticles on oxidative damage and antioxidant defense system in plants. In *Molecular plant abiotic stress: Biology and biotechnology* (pp. 315–333). https://doi.org/10.1002/9781119463665.ch17

Shi, X., Zhou, K., Huang, F., & Wang, C. (2017). Interaction of hydroxyapatite nanoparticles with endothelial cells: Internalization and inhibition of angiogenesis in vitro through the PI3K/Akt pathway. *International Journal of Nanomedicine*, 549(7673):523–527

Shirasuna, K., Karasawa, T., & Takahashi, M. (2019). Exogenous nanoparticles and endogenous crystalline molecules as danger signals for the NLRP3 inflammasomes. *Journal of Cellular Physiology, 234*(5), 5436–5450.

Simonin, M., Martins, J. M., Le Roux, X., Uzu, G., Calas, A., & Richaume, A. (2017). Toxicity of TiO2 nanoparticles on soil nitrification at environmentally relevant concentrations: Lack of classical dose–response relationships. *Nanotoxicology, 11*(2), 247–255.

Somade, O. T., Ajayi, B. O., Safiriyu, O. A., Oyabunmi, O. S., & Akamo, A. J. (2019). Renal and testicular up-regulation of pro-inflammatory chemokines (RANTES and CCL2) and cytokines (TNF-α, IL-1β, IL-6) following acute edible camphor administration is through activation of NF-kB in rats. *Toxicology Reports, 6*, 759–767.

Souri, M., Soltani, M., Kashkooli, F. M., Shahvandi, M. K., Chiani, M., Shariati, F. S., et al. (2022). Towards principled design of cancer nanomedicine to accelerate clinical translation. *Materials Today Bio, 13*, 100208.

Stern, S. T., Adiseshaiah, P. P., & Crist, R. M. (2012). Autophagy and lysosomal dysfunction as emerging mechanisms of nanomaterial toxicity. *Particle and Fibre Toxicology, 9*, 1–17.

Sun, Q., Li, Y., Shi, L., Hussain, R., Mehmood, K., Tang, Z., & Zhang, H. (2022). Heavy metals induced mitochondrial dysfunction in animals: Molecular mechanism of toxicity. *Toxicology, 469*, 153136.

Weng, D. S., Zhou, J., Zhou, Q. M., Zhao, M., Wang, Q. J., Huang, L. X., et al. (2008). Minimally invasive treatment combined with cytokine-induced killer cells therapy lower the short-term recurrence rates of hepatocellular carcinomas. *Journal of Immunotherapy, 31*(1), 63–71.

Wijayanti, H. B., Bansal, N., & Deeth, H. C. (2014). Stability of whey proteins during thermal processing: A review. *Comprehensive Reviews in Food Science and Food Safety, 13*(6), 1235–1251.

Xia, Q., Fu, P. P., Hwang, H. M., Ray, P. C., & Yu, H. (2014). Mechanisms of nanotoxicity: Generation of reactive oxygen species. *Journal of Food and Drug Analysis, 22*(1), 64–75.

Yankovsky, I., Bastien, E., Yakavets, I., Khludeyev, I., Lassalle, H. P., Gräfe, S., & Zorin, V. (2016). Inclusion complexation with β-cyclodextrin derivatives alters photodynamic activity and biodistribution of meta-tetra (hydroxyphenyl) chlorin. *European Journal of Pharmaceutical Sciences, 91*, 172–182.

Yu, M., & Zheng, J. (2015). Clearance pathways and tumor targeting of imaging nanoparticles. *ACS Nano, 9*(7), 6655–6674.

Zhang, Y. N., Poon, W., Tavares, A. J., McGilvray, I. D., & Chan, W. C. W. (2016). Nanoparticle-liver interactions: Cellular uptake and hepatobiliary elimination. *Journal of Controlled Release, 240*, 332–348.

Zhao, H., Wu, L., Yan, G., Chen, Y., Zhou, M., Wu, Y., & Li, Y. (2021). Inflammation and tumor progression: Signaling pathways and targeted intervention. *Signal Transduction and Targeted Therapy, 6*(1), 263.

Chapter 7
Oxidative Stress and Inflammation Induced by Nanoparticles

Abdulmajid Musa Maku, Adamu Binta Buba, Oluwafemi Adebayo Oyewole, Alhassan Mohammed Alhassan, and Patrick Omoregie Isibor

7.1 Introduction

Nanotechnology is a recent field that concentrates on designing, characterizing, and producing systems and devices by changing their size and shape at the nanoscale. Recent advances in nanotechnology and nanoscience have uses in various industries, including aerospace engineering, food technology, medicine, transportation, drug delivery, cosmetics, biotechnology, and electronics (Samrot & Prakash, 2023). Commercial interest in manufactured nanomaterials (NM) has grown in a number of consumer goods. Their unique physicochemical, thermal, and electrical qualities make it easier for them to be used in clothing, cosmetics, and medicine, increasing the likelihood that people and the environment will come into contact with these nanomaterials (Manke et al., 2013). Engineered nanoparticles differ greatly from bigger particles of the same composition in their form, crystallinity, surface charge, reactive surface groups, dissolving rate, small size, high surface area-to-mass ratio, and state of aggregation and dispersal, despite the fact that larger particles have been shown to have a stronger mass-dependent biological effect and particle-mediated toxicity than nanoparticles (Nemmar et al., 2016). Although there may be beneficial uses for nanoparticles in many different industries, multiple research have disclosed that coming in contact with nanoparticles can have negative consequences on animals and cells (Yu et al., 2020). The study of particles with a size of below 100 nm,

A. M. Maku (✉) · O. A. Oyewole · A. M. Alhassan
Department of Microbiology, Federal University of Technology, Minna, Nigeria
e-mail: abdulmajeedmoosa3@gmail.com

A. B. Buba
National Biotechnology Development Agency, Abuja, Nigeria

P. O. Isibor
Department of Biological Sciences, Covenant University, Ota, Ogun State, Nigeria
e-mail: patrick.isibor@covenantuniversity.edu.ng

P. O. Isibor et al. (eds.), *Environmental Nanotoxicology*,
https://doi.org/10.1007/978-3-031-54154-4_7

commonly referred to as ultrafine particles, has been going on for years even though nanotechnology may be a relatively new field. Nanoparticles are produced in a variety of ways, including naturally occurring processes in the environment (such as forest fires, viruses, and volcanoes), as byproducts of industrial or burning processes and on purpose for application in both commercial and consumer products (such as pigments and chemical catalysts) (Madl et al., 2014). Nanoparticles are pervasive in the environment and can penetrate the body by eating, absorption through the skin, intravenous injection, and inhalation when employed for medical purposes. When nanoparticles are released into the environment, they can easily be absorbed in the cells through free penetration or receptor-mediated endocytosis, where they can interact with proteins, lipids, and genomic DNA (Habas et al., 2021). As a result, reactive oxidative species (ROS), which are the main cause of nanotoxicity, produce oxidative stress. Although some nanoparticles, including those made of gold, silver, copper, and iron, among others, have antioxidant properties, the majority of them contribute to the development of intracellular reactive oxidative stress because they are taken up by cells, react inside of them, and release intracellular metal ions. They are also made during various cellular signaling processes and are a component of the immune system's defense mechanism (Samrot & Prakash, 2023). Damage to biological macromolecules including proteins, lipids, and DNA caused by reactive oxidative stress (ROS) has an adverse impact on cells and leads to mitochondrial malfunction. Numerous physiopathological outcomes, including apoptosis, necrosis, hypertrophy, genotoxicity, inflammation, fibrosis, and even cancer, can be caused by a high amount of reactive oxidative stress (Shields et al., 2021). Additionally, it stimulates the activity of inflammatory cells like macrophages and raises the creation of pro-inflammatory cytokines, which enhances the reactive oxidative stress (Yu et al., 2020). In fact, because of their size and makeup, certain nanoparticles have been discovered to permanently harm cells through organ injury and oxidative stress. Researchers such as Magrez et al. evaluated the cytotoxicity of carbon nanomaterials on lung cancer cells in their study, and the results point to size-dependent cytotoxicity being caused by carbon-based nanoparticles. According to certain theories, a nanoparticle's level of toxicity might vary depending on its size, surface functionality, crystallinity, composition, and aggregation. Additionally, a person's genetic makeup, which is influenced by their capacity to respond and adapt to harmful compounds, determines the toxicity of a nanoparticle in that person (Kumah et al., 2023).

7.2 Types of Nanoparticles

The four types of nanoparticles that can be used for different applications include those with an inorganic basis, those with a carbon basis, those with an organic or polymer basis, and those with a composite basis (Shin et al., 2016; Khan et al., 2019).

7.2.1 Inorganic-Based Nanoparticles

Inorganic-based nanomaterials are made up of various metals and metal oxides. Metal-based inorganic nanomaterials include those made of aluminum, silver, gold, zinc, lead, iron, cadmium, and copper (Kumah et al:, 2023). Zinc oxide (ZnO), copper oxide (CuO), magnesium aluminum oxide ($MgAl_2O_4$), titanium dioxide (TiO_2), cerium oxide (CeO_2), iron oxide (Fe_2O_3), silica (SiO_2), and iron oxide (Fe_3O_4) are examples of metal oxide-based inorganic nanomaterials (Mahji & Yadav, 2020). Metallic nanomaterials, which are solely composed of metal ingredients, may be monometallic, bimetallic, or polymetallic. Bimetallic nanoparticles can be made from alloys or may be constructed up in sheets (coreshell). These nanoparticles have distinctive optical and electrical properties thanks to the features of localized surface plasmon resonance. Additionally, several metal nanoparticles display unusual biological, magnetic, and thermo characteristics. They are playing an increasingly significant role as components for nanomaterial with a range of physical, chemical, biological, biomedical, and pharmaceutical applications (Joudeh & Linked, 2022).

7.2.2 Carbon-Based Nanoparticles

This class of nanoparticles is made up of those that are only composed of carbon atoms. Fullerenes and carbon nanotubes are well-known representatives of this class (Joudeh & Linked, 2022). Allotropic forms of carbon are present in the globular void cage-based nanomaterial known as fullerenes. Due to their conductivity in electricity, excellent strength, organization, electron attraction, and adaptability, they have generated a great deal of economic interest. These substances consist of sp2 hybridized five and six carbon units that are organized (Khan et al., 2019). The cylinder-shaped carbon structures known as carbon nanotubes (CNTs) have unique characteristics. Depending on the number of layers, they can be single-walled (SWCNTs) or multiwalled (MWCNTs). CNTs exhibit outstanding mechanical toughness, electrical conductivity, and thermal characteristics. They have potential usefulness in the domains of medicine, electronics, materials research, and nanotechnology (Zhang et al., 2021).

7.2.3 Organic Nanoparticles

Nanoparticles composed of polymers, lipids, proteins, carbohydrates, and other organic materials fall under this group. The most well-known types are ferritin, dendrimers, liposomes, micelles, and protein complexes. These nanoparticles occasionally have hollow cores like liposomes and are typically nontoxic and biodegradable. Heat, light, and other forms of electromagnetic and thermal radiation can damage organic nanoparticles. They typically come about as a result of non-covalent intermolecular interactions as well that allow them to be more labile and also gives a way for excretion from the body. The composition, surface shape, stability, and carrying capacity of organic nanoparticles are some of the factors that can affect their potential applications (Joudeh & Linked, 2022).

7.2.4 Composite Nanoparticles

Nanoparticles made up of two or more different materials or components are known as composite nanoparticles. They are created using mixtures of nanoparticles made of metal oxide, metal, organic, and/or carbon (Kumah et al., 2023). Through interfaces, the various parts can talk to one another. At the nanoscale, the components' varying capacities interact with one another in a major and powerful way. As a result, other nanoparticles besides composite ones also incorporate these effects. The constraints of single-component attributes are overcome by the composite nanodevices, which also exhibit a variety of distinctive properties and considerably improve intrinsic performance. The development of novel useful materials, the effective utilization of novel energy, wastewater management, and medicine are just a few of the significant domains where composite nanoparticles have shown great potential (Luo et al., 2014).

7.3 Human Exposure to Nanoparticles

The safeness of these elements in the body of human is unaffected by the wide variety of produced nanoparticles used in therapies created from diverse materials. Since it is crucial to comprehend their potentials and toxicological profile following delivery, knowing the process of nanoparticle-induced toxicity is crucial. Due to their unrestrained use and waste release, nanomaterials could be hazardous to the environment's health. The potential for interactions between nanomaterials and aquatic and terrestrial environments is increased by the commercial usage of nanomaterials in products including pharmaceuticals, cosmetics, coatings, electronics, and textiles. Additionally, the use of nanopesticides and fertilizers has increased agricultural exposure, posing unknown dangers to plants and animals (Ferdous & Nemmar, 2020).

Due to repeated exposure over extended periods of time and low excretion levels, metal, metal oxide, and carbon-based nanomaterials are the most significant nanostructures linked to human exposure. These nanomaterials include sheets of graphene, CNTs, silicon nanodevices already used in commercial equipment, tin and titanium oxide nanostructures used as electrode substances in lithium-ion batteries, and silver and gold nanoparticles found in fertilizers, insecticides, and antimicrobials. Nanomaterials have a wide range of uses, from everyday items like cosmetics to building materials, analytical software, and electronic equipment (Ma et al., 2023). Nanoparticles can enter the human body through the skin, ingestion, or inhalation (Matteis, 2017).

7.3.1 *Ingestion*

When eaten elements get access to the human gastrointestinal tract, intestine, and circulatory system, exposure to nanomaterials via ingestion occurs. The particles can contact other systems like the cardiovascular and central neurological systems due to their distribution in the systemic bloodstream (Geiser & Kreyling, 2010). The early pH range of the stomach is 1.22.0, indicating an acidic environment. However, the pH rises to around 5.0 when the bolus is formed, followed by a gradual acidification. The continued destruction of nanoparticles in the digestive juices is encouraged by their solubility at a low pH. Gastric fluids showed signs of degradation and ion release, despite the fact that the stability of nanoparticles in vivo is yet unclear. It has been shown that in ordinary epithelial cells, endocytosis accounts for the majority of the absorption of nanoparticles with a size below 100 nm. The small intestine, which typically has a pH concentration of between 6 and 7 at the duodenum site, is responsible for digestion and absorption activities. Villi and enterocytes are responsible for improving the absorptive surface. Nanoparticles can cause DNA damage, oxidative stress, and inflammations in intestinal cells (Matteis, 2017).

7.3.2 *Inhalation*

The respiratory system serves as the most frequent pathway for the deposit of inhaled nanomaterials since inhalation is the major way that people are exposed to nanomaterials. Multiple worries concerning their possible toxicity to customers and industrial employees during the various procedures have been raised as a result of this. This contact with nanomaterials may result from an unintentional leak that develops during the production, handling, transportation, storage, or disposal of waste. Once they have gotten past the immune system and into the bloodstream, nanoparticles can migrate to other organs. These nanoparticles are transported into cellular compartments, where they have negative short- and long-term effects and raise the risk of bioaccumulation (Liu & Tang, 2020).

About 300,106 alveoli make up the interior surface area of a human lung, which ranges from 75 to 140 m^2. Since the lungs are linked with the outside environment, they serve as the body's principal entry point for nanoparticles. They are also seen as crucial and the main area to focus on while researching the impact of nanostructures (Pacurari et al., 2016). Oxygen is taken in through the nose and mouth, moves through the pharynx and tracheobronchial tree, and then arrives at the alveoli. The Weibel bifurcating tubes model, which is divided into the conducting area and breathing region, as already covered in depth, can be used to depict particle transport, which is reliant on the structure of the respiratory airway. Nanomaterials can disperse and accumulate in the alveolar region because of their small size and longer retention times. Some of these particles may also get through the capillary endothelial cells and the alveolar epithelium and get into the cardiovascular system and other structures within the body (Qiao et al., 2015).

7.3.3 *Skin Penetration*

Skin penetration from handling nanomaterials or skin contact with contaminated surfaces are two additional potential exposure routes. Additionally, this exposure could happen during therapeutic or diagnostic procedures involving subcutaneous, intravenous, or intramuscular injection (Boushehri, 2020). Nanomaterials are evolving, and their use in the cosmetics industry is expanding. Cosmetics incorporating nanoparticles have shown a significant spread since 2006. According to estimates, TiO_2NPs (70–80%), ZnONPs (70%), and AgNPs (20%) are present in cosmetic items. Nanoparticles cannot cross the epidermal barrier, according to some studies. In contrast, some investigations demonstrated the entry of metal nanoparticles, such as iron nanoparticles, into the basal and spinous layers of hair follicles. Size has an impact on how nanoparticles enter the body through the skin in this situation as well. According to the available data, nanoparticles with a size of about 4 nm can get access to healthy skin, but it has been discovered that particles larger than 45 nm enter only a compromised skin (Matteis, 2017).

7.4 Oxidative Stress

Reactive oxygen species (ROS), which are common in the body and can lead to potentially damaging biological reactions, are what cause oxidative stress. ROS imbalance inhibits body's systems from quickly detoxifying reactive intermediates or undoing damage by activating antioxidant enzymes and nonenzymatic antioxidants, which causes of the problem. When ROS level climbs above the threshold as a result of nanoparticle penetration, further negative effects, such as inflammation, genotoxicity, and fibrosis, ensue from activated immune cells and activated cell signaling pathways (Samrot & Prakash, 2023).

Reactive oxygen species (ROS) excessive creation can cause oxidative stress, which makes it difficult for cells to maintain typical physiological redox-regulated processes. The oxidative alteration of proteins to produce protein radicals, the beginning of lipid peroxidation, DNA strand separation, modifications to nucleotides, the alteration of genome transcription by the stimulation of redox-sensitive transcription elements, and the alteration of autoimmune reactions by transmitted signals are examples of damage to cell function and development that may result in cell destruction and genotoxic reactions. To counteract an overactive ROS response, cells can activate both enzymatic and nonenzymatic antioxidant systems (Fu et al., 2014). The hierarchy system of oxidative stress was suggested as the basis for nanoparticle-mediated oxidative stress. This idea states that antioxidant enzyme systems in cells and organs react to high degrees of oxidative stress when exposed to nanoparticles. Under mild oxidative stress, nuclear element stimulation causes transcriptional stimulation of phase II antioxidant enzymes (Manke et al., 2013).

Oxidative stress, which is brought on by an disparity involving antioxidants and free radicals in the body, is the principal effect of nano-toxicity. As a result, a disparity in the respiratory tract is brought on by additional free radicals that unintentionally interact with other molecules and have an unbalanced number of electrons (Sonwani et al., 2021). When the body is exposed to nanoparticles, oxidative stress occurs, causing alterations in calcium homeostasis, cellular respiration, cell death, stimulation of the NADPH, and a depletion of antioxidant enzymes (Habas et al., 2021). Numerous studies show the importance of reactive particle surface in the formulation of ROS. When these free radicals and oxidants are bonded to the outer layer of the nanoparticle, free radicals are produced from that surface. ROS like OH is created from surface-bound radicals like SiO that are present on quartz particles. Moreover, outside elements like ozone and nitrogen dioxide (NO_2) that adhere to particle surfaces have the ability to cause oxidative damage (Manke et al., 2013).

It is predicted that nanoparticle intrinsic properties and extrinsic nanomaterial-cell associations may result to oxidative stress caused by nanomaterials in a number of different ways. It is possible for nanoparticles to generate oxidants. Through Fenton-like processes, transition metals can encourage the creation of OH- radical from H_2O_2, which bring about lipid peroxidation. Transition metals can be large or minute pollutants of unintentional or deliberate nanomaterials. Fixed free radical precursors on the layers' active particles, as in quartz, and redox-active classes, such as quinones, on modified nanomaterials, can both be sources of intrinsic particle oxidants. ROS can be produced in significant quantities by cells, especially neutrophils and macrophages, in response to interactions with nanomaterials. ROS are primarily created by mitochondria through normal methods of cellular respiration; nevertheless, oxidant stress can result if there is an imbalance involving oxidant formation and antioxidant enzymes and proteins (Madl et al., 2014).

In addition to being self-oxidative in origin, nanoparticles interact with biological systems and causes their prooxidant reaction by cellular reactive oxygen species creation that involves cellular respiration and stimulation of NADPH-like enzymes. Nanoparticles may excite the cell's redox system, especially in the lower respiratory tract where defensive cells like alveolar macrophages and neutrophils serve as major reactive oxygen species initiators. Neutrophils and alveolar macrophages are examples of immune system cells known as professional phagocytic cells. When these cells internalize a nanoparticle, they produce a considerable amount of ROS via the NADPH oxidase enzyme system (Pacurari et al., 2016).

7.5 Inflammation

Inflammation is a natural defensive reaction to stimuli including infection, damage, and poisons. However, it has been shown that excessive and unchecked inflammation can lead to a variety of illnesses, including heart disease, hepatitis, nephritis, and sluggish wound healing (Liu et al., 2020).

Our immune system is made up of natural and acquired defense mechanisms. The natural defense mechanism acts as the body's initial line of immunity when an outside substance enters the human system. If the natural defense mechanism is unable to successfully eliminate the harmful substances (antigen) personally, then the acquired defense mechanism, which is more advanced and potent, is activated. This activity is done by the body's protective cells. Due to their foreign nature, nanoparticles also trigger the activation of dendritic cells, which in turn trigger the activation of naive T cells, various inflammasomes, reactive oxygen species (ROS), cytokines, and chemokines (Sonwani et al., 2021). Due to its ability to perpetuate chronic inflammation, oxidative response is closely related to inflammatory response. The immune system's macrophages are recognized for producing pro-inflammatory hormones as well as free radicals, which are the root of oxidative stress. Proteins that are created as a result of the activation and synthesis of transcription elements via the mitogen-activated protein kinase (MAPK) pathway causes inflammation, cell stress, and damage within the cell (Lu et al., 2014).

A nanoparticle's zeta potential (P) is a well-known sign of how likely it is to cause inflammation. It is associated with a charged group on a particle's surface that interacts with the suspension medium to produce an electric potential (P). More positively charged groups on the surface will boost a particle's solubility there since the medium in the human body is acidic, which will increase its contact with macrophages and trigger inflammation (Schins, 2013).

The quantity of white blood cells (WBCs) in the blood is another indicator of whether inflammation is occurring in the body. Inflammation is determined based on the increase in WBC count over the normal level, which shows a failing immune system. Alveolar macrophages, the body's first line of defense, interact with inhaled nanoparticles and generate pro-inflammatory hormones as a result of their buildup in the lungs. As a result, a number of pro-inflammatory cytokines – including those from the IL-1 family, IL-6, IL-8, and the synthesis of the pro-inflammatory protein i-CAM – move toward the injured location once the dormant macrophages are prompted to become active (Clift et al., 2011).

7.6 Roles of Antioxidant Enzymes in the Prevention of Nanoparticle-Induced Toxicities

The human body uses endogenous antioxidant molecules such as superoxide dismutase (SOD), glutathione peroxidase (GPx), catalase (CAT), lipoic acid, glutathione, and coenzyme Q10 to fight the effects of free radicals and oxidative stress. A variety of exogenous antioxidant compounds of plant or animal origin are also present; these molecules are typically consumed through diet (Pizzino et al., 2017). They are molecules capable of protecting many components of biological systems from oxidative damage. In addition to repairing new antioxidants to counteract their effects and stop the oxidation of biologically relevant components, they can prevent injury by scavenging and neutralizing free radicals, reactive oxygen and nitrogen species, and free radicals from causing damage (Mihailovic et al., 2021).

Reactive oxygen species (ROS) damage can often be prevented or at least limited with the help of antioxidants. The hydroxyl radical is strongly reactive with practically every form of biomolecule, including proteins and nucleic acids, and it has the highest one-electron reduction potential of all physiologically relevant reactive oxygen species (ROS) (Evans et al., 2004). The hydroxyl radical cannot be removed by an enzymatic reaction in vivo, as far as is known. Antioxidants are the sole known means of protection against hydroxyl radicals. Antioxidants participate in redox reactions by giving electrons or hydrogen atoms; they are essentially reducing agents. As long as certain restrictions are followed, this process enables cells to run normally and prevents the oxidation of structural and other essential components. This lessens the likelihood of early damage due to errors (Fu et al., 2014). Exogenous nonenzymatic antioxidants, such as dietary supplements, vitamins, minerals, or plant components, are a significant source of compounds for the human antioxidant defense system in the prevention and mitigation of organism damage brought on by oxidative stress (Pisoschi et al., 2021).

The subclasses of antioxidants are diverse. Depending on how they work, they can be classified as enzyme-based or nonenzymatic antioxidants. Enzymes are used by antioxidants to counteract and get rid of free radicals. When substrates like copper, zinc, manganese, and iron are available, the antioxidant enzymes transform damaging oxidative molecules into H_2O_2 and then into H_2O in a complex process. Antioxidants that are not enzymes work by stopping the link interactions of free radicals. Glutathione, vitamin C, vitamin E, plant polyphenols, and carotenoids are a few antioxidants that are not enzyme-based (Nimse & Pal, 2015).

The body uses enzymatic antioxidant processes to defend itself against reactive oxygen species (ROS) that are based on inorganic materials. The antioxidant enzymes play a crucial part in preventing lipid peroxidation and preserving the integrity of cell membranes by lowering the amounts of lipid hydroperoxide and H_2O_2. Superoxide dismutase (SOD), glutathione peroxidase (GPx), and catalase (CAT) are a few examples of the enzymatic antioxidants (Mihailovic et al., 2021). In the presence of metal ion cofactors like copper (Cu), zinc (Zn), or manganese (Mn), superoxide dismutase (SOD), which is found in the cytoplasm and mitochondria, helps to convert atmospheric oxygen into oxygen and hydrogen peroxide. The peroxisome's CAT enzyme breaks down H_2O_2 into water and oxygen. Nearly all human tissues contain GSHPx, which can be found both intracellularly and extracellularly. H_2O_2 is converted into water by GSHPx. The GSHPx enzyme reacts vigorously with both H_2O_2 and fatty acid hydroperoxides. Peroxynitrite (ONOO-), organic hydroperoxides, and H_2O_2 are all reduced by the enzyme peroxiredoxin (Nimse & Pal, 2015).

7.6.1 Superoxide Dismutase (SOD)

The first line of protection against damage induced by ROS is superoxide dismutase. These proteins catalyze the dismutation of superoxide anion free radical (O_2-), which when present in excess damages cells, into O_2 and H_2O_2. This lowers the percentage molecular oxygen, which is harmful when present. In response to this action, metal ions present in the reactive part of SODs alternately undergo redox reaction (Younus, 2018). Depending on the metal substrates available at the reactive part, SODs can be categorized into four groups: copper-zinc-SOD, iron SOD, manganese SOD, and nickel SOD. The body's antioxidant defense against oxidative stress depends heavily on SODs. The enzyme can inhibit the emergence of precancerous cell changes and reduce inflammation (Younus, 2018).

7.6.2 The Glutathione Peroxidase

In mammals, the glutathione peroxidase family (GPx) of antioxidant enzymes is a crucial class of selenoenzymes. They belong to the same class of heme-free thiol peroxidases as peroxidases and reduce the toxicity of organic hydroperoxides by catalyzing their conversion to water or matching alcohols (Pei et al., 2023). When reacting with hydrogen peroxide or lipid peroxides, glutathione peroxidase (GPx) catalyzes the process and aids in the detoxification of these compounds by forming a glutathione bridge with another glutathione molecule (GSSG) form. Catalase and glutathione peroxidase detoxify H_2O_2. GSH is well known for its variety of physiological roles in the detoxification of xenobiotic substances as an antioxidant against ROS and free radicals (Sarikaya & Dogan, 2020).

7.6.3 Catalase

Hydrogen peroxide, a nonradical reactive oxygen species (ROS), serves as the material for catalase. Catalase is in charge of counteracting H_2O_2 through disintegration, so preserving an usual percentage of the substances in the cell, that is crucial for cell signal functions (Nandi et al., 2019). The most crucial antioxidant enzyme is catalase, which is present in all aerobes. Two molecules of H_2O_2 are disintegrated by catalase to form one O_2 and two H_2O molecules in a double-step reaction. One hydrogen H_2O_2 molecule is reduced to a through spectroscopy unique between chemical I, a conjugated oxyferryl species (FeIVO), with a porphyrin-cation radical, while the other H_2O_2 molecule functions as an electron donor in the second-step process, which decreases chemical I by oxidation-reduction reactions to create free enzyme, oxygen, and water (Nandi et al., 2019).

7.7 Conclusion

In conclusion, the review highlights the important role that nanoparticles play in inducing oxidative stress and inflammation in biological systems. The combined data indicates that nanoparticle exposure can be a result of ROS formation and cause inflammatory response in different tissues and cell types. In order to discover practical methods for reducing the negative consequences of oxidative stress and inflammation brought on by nanomaterials, more investigation is required.

References

Boushehri, S. (2020). Nanotechnology as a platform for the development of injectable parenteral formulations. *Pharmaceutics, 12*, 510.

Clift, M. J., Gehr, P., & Rothen-Rutishauser, B. (2011). Nanotoxicology: A perspective and discussion of whether or not in vitro testing is a valid alternative. *Archives of Toxicology, 85*, 723–731.

Evans, M. D., Dizdaroglu, M., & Cooke, M. S. (2004). Oxidative DNA damage and disease: Induction, repair and significance. *Reviews in Mutation Research, 567*(1), 1–61.

Ferdous, Z., & Nemmar, A. (2020). Health impact of silver nanoparticles: A review of the biodistribution and toxicity following various routes of exposure. *International Journal of Molecular Sciences, 21*(7), 2375.

Fu, P. P., Xia, Q., Hwang, H., Ray, P. C., & Yu, H. (2014). Mechanisms of nanotoxicity: Generation of reactive oxygen species. *Journal of Food and Drug Analysis, 22*(1), 64–75.

Geiser, M., & Kreyling, W. G. (2010). Deposition and biokinetics of inhaled nanoparticles. *Particle and Fibre Toxicology, 7*(1), 1–17.

Habas, K., Demir, E., Guo, C., Brinkworth, M. H., & Anderson, D. (2021). Toxicity mechanisms of nanoparticles in the male reproductive system. *Drug Metabolism Reviews, 53*(4), 604–617.

Joudeh, N., & Linked, D. (2022). Nanoparticle classification, physicochemical properties, characterization, and applications: A comprehensive review for biologists. *Journal of Nanotechnology, 20*, 626.

Khan, L., Saeed, K., & Khan, I. (2019). Nanoparticles: Properties, applications and toxicities. *Arabian Journal of Chemistry, 12*(7), 908–931.

Kumah, E. A., Fopa, R. D., Harati, S., Boadu, P., Zohoori, F. V., & Pak, T. (2023). Human and environmental impacts of nanoparticles: A scoping review of the current literature. *BMC Public Health, 23*, 1059.

Liu, N., & Tang, M. (2020). Toxic effects and involved molecular pathways of nanoparticles on cells and subcellular organelles. *Journal of Applied Toxicology, 40*(1), 16–36.

Liu, T., Xiao, B., & Deng, J. (2020). Ultrasmall copper-based nanoparticles for reactive oxygen species scavenging and alleviation of inflammation related diseases. *Nature Communications, 11*, 2788.

Lu, X., Zhu, T., Chen, C., & Liu, Y. (2014). Right or left: The role of nanoparticles in pulmonary diseases. *International Journal of Molecular Science, 15*, 17577–17600.

Luo, G., Du, L., Wang, Y., & Wang, K. (2014). Composite nanoparticles. In *Encyclopedia of microfluidics and nanofluidics* (pp. 1–9). Springer.

Ma, N. L., Zhang, L., Yong, W. T., Misbah, S., Hashim, F., Soon, C. F., Lim, G. P., Peng, W., & Sonne, C. (2023). Use, exposure and omics characterisation of potential hazard in nanomaterials. *Materials Today Advances, 17*, 100341.

Madl, A. K., Plummer, L. E., Carosino, C., & Pinkerton, K. E. (2014). Nanoparticles, lung injury, and the role of oxidant stress. *Annual Review of Physiology, 76*, 447–465.

Mahji, K. C., & Yadav, M. (2020). Synthesis of inorganic nanomaterials using carbohydrates. *Green Inorganic Synthesis, 27*, 109–135.

Manke, A., Wang, L., & Rojanasakul, Y. (2013). Mechanisms of nanoparticle-induced oxidative stress and toxicity. *BioMed Research International, 942916*, 15.

Matteis, V. D. (2017). Exposure to inorganic nanoparticles: Routes of entry, immune response, biodistribution and in vitro/in vivo toxicity evaluation. *Toxics, 5*(4), 29.

Mihailovic, V., Stankovic, S. J., Selakovic, D., & Rosi, G. (2021). An overview of the beneficial role of antioxidants in the treatment of nanoparticle-induced toxicities. *Oxidative Medicine and Cellular Longevity, 10*, 1155.

Nandi, A., Yan, L. J., Jana, C. K., & Dad, N. (2019). Role of catalase in oxidative stress- and age-associated degenerative diseases. *Oxidative Medicine and Cellular Longevity, 10*, 1155.

Nemmar, A., Yuvaraju, P., Beegam, S., Pathan, J., Kazzam, E. E., & Ali, B. H. (2016). Oxidative stress, inflammation, and DNA damage in multiple organs of mice acutely exposed to amorphous silica nanoparticles. *International Journal of Nanomedicine, 11*(1), P919–P928.

Nimse, S. B., & Pal, D. (2015). Free radicals, natural antioxidants, and their reaction mechanisms. *Royal Society of Chemistry, 5*, 27986–28006.

Pacurari, M., Lowe, K., Tchounwou, P. B., & Kafoury, R. (2016). A review on the respiratory system toxicity of carbon nanoparticles. *International Journal of Environmental Research and Public Health, 13*(3), 325.

Pei, J., Pan, X., Wei, G., & Hua, Y. (2023). Research progress of glutathione peroxidase family (GPX) in redoxidation. *Frontiers Pharmacology, 14*, 1147414.

Pisoschi, A. M., Pop, A., Iordache, F., Stanca, L., Predoi, G., & Serban, A. I. (2021). Oxidative stress mitigation by antioxidants - An overview on their chemistry and influences on health status. *European Journal of Medicinal Chemistry, 209*, 112891.

Pizzino, G., Irrera, N., Cucinotta, M., Pallio, G., Mannino, F., Arcoraci, V., Squardrito, F., Altavilla, D., & Bitto, A. (2017). Oxidative stress: Harms and benefits for human health. *Oxidative Medicine and Cellular Longevity, 13*.

Qiao, H., Liu, W., Gu, H., Wang, D., & Wang, Y. (2015). The transport and deposition of nanoparticles in respiratory system by inhalation. *Journal of Nanomaterials*, 2015: 1687–4110.

Samrot, A. V., & Prakash, L. N. (2023). Nanoparticles induced oxidative damage in reproductive system and role of antioxidants on the induced toxicity. *Life, 13*(3), 767.

Sarikaya, E., & Dogan, S. (2020). Glutathione peroxidase in health and diseases. In *Glutathione system and oxidative stress in health and disease* (Vol. 10, p. 5772). Chapter Metrics Overview.

Schins, R. (2013). Genotoxicity of nanoparticles. *Nanomaterials*, 60–64.

Sheilds, H. J., Traa, A., & Van Raamsdunk, J. M. (2021). Beneficial and detrimental effects of reactive oxygen species on lifespan: A comprehensive review of comparative and experimental studies. *Frontiers in Cell and Developmental Biology, 9*(628157): 1–27.

Shin, W. K., Cho, J., Kannan, A. G., Lee, Y. S., & Kim, D. W. (2016). Cross-linked composite gel polymer electrolyte using mesoporous methacrylate-functionalized SiO2 nanoparticles for lithium-ion polymer batteries. *Scientific Reports, 6*, 26332.

Sonwani, S., Madaan, S., Arora, J., Suryanarayan, S., Rangra, D., Mongia, N., Vats, T., & Saxena, P. (2021). Inhalation exposure to atmospheric nanoparticles and its associated impacts on human health: A review. *Frontiers Sustainable Cities, 3*, 690444.

Younus, H. (2018). Therapeutic potentials of superoxide dismutase. *International Journal of Health Science, 12*(3), 88–93.

Yu, Z., Li, Q., Wang, J., Yu, Y., Wang, Y., Zhou, Q., & Li, P. (2020). Reactive oxygen species-related nanoparticle toxicity in the biomedical field. *Nanoscale Research Letters, 15*, 115.

Zhang, C., Wu, L., Perrot, M., & Zhao, X. (2021). Carbon nanotubes: A summary of beneficial and dangerous aspects of an increasingly popular group of nanomaterials. *Frontiers Oncology, 11*, 693814.

Chapter 8
Aquatic Ecotoxicity of Nanoparticles

Patrick Omoregie Isibor, Ifeoluwa Ihotu Kayode-Edwards, Oluwafemi Adebayo Oyewole, Clement Shina Olusanya, Tsado Priscilla Yetu, Samuel Adeniyi Oyegbade, David Osagie Agbontaen, Ogochukwu Oritseweyinmi Onwaeze, and Hephzibah Adeniji

8.1 Introduction

Nanoparticles (NPs), with their minuscule size, exhibit unique physicochemical properties that can be tailored for diverse purposes. Due to their remarkable properties and wide-ranging applications, these particles can be utilized in numerous fields like in agriculture, in aquaculture, in medicine for targeted drug delivery and diagnostics, in electronics for miniaturization and enhanced performance, in environmental science for pollution remediation, and in materials science for improving the strength, durability, and functionality of materials (Nirmal, 2019; Bai et al., 2020; Kraśniewska et al., 2020; Ma et al., 2020; Paul & Sharma, 2020; Ren et al., 2020; Amreen & Goel, 2021; Khan et al., 2021; Luo et al., 2021). Zhou et al. (2023) report that the global market value of nanomaterials is expected to grow at a compound annual growth rate (CAGR) of 14.8% from 2023 to 2030. Despite the multifaceted nature and capabilities of nanoparticles, their interactions with other substances have raised questions about their ecotoxicological effects on aquatic organisms (Yamini et al., 2023). This area of study, known as the ecotoxicity of nanoparticles, explores how these minuscule particles can adversely affect the delicate balance of aquatic ecosystems (Rana & Kalaichelvan, 2013). Aquatic ecosystems, encompassing diverse environments such as freshwater lakes, bogs, rivers, wetlands, ponds, streams, and springs (Balasubramanian, 2011), and marine open oceans, deep-sea oceans, coral reefs, estuaries, mangrove forests, and salt marshes (Williamson &

P. O. Isibor (✉) · I. I. Kayode-Edwards · S. A. Oyegbade · O. O. Onwaeze · H. Adeniji
Department of Biological Sciences, Covenant University, Ota, Ogun State, Nigeria
e-mail: patrick.isibor@covenantuniversity.edu.ng

O. A. Oyewole · C. S. Olusanya · T. P. Yetu
Department of Microbiology, Federal University of Technology, Minna, Nigeria

D. O. Agbontaen
Department of Public Health, University of South Wales, Pontypridd, UK

P. O. Isibor et al. (eds.), *Environmental Nanotoxicology*,
https://doi.org/10.1007/978-3-031-54154-4_8

Guinder, 2021) are vital for biodiversity and provide essential services to humans. They serve as habitats for a wide range of organisms, forming intricate food webs and supporting the livelihoods of millions of aquatic organisms. However, the introduction of nanoparticles into these ecosystems poses significant challenges (Kurwadkar et al., 2015). Engineered nanoparticles from various point sources, including industrial processes and consumer products, and nonpoint sources, including stormwater runoff and atmospheric deposition, can find their way into water bodies through wastewater discharges, runoff, or direct release, leading to the potential exposure of aquatic organisms (Kurwadkar et al., 2015; Topuz & van Gestel, 2016). Over 90% of the waste generated in developing countries is openly managed with the possibility of finding its way into the hydrosphere and atmosphere. Dube and Okuthe (2023) identify examples in the case of nanoparticles like titanium dioxide nanoparticles (TiO_2 NPs) in sunscreens that can eventually be washed down the drain and silver nanoparticles (AgNPs) in fabrics that could leach into aquatic systems, such as streams and rivers, during washing.

An analysis of the effects of nanomaterial exposure on the aquatic environment is essential because it is the final recipient of erosion and wastewater. This is especially true considering that there are a growing number of nanoscale applications and that the use and development of nanomaterials is expected to continue to grow in the future). A major concern regarding the ecotoxicity of nanoparticles is their ability to enter the bodies of aquatic organisms through processes like ingestion, inhalation, or dermal penetration (Prajitha et al., 2019). Once inside an organism, nanoparticles can translocate to different organs, tissues, and cells, potentially causing adverse effects. The unique surface properties of nanoparticles can facilitate interactions with cell membranes, proteins, and other biomolecules, leading to cellular stress, inflammation, and even cell death (Khanna et al., 2015; Tee et al., 2016). Furthermore, nanoparticles can accumulate in organisms over time, leading to bioaccumulation and biomagnification through the food chain (Uddin et al., 2020), with potential consequences for higher trophic organisms and human health when contaminated organisms are consumed (Carbery et al., 2018). Fish are highly sensitive to the presence of natural water contaminants and are therefore used as biomarkers of aquatoxicity (Rana & Kalaichelvan, 2013; Khan et al., 2019; Sibiya et al., 2022; Dube & Okuthe, 2023).

The ecotoxicological effects of nanoparticles on aquatic organisms are multifaceted. They can interfere with various physiological processes, including metabolism, reproduction, and immune responses. For example, Wu and Zhou (2013) reported that silver nanoparticles (AgNPs) can negatively affect the cell lines and embryos of freshwater fish, and Kaya et al. (2016) reported that zinc oxide nanoparticles (ZnO NPs) caused damage to the gills of *Oreochromis niloticus* (Nile tilapia), affecting their respiratory functions and osmoregulation. Similarly, Purushothaman et al. (2014) reported that titanium dioxide nanoparticles (TiO_2 NPs) induced oxidative stress in *Danio rerio* (zebra fish), leading to respiratory difficulties. Furthermore, nanoparticles can affect the behavior and ecological interactions of aquatic organisms. For instance, nanoparticles may alter the feeding behavior of zooplankton (Li et al., 2023) and of fish on nanoparticle-containing zooplankton (Land, 2015), disrupting the balance of the entire aquatic food web. Additionally, nanoparticles can influence the development and behavior of aquatic larvae leading to abnormalities, with potential repercussions for population dynamics and community structure (Gambardella et al., 2013; Pitt et al., 2018; Vogt et al., 2022).

Understanding the ecotoxicity of nanoparticles on aquatic organisms is essential for safeguarding the health of the aquatic ecosystems and also for ensuring the well-being of human populations that rely on them. Consequently, rigorous scientific research and environmental risk assessments (ERA) are imperative to establish guidelines and regulations governing the production, use, and disposal of nanoparticles, mitigating their potential adverse effects on aquatic environments (Singh & Gurjar, 2022).

8.2 Sources of Nanoparticles into Aquatic Systems

Aquatic organisms can be exposed to nanoparticles through various routes, both anthropogenic and natural. The entry of nanoparticles into water bodies is a concern because it can lead to potential ecological risks.

Nanoparticles can enter water bodies through direct discharge from industrial effluents (Batley et al., 2013), municipal wastewater (Li et al., 2013), sewage treatment systems (Zhang et al., 2016a), wastewater treatment plants (Shi et al., 2020), shipping and maritime activities, biosolids application, and rainwater runoff (Batley et al., 2013). As illustrated in Fig. 8.1, natural processes like weathering and erosion of rocks containing nanoparticles, volcanic eruptions, and atmospheric deposition can also introduce nanoparticles into aquatic ecosystems (Hochella et al., 2012; Buzea & Pacheco, 2017). In some cases, nanoparticles may be unintentionally released directly into the aquatic environment. For example, nanoparticles used in sunscreen lotions and cosmetics can be washed off and enter water bodies during recreational activities (Gondikas et al., 2018). Nanoparticles from agricultural products like pesticides and fertilizers can be carried into water bodies through runoff or soil erosion (Sarkar et al., 2021).

Nanoparticles that settle and accumulate in sediment can pose a risk to benthic (bottom-dwelling) organisms (Ciacci et al., 2019). These organisms may come in contact with or ingest nanoparticles when they burrow into or feed on sediment (Pang et al., 2012; Baker et al., 2014). Some aquatic species can transfer nanoparticles to their offspring through reproductive processes. This can occur when nanoparticles are present in the parental organism's reproductive organs or the reproductive fluids (Yang et al., 2023).

Aquatic organisms can ingest nanoparticles directly from the water or through their food. Once inside the organisms, nanoparticles can be absorbed and distributed within their tissues. This can lead to bioaccumulation and biomagnification, where nanoparticles are concentrated as they move up the food chain, posing potential risks to higher trophic levels, including humans who consume contaminated aquatic organisms (Shi et al., 2020; Uddin et al., 2020; Rakib et al., 2023). Some

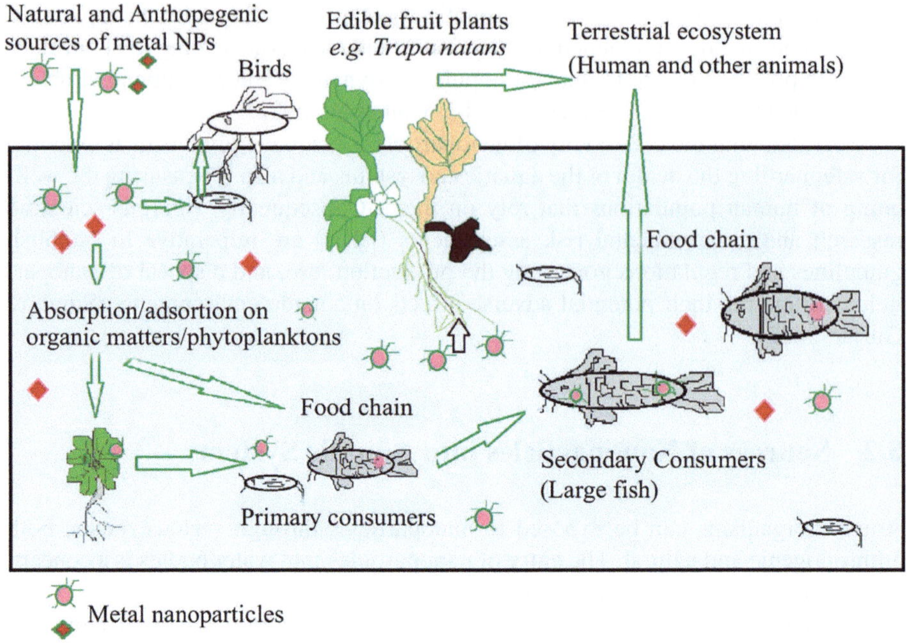

Fig. 8.1 The release of metal nanoparticle in the aquatic food chain

aquaculture practices involve the use of nanoparticle-enhanced feeds, which can be ingested by farmed aquatic organisms, leading to potential exposure and bioaccumulation within these organisms (Handy et al., 2022).

The entry of nanoparticles into water bodies can have environmental implications, including the potential for ecotoxicity, and long-term impacts on aquatic ecosystems (Sajid et al., 2015). As a result, there are ongoing research and regulatory efforts to better understand and manage the release of nanoparticles into the environment to mitigate potential risks (Gupta & Xie, 2018).

8.3 Entry Routes of Nanoparticles into Aquatic Organisms

Nanoparticles can enter aquatic organisms through various routes, and their entry can have potential ecological and toxicological implications. The routes of entry include direct ingestion, exposure by inhalation, and dermal penetration (Brandelli, 2020).

Nanoparticles suspended in the water column can adhere to the surfaces of aquatic organisms through dermal contact. This contact can occur when nanoparticles are present in the water and come into direct contact with the skin, scales, or shells of aquatic species (Ma & Lin, 2013). Alternatively, some nanoparticles can release ions or compounds that are soluble in water, which can also be absorbed by organisms through their gills or body surfaces (Scown et al., 2010).

Another route of entry of nanoparticles into aquatic organisms is by consumption of nanoparticle-contaminated food. Aquatic organisms, particularly filter-feeding organisms like mussels and zooplankton, can directly ingest nanoparticles suspended in water (Duan et al., 2023). On the other hand, aquatic organisms can ingest nanoparticles when they consume other organisms that have already accumulated nanoparticles in their cells and tissues (Werlin et al., 2011; Uddin et al., 2021). Nanoparticles ingested by aquatic organisms can be absorbed through the gastrointestinal tract, entering the organism's circulatory system and potentially accumulating in various tissues (Bergin & Witzmann, 2013; Wright et al., 2013; Baker et al., 2014). Organisms that have accumulated nanoparticles in their tissues can transfer them to predators when they are consumed as prey. This can lead to higher concentrations of nanoparticles in organisms higher up the food chain (Mattsson et al., 2017).

The third route of entry of nanoparticles is through inhalation of airborne nanoparticles. Some aquatic organisms, especially those with gills, can potentially be exposed to nanoparticles through inhalation (Prüst et al., 2020). Nanoparticles present in the water can be taken up by organisms when they respire, which may lead to nanoparticle accumulation in respiratory organs and consequent inflammation (Vlachogianni et al., 2013).

Once nanoparticles are ingested or enter the body of an aquatic organism, they may be taken up by cells through various mechanisms such as endocytosis or passive diffusion (Beddoes et al., 2015).

8.4 Toxicity of Nanoparticles to Aquatic Organisms

Nanoparticles have gained increasing attention in various industries and applications due to their unique properties. However, their introduction into aquatic ecosystems can have adverse effects on these ecosystems as they can be toxic to aquatic organisms, including fish, invertebrates, bacteria, and other microorganisms (Rocha et al., 2017).

At high concentrations from accumulation or exposure, nanoparticles can result in reduced biomass and loss of viability, growth inhibition, oxidative stress, mortality, gene alterations to target organs such as fish gills, morphological and physiological abnormalities such as ascites and color change, decreased muscle activity, decreased chlorophyll content, genotoxic effects, cytotoxicity cell membrane damage, and skeletal anomalies (Manke et al., 2013; Zhu et al., 2019; Khoshnamvand et al., 2020; Ale et al., 2021). The toxicity of nanoparticles typically causes damage to the skin and respiratory systems of multicellular organisms, but in unicellular organisms, the effects of these particles are usually pervasive (Khoshnamvand et al., 2020). This prevalent toxicity of nanoparticles to unicellular organisms, such as amoeba and microalga, can be attributed to a variety of mechanisms, including membrane adhesion, denaturation of ribosomes, increased cellular porosity, inactivation of proteins and enzymes, malfunctioning proton pumps, inhibition of DNA synthesis, production of reactive oxygen species (ROS), DNA damage, and degradation of lipopolysaccharide molecules (Moreno-Garrido et al., 2015; Hu et al., 2018; Xiang et al., 2018).

Li et al. (2023) reported on the toxic effects of silver nanoparticles (AgNPs) on zooplankton, and Khoshnamvand et al. (2020) reported that on entry into living cells of *Daphnia magna* (water flea), silver nanoparticles (AgNPs) can produce oxidative stress and ROS, which can have detrimental effects such as DNA damage, mitochondrial damage, membrane lipid peroxidation, and consequently cell apoptosis. In the same report, the observation was stated that *D. magna* is located at lower trophic levels and serves as food for higher organisms; hence, AgNPs might be transferred to organisms at higher trophic levels. Ale et al. (2021) identified the main toxic effects of silica nanoparticles (SiNPs) on various aquatic organisms summarized in Fig. 8.1 and their toxicity mechanisms and the parameters that affect their toxicity as schematized in Fig. 8.2. Zhang et al. (2016b) investigated the toxicity of magnetic iron oxide nanoparticles (IONPs) on *Raphidocelis subcapitata*, *Lemna minor*, and *D. magna*. Although the IONPs were nontoxic to *L. minor*, growth inhibition was observed in *R. subcapitata* and *D. magna*, while acute immobilization and moulting disruption were observed in *D. magna* on exposure to uncoated IONPs. Flocculation, which interrupts gas exchange and impedes access to light and nutrients (Navarro et al., 2008), was also reported in *R. subcapitata*. Titanium dioxide nanoparticles (TiO_2 NPs) have also been reported to cause flocculation in *R. subcapitata* (Aruoja et al., 2009) and molting disruption in *D. magna*. Srinivasan et al. (2019) evaluated the toxicity of TiO_2 NPs on *Danio rerio* (zebra fish) embryos where malformations like tail curvature, egg coagulation, bent spine, and delayed hatching were observed at high concentrations of the nanoparticles.

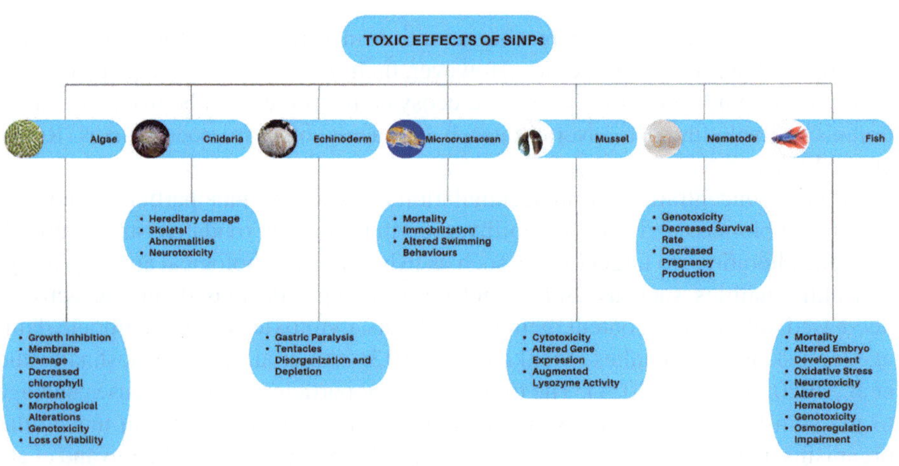

Fig. 8.2 Main toxic effects of silica nanoparticles (SiNPs) on various aquatic organisms (Ale et al., 2021)

Davarpanah and Guilhermino (2019) compiled the toxic effects of gold nanoparticles (AuNP) including cytotoxicity (Iswarya et al., 2016), reduction of metabolic activity (Perreault et al., 2014), growth inhibition (Nam et al., 2015), mortality (Renault et al., 2008), chlorophyll reduction (Moreno-Garrido et al., 2015), and oxidative stress.

Engineered nanoparticles (ENPs) such as quantum dots (QDs) are also toxic to aquatic organisms. Rocha et al. (2017) compiled a report on the ecotoxicity of QDs on numerous strains of bacteria (gram-positive and gram-negative bacteria), microorganisms (algae, protozoa, fungi), aquatic invertebrates (crustacean, annalida, platyhelminthes, nematoda, bivale, and gastropoda), and fish species in which growth inhibition, genotoxic effects, colony-forming unit (CFU) inhibition, production of reactive oxygen species (ROS) in culture medium, increased oxidative stress, embryonic malformations, decreased heart rate, decreased fecundity and hatching rate in fish, decreased male reproductive fitness and male building nests in fish, carbon dioxide (CO_2) depletion and impeded oxygen (O_2) production in photosynthetic microorganisms, reduced cell viability, morphological changes, reduced mitochondrial membrane potential, reduced thickness and density of cell wall, trophic transfer, reduced life span and brood size, mortality, vascular damage, hampered gene expression, and DNA damage were major effects.

Other nanoparticles (NPs), including metal nano-powders (such as holmium oxide NPs, zinc oxide NPs, yttrium iron oxide NPs, erbium oxide NPs, copper oxide NPs, nickel-zinc iron oxide NPs, indium tin oxide NPs, copper zinc iron oxide NPs, samarium oxide NPs, and strontium ferrite NPs), and organic nanoparticles such as single-walled carbon nanotube (SWCNT), multi-walled carbon nanotubes (MWCNT), and fullerene-C60 have also been reported to have toxic effects on aquatic organisms (Blaise et al., 2008; Canesi et al., 2010; Bondarenko et al., 2013; Jackson et al., 2013; Fard et al., 2015; Suman et al., 2015; Chen et al., 2019). Additionally, hereditary disorders have been proposed to be caused by SiNPs, which could be a threat to the fate of populations (Ale et al., 2021).

Although nanoparticles synthesized biologically are regarded as green products, that have been designed to have minimal impact on the environment by reducing waste and maximizing resource efficiency, research on their toxicity in the aquatic ecosystem requires more attention (Khoshnamvand et al., 2020). In addition, more comprehensive research encompassing the toxicological effects of nanoparticles on a wider range of aquatic species located in different ecosystems is of importance in order to provide early indications and warning signs to control and prevent environmental degradation (Ale et al., 2021).

8.5 Toxicity Mechanisms of Nanoparticles

Nanoparticles exhibit unique properties that make them invaluable for various applications. However, their potential toxicity raises concerns. When nanoparticles enter biological systems, they can interact with cells and tissues, triggering diverse toxic responses. Toxicity mechanisms involve reactive oxygen species (ROS) formation, oxidative stress, biotransformation and chemical transformation, bioaccumulation, endocrine disruption, cytotoxicity, inflammation, genotoxicity, and disruption of cellular processes. Factors like the size, shape, surface charge, and chemical composition of nanoparticles play pivotal roles in determining the extent and nature of nanoparticle toxicity.

8.5.1 Reactive Oxygen Species

Reactive oxygen species (ROS) formation is one toxicity mechanism of nanoparticles. Reactive oxygen species are highly reactive molecules or ions that contain oxygen, such as superoxide radicals, hydrogen peroxide, and hydroxyl radicals (Čapek & Roušar, 2021). They are produced as natural byproducts of cellular metabolism, particularly during processes like respiration and energy production. However, their control is of great importance as an imbalance in their influx and efflux could lead to pathophysiological damages (Manke et al., 2013). As a result of their minuscule size and increased surface area to volume ratio, these nanoparticles have higher chemical reactivity, which results in the generation of reactive oxygen species (ROS) (Fard et al., 2015). ROS formation serves as a preliminary for other toxicity mechanisms including inflammation, oxidative stress, genotoxicity, immunotoxicity, and damage to DNA, cell membranes and proteins (Fard et al., 2015). Nanoparticles, such as TiO_2 NPs, do not release toxic ions, so their toxicity can be attributed to their size-dependent interactions with intracellular biomolecules that get adsorbed onto their surface (Xiong et al., 2013). These interactions lead to ROS generation, cytokine release, leakage of the plasma membrane, mitochondrial depolarization, and intracellular calcium influx (Ghosh et al., 2013; Fard et al., 2015). Manke et al. (2013) report that it is inaccurate to assume that ROS formation is a prerequisite to all nanoparticle-induced toxicity as some researchers have reported direct toxicity of NPs without causing ROS.

8.5.2 Oxidative Stress

Oxidative stress is another toxicity mechanism of nanoparticles in aquatic organisms. Oxidative stress is a physiological imbalance between the production of reactive oxygen species (ROS) or free radicals and the body's ability to counteract or detoxify their harmful effects through neutralization by antioxidants (Walke et al., 2023). The oxidative stress induced by nanoparticles, such as TiO_2 NPs, zinc oxide NPs, (ZnO NPs), cerium oxide NPs (CeO_2 NPs), and AgNPs, occurs as a result of abiotic factors including particle composition, size, surface, and the presence of metals (Manke et al., 2013). ROS-mediated damage is then caused by cellular responses including immune cell activation, NP-cell interaction, and mitochondrial respiration (Manke et al., 2013). Excessive ROS can damage cellular components, including DNA, proteins, and lipids, leading to various health issues, such as tissue inflammation, impaired growth, and decreased reproductive success (Fard et al., 2015). Oxidative stress also serves as a prelude for other toxicity mechanisms including fibrosis, inflammation, and genotoxicity. Oxidative stress caused by nanoparticles is a concern for aquatic ecosystems, as it can disrupt the health of aquatic organisms and the overall balance of the ecosystem, potentially affecting water quality and biodiversity.

8.5.3 Biotransformation and Chemical Transformations

Biotransformation and chemical transformations are also toxicity mechanisms of nanoparticles in aquatic organisms. Transformations refer to the metabolic processes through which these particles are chemically modified or processed within an organism's body (Spurgeon et al., 2020). When aquatic organisms are exposed to nanoparticles, they may attempt to metabolize or detoxify them, typically in organs like the liver or gills (Kumar et al., 2023). This biotransformation can lead to the formation of new compounds or the modification of the nanoparticle surfaces (Dabrowski et al., 2022). Depending on the organism's capability and the nanoparticle's properties, these processes can either reduce or enhance toxicity.

Chemical transformations caused by nanoparticles, such as ENPs, occur due to the inherent properties of these NPs. They include processes such as surface coating, agglomeration, adsorption, and redox reactions as schematized in Fig. 8.3. These chemical transformations influence the uptake, toxic effects, bioavailability, transport, and persistence of the NPs (Dube & Okuthe, 2023).

The chemical transformation of nanoparticles through agglomeration and aggregation refers to the process in which individual nanoparticles clump together into a mass due to attractive forces, altering their physical and chemical properties (Dube & Okuthe, 2023). Agglomeration involves loosely bound clusters, while aggregation results in tighter, more stable particle associations (Singer et al., 2019). These transformations can change nanoparticle size, surface area, and reactivity, impacting their functionality and behavior and potentially reducing the number of free NPs in the media (Markus et al., 2015; Khan et al., 2019).

Transformation through dissolution involves the process of NPs changing their chemical composition or structure by the discharge of materials from their surface due to their interaction with their environment, resulting in the dispersion of the released materials into the exposure media (Dube & Okuthe, 2023). The dissolution and dispersion and aggregates and agglomerates formation patterns of NPs are influenced by their inherent properties and the physicochemical properties of the aquatic environment (Dube & Okuthe, 2023).

Transformation through redox reactions involves the exchange of electrons between nanoparticles and other chemical entities and substances, such as functional groups and adsorbed pollutants, and oxidation-reduction processes. This

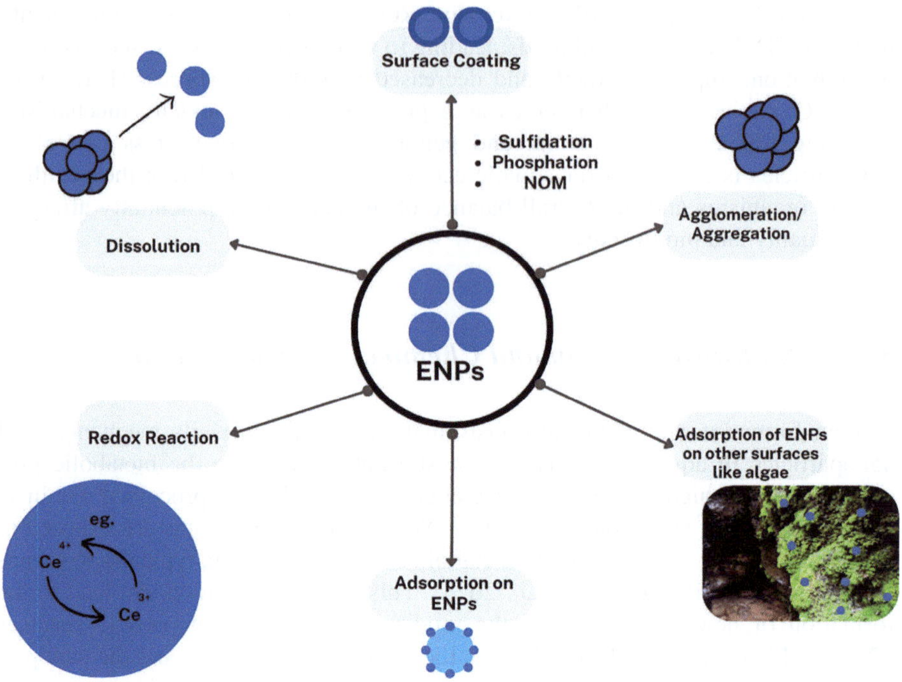

Fig. 8.3 Illustration of ENP transformation in aquatic systems (Dube & Okuthe, 2023)

leads to alterations in their chemical composition and properties. In these reactions, nanoparticles may donate or accept electrons, resulting in the oxidation (loss of electrons) or reduction (gain of electrons) of their constituent materials. This process can modify nanoparticle size, surface charge, and reactivity. Redox transformations are fundamental in various bioprocesses, such as catalysis and energy storage, where nanoparticles play a critical role in facilitating electron transfer reactions. The redox reaction pattern of NPs is influenced by the presence of a reducing or oxidizing agent and conditions in the aquatic environment such as pH (Dube & Okuthe, 2023).

The chemical transformation of nanoparticles through adsorption or desorption involves the binding of substances from the aquatic media onto the surface of nanoparticles (sorbent) or the binding of NPs onto solid surfaces in the aquatic systems (sorbate) (Dube & Okuthe, 2023). This leads to changes in their properties and functionality as the transformation can alter the surface chemistry, reactivity, and stability of nanoparticles (Fig. 8.3). Chemical transformations through alterations in surface coatings involve modifying the outer layer of nanoparticles, typically by changing or replacing the molecules or materials that coat their surfaces. Additionally, the NPs can adsorb natural organic matter (NOM), thereby increasing their coating (Dube & Okuthe, 2023). These surface coatings can impact nanoparticle stability, reactivity, and interactions with their environment (Nel et al., 2009; Connolly et al., 2022). Transformations through interaction with phosphate (phosphation) and sulfur (sulfidation) result in changes to the NPs' composition and properties. Phosphate and sulfur tend to form strong bonds with certain metal nanoparticles (Pentsak et al., 2021), leading to surface modifications or the formation of new materials. This interaction can impact nanoparticle stability, catalytic activity, and surface chemistry (Dube & Okuthe, 2023).

8.5.4 Bioaccumulation

Bioaccumulation is another toxicity mechanism of nanoparticles on aquatic organisms. Bioaccumulation is the gradual buildup of substances, such as nanoparticles, within the tissues of organisms in a food chain or ecosystem (Uddin et al., 2021). This can occur when organisms ingest or absorb nanoparticles from the surrounding water or sediments or consume other organisms that have ingested nanoparticles. Due to their minute size and unique properties, nanoparticles can be absorbed and retained in organisms more readily than larger particles (Ali et al., 2021), potentially leading to higher concentrations as they move up the food chain, with apex predators or tertiary consumers often harboring the highest levels. This is because organisms that feed on nanoparticle-contaminated prey may accumulate higher concentrations of nanoparticles in their tissues, leading to biomagnification of nanoparticles in higher trophic levels (Werlin et al., 2011; Uddin et al., 2021). As a result, the concentration of these particles increases within food webs. Bioaccumulation also leads to bioconcentration, whereby the concentration of the toxins in the organism surpasses that in the environment (Uddin et al., 2021). Dong et al. (2023) report that the bioaccumulation of specific nanoparticles was intrinsically determined by the mass equilibrium between the inflow and the outflow of the suspending nanoparticles.

8.5.5 Inflammation

Another toxicity mechanism is inflammation, which is a biological response triggered by the immune system of aquatic organisms from exposure to these tiny particles in aquatic environments (Jovanović & Palić, 2012; Brun et al., 2018). This response is triggered when the immune system recognizes the nanoparticles as "foreign objects" and attempts to isolate and eradicate them (Williams, 2017). This inflammatory response can lead to localized or systemic inflammation, potentially causing health issues. Chronic inflammation induced by nanoparticles and their interactions with immune cells has been associated with cell death as summarized in Fig. 8.4 (Boraschi et al., 2017).

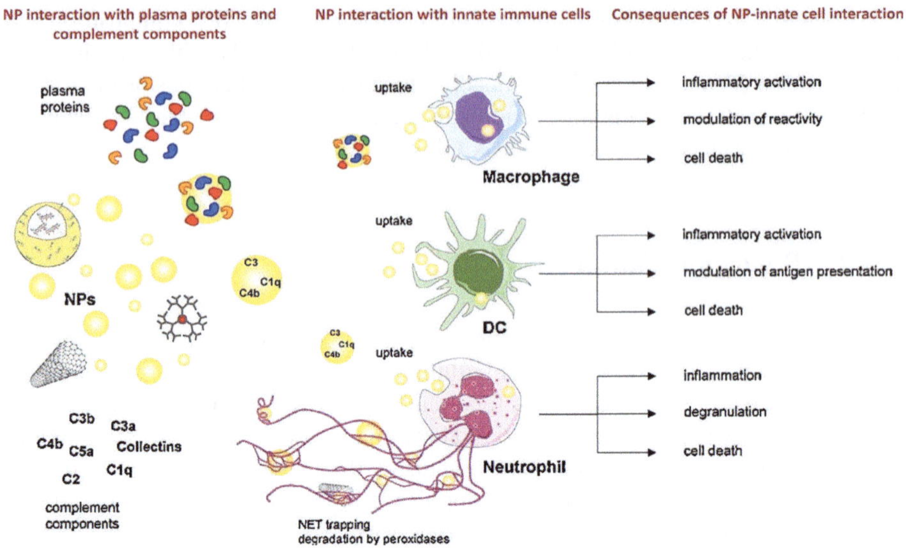

Fig. 8.4 NP interactions and consequences of interactions (Boraschi et al., 2017)

8.5.6 *Genotoxicity*

Genotoxicity of nanoparticles is also a toxicity mechanism, and it refers to the ability of nanoparticles to cause damage to an organism's genetic material (Yao et al., 2022). Due to their small size and unique properties, nanoparticles can interact with DNA and cellular structures, potentially leading to genetic mutations and chromosomal abnormalities through clastogenic effects (Magdolenova et al., 2014; Benameur et al., 2015; Shukla et al., 2021). This genotoxic effect is a significant concern as it can contribute to the development of diseases, including cancer, and hereditary disorders (Choudhuri et al., 2021). The mechanisms underlying nanoparticle genotoxicity involve direct DNA interaction, oxidative stress, or the release of toxic ions (Manke et al., 2013; Karlsson et al., 2014; Benameur et al., 2015).

8.5.7 *Interactions with Membranes*

The interaction of nanoparticles with membranes in aquatic organisms may lead to changes in membrane integrity, fluidity, or permeability, affecting the normal functioning of cells (Rawat et al., 2021; Tian et al., 2021), hence making it another toxicity mechanism. Nanoparticles can interact and penetrate with the lipid bilayers that compose cell membranes and this can have a range of consequences, including cellular stress, damage, or even cell death (Wang et al., 2022). For example, nanoparticles that enter the gills of aquatic organisms can hinder the exchange of gases like oxygen and carbon dioxide, which are essential for respiration, and this can result in respiratory distress and reduced oxygen uptake (do Carmo et al., 2018).

8.5.8 Endocrine Disruption

Endocrine disruption, another toxicity mechanism of NPs, refers to the interference of these particles with the endocrine system, which regulates hormones and various physiological processes (Knight, 2021). Nanoparticles, especially those containing certain chemicals or metals, can disrupt endocrine function when absorbed by aquatic organisms as they may mimic or interfere with natural hormones, leading to hormonal imbalances (Dagar & Bagchi, 2020; Ismanto et al., 2022). This disruption can result in adverse effects on reproduction, development, and immune health in aquatic species (Zhou et al., 2010). Moreover, endocrine disruption in these organisms can lead to impaired reproduction and can have ecological repercussions, affecting population dynamics and reproductive success (Kar et al., 2021; Marana et al., 2022). This can also lead to hereditary developmental anomalies in offspring (Sharma et al., 2020).

8.5.9 Cellular Damage

Another toxicity mechanism of NPs is cellular damage. Nanoparticles, especially those that are highly reactive or have sharp structural edges, can cause physical damage to cells and tissues when they come into contact with cellular structures (Augustine et al., 2020; Rhazouani et al., 2021). This damage may disrupt cellular membranes and organelles, leading to impaired cellular function.

8.5.10 Altered Behavior and Feeding

A final toxicity mechanism of NPs is altered behavior and feeding. Nanoparticles can interfere with an organism's ability to evade predators and feed (Justice & Bernot, 2014; Bergami et al., 2016). For example, they may disrupt the swimming behavior of aquatic organisms, impairing their ability to evade predators or find food (Wright et al., 2013; Carmo et al., 2019).

Chronic exposure to nanoparticles over time can lead to cumulative and long-term effects on aquatic organisms, potentially causing sublethal effects that may not be immediately apparent (Baker et al., 2014; Stara et al., 2021). Researchers continue to study the ecotoxicological impacts of nanoparticles in aquatic ecosystems to better understand the risks they pose and develop strategies to mitigate their adverse effects. Regulatory agencies may also establish guidelines and regulations to manage nanoparticle pollution in aquatic environments to protect aquatic life and ecosystem health.

8.6 Fate of Nanoparticles in Aquatic Environment

Environmental factors, such as water chemistry (pH, salinity), temperature, and dissolved organic matter (DOM), significantly influence nanoparticle behavior in aquatic ecosystems (Yu et al., 2018, 2019). Changes in pH can alter nanoparticle surface charges, affecting their stability and interactions with organisms (Handy et al., 2008). Temperature variations impact nanoparticle aggregation and dissolution rates, potentially intensifying toxicity (Zhang et al., 2018). Salinity can influence nanoparticle dispersion and bioavailability, varying across freshwater and marine systems (Wang & Wang, 2014). Dissolved organic matter can form coatings on nanoparticles, altering their reactivity and bioavailability (Yu et al., 2018).

The most important factors affecting the destiny of nanoparticles in aquatic systems are agglomeration and aggregation, dissolution, redox reactions, and transformation into new solid phases. When nanoparticles aggregate, larger particles are created that can be transported to the sediments and removed from the water column. Agglomeration very weakly binds nanoparticles, whereas aggregation creates chemical bonds between the particles. For bigger agglomerated particles, the bioavailability of NP and their dissolving behavior may also alter (Fig. 8.5). The nature of the medium, particularly its ionic strength, pH, and natural organic matter concentration, all affects agglomeration.

Particles of elemental MeNP are maintained in suspension by their surface coating, which is either stabilizing or negatively or positively charged depending on steric parameters. Examples include carbonate-, citrate-, or charged polymer-stabilized AgNP solutions. AgNP is stabilized by steric effects in other polymeric coverings, such as polyvinylpyrrolidone PVP. The effects of pH and ionic strength on the agglomeration of AgNP suspensions have been researched for various coatings. Aggregation has been observed, particularly at low pH values when the coating's negative surface charge is neutralized and at high ionic strengths. According to predictions, humic acids will interact with surfaces and have steric effects that will alter how stable the suspension is. The water composition, including pH, ionic strength, the concentration of divalent cations, and the concentrations of fulvic and humic acids, has an impact on the colloidal stability of AgNP suspensions in natural fluids. If not functionalized or covered in organic molecules, metal oxide nanoparticles like TiO_2, Al_2O_3, and CeO_2 behave as oxides with a pH-dependent surface charge. At neutral surface charge in the pH range of the zero point of charge, each of the metal oxides has a distinct characteristic that leads to agglomeration. For instance, CeO_2 NP exhibit zero point of charge at the pH 7–8 neutral range. Around pH 7, such as in media used for aquatic organism development and toxicity testing, CeO_2 NP agglomeration has been seen. Similar behavior is displayed by TiO_2, which exhibits a zero point of charge in the pH-neutral range. It has been shown that

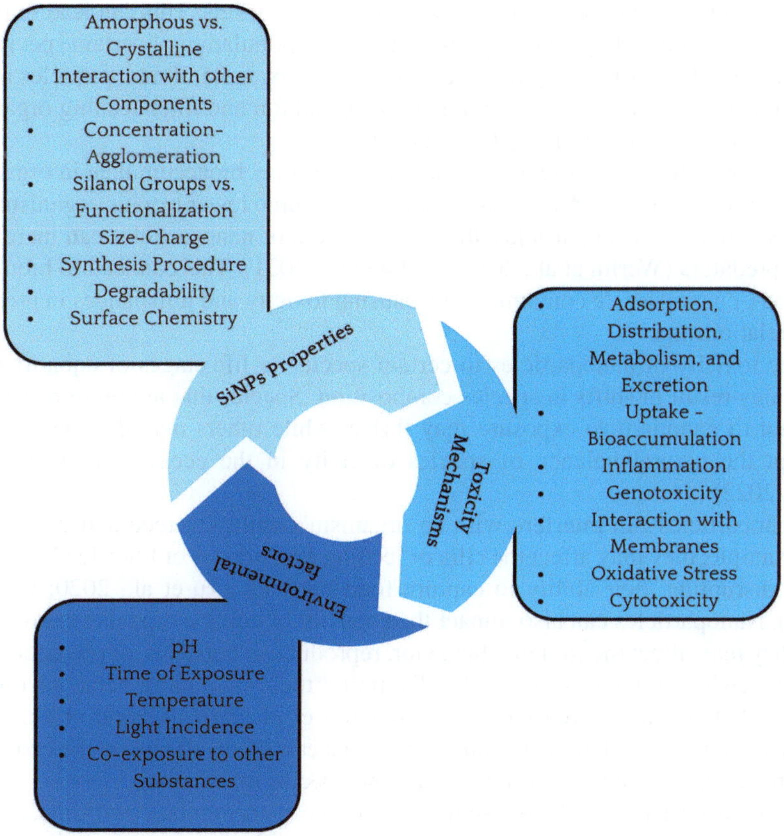

Fig. 8.5 Main parameters affecting the toxicity mechanisms of silica nanoparticles (SiNPs) (Ale et al., 2021)

humic acids can stabilize TiO_2 nanoparticles. Toxic metal ions, such as Ag^+ from AgNP, Cu^{2+} from elemental Cu or from CuO NP, Zn^{2+} from ZnO, and Cd^{2+} from quantum dots, may be liberated from the nanoparticles during their dissolution reactions. These ions' well-known toxicity to aquatic life depends heavily on their speciation in solution (Schirmer et al., 2013).

8.7 Ecological Imbalances Caused by Nanoparticles Affecting Aquatic Species and Ecosystems

Nanoparticles can lead to ecological imbalances in aquatic ecosystems by disrupting the delicate interactions and processes that govern these ecosystems, as well as having negative effects on biodiversity. Nanoparticles can alter the abundance and composition of primary producers (such as phytoplankton and algae) (Khan, 2020) and disrupt the availability of prey for higher trophic levels. This can lead to imbalances in the food web, potentially resulting in overpopulation of certain species and declines in others (Maurer-Jones et al., 2013). For example, if nanoparticles reduce phytoplankton populations, it can impact zooplankton and filter-feeding organisms that rely on phytoplankton as a food source.

On entry into aquatic organisms, nanoparticles may bioaccumulate in organisms at higher trophic levels. As predatory species consume lower trophic organisms that have accumulated nanoparticles, the concentration of nanoparticles can increase in these predators (Werlin et al., 2011; Uddin et al., 2021). This can lead to biomagnification of nanoparticle contamination, causing toxicity and imbalances in predator-prey relationships.

The toxicity of nanoparticles to certain species or life stages of aquatic organisms may result in shifts in species composition. Species that are more resistant or tolerant to nanoparticle exposure may thrive, while others may decline. This can disrupt the natural balance of species diversity in the ecosystem (Muhammad et al., 2022).

Nanoparticles can interfere with an organism's ability to feed and digest food. For example, they may affect the cilia or feeding appendages of filter-feeding organisms, disrupting their ability to capture food particles (Yu et al., 2020; Li et al., 2023). Nanoparticles can also impact the behavior of aquatic organisms. For example, they may affect the foraging behavior, reproductive behavior, or predator avoidance behavior of aquatic species by disrupting their swimming behavior (Wright et al., 2013; Justice & Bernot, 2014; Bergami et al., 2016; Carmo et al., 2019). These feeding and behavioral changes can have cascading effects on the ecosystem by influencing the success of individuals and species interactions.

Nanoparticles that settle into sediment layers can accumulate over time, and this can lead to contamination of benthic ecosystems, affecting organisms living in or near sediments (Ciacci et al., 2019). Sediment-dwelling species may experience direct toxicity or alterations in their habitat, leading to imbalances in benthic communities (Pang et al., 2012; Baker et al., 2014).

The interactions of nanoparticles with other pollutants in the aquatic environment can potentially amplify the toxicity of co-occurring contaminants (Bundschuh et al., 2018). These interactions can result in unforeseen ecological consequences and magnify the impact of pollution on aquatic species and ecosystems (Bundschuh et al., 2018). Additionally, nanoparticles, such as AgNPs, can influence nutrient cycling in aquatic ecosystems (McGee, 2020). For example, they may alter the availability of nutrients like nitrogen and phosphorus, which are essential for primary production by binding to them (Das et al., 2014; Bao et al., 2022). Changes in nutrient dynamics can affect the growth and distribution of aquatic plants and algae, with consequences for other species dependent on these resources.

The cumulative and chronic effects of nanoparticles on aquatic ecosystems may not become apparent immediately (Baker et al., 2014; Stara et al., 2021). Over time, persistent nanoparticle contamination can lead to gradual ecological changes, including population declines and shifts in community structure (Gambardella et al., 2013; Pitt et al., 2018; Vogt et al., 2022). Ecological imbalances caused by nanoparticle exposure can reduce the resilience of aquatic ecosystems. Resilient ecosystems can better withstand environmental stressors and recover from disturbances (Côté & Darling, 2010). Hence, when ecological imbalances occur, ecosystems may become less capable of adapting to changing conditions.

8.8 Conclusion

With the usage of nanoparticles in sunscreen, cosmetics, paint, medicine, and other products, they have become an essential component of our daily lives. Due to their impact on the food chain and disruption of the ecological balance, they have grown to be a significant worry in the modern world. When discharged into the aquatic environment, their incredibly small size has proven to be fatal to the developmental stages in the embryos of aquatic organism, which is mostly owing to their chemical composition. Understanding how these NPs affect the organisms is the subject of extensive investigation.

Overall, the introduction of nanoparticles into aquatic ecosystems can disrupt the ecological dynamics and functions that support the health and stability of these ecosystems. To mitigate the potential for ecological imbalances, it is essential to understand the risks associated with nanoparticle pollution and implement measures to minimize their release and exposure in aquatic environments.

References

Ale, A., Gutierrez, M. F., Rossi, A. S., Bacchetta, C., Desimone, M. F., & Cazenave, J. (2021). Ecotoxicity of silica nanoparticles in aquatic organisms: An updated review. *Environmental Toxicology and Pharmacology, 87*, 103689.

Ali, S., Mehmood, A., & Khan, N. (2021). Uptake, translocation, and consequences of nanomaterials on plant growth and stress adaptation. *Journal of Nanomaterials, 2021*, 1–17.

Amde, M., Liu, J. F., Tan, Z. Q., & Bekana, D. (2017). Transformation and bioavailability of metal oxide nanoparticles in aquatic and terrestrial environments. A review. *Environmental Pollution, 230*, 250–267.

Amreen, K., & Goel, S. (2021). Miniaturized and microfluidic devices for automated nanoparticle synthesis. *ECS Journal of Solid State Science and Technology, 10*(1), 017002.

Aruoja, V., Dubourguier, H. C., Kasemets, K., & Kahru, A. (2009). Toxicity of nanoparticles of CuO, ZnO and TiO_2 to microalgae *Pseudokirchneriella subcapitata. Science of the Total Environment, 407*(4), 1461–1468.

Augustine, R., Hasan, A., Primavera, R., Wilson, R. J., Thakor, A. S., & Kevadiya, B. D. (2020). Cellular uptake and retention of nanoparticles: Insights on particle properties and interaction with cellular components. *Materials Today Communications, 25*, 101692.

Bai, X., Wang, Y., Song, Z., Feng, Y., Chen, Y., Zhang, D., & Feng, L. (2020). The basic properties of gold nanoparticles and their applications in tumor diagnosis and treatment. *International Journal of Molecular Sciences, 21*(7), 2480.

Baker, T. J., Tyler, C. R., & Galloway, T. S. (2014). Impacts of metal and metal oxide nanoparticles on marine organisms. *Environmental Pollution, 186*, 257–271.

Balasubramanian, A. (2011). *Aquatic ecosystems-freshwater types*. University of Mysore.

Bao, S., Xiang, D., Xue, L., Xian, B., Tang, W., & Fang, T. (2022). Pristine and sulfidized ZnO nanoparticles alter microbial community structure and nitrogen cycling in freshwater lakes. *Environmental Pollution, 294*, 118661.

Batley, G. E., Kirby, J. K., & McLaughlin, M. J. (2013). Fate and risks of nanomaterials in aquatic and terrestrial environments. *Accounts of Chemical Research, 46*(3), 854–862.

Beddoes, C. M., Case, C. P., & Briscoe, W. H. (2015). Understanding nanoparticle cellular entry: A physicochemical perspective. *Advances in Colloid and Interface Science, 218*, 48–68.

Benameur, L., Auffan, M., Cassien, M., Liu, W., Culcasi, M., Rahmouni, H., Stocker, P., Tassistro, V., Bottero, J. Y., Rose, J., & Botta, A. (2015). DNA damage and oxidative stress induced by CeO_2 nanoparticles in human dermal fibroblasts: Evidence of a clastogenic effect as a mechanism of genotoxicity. *Nanotoxicology, 9*(6), 696–705.

Bergami, E., Bocci, E., Vannuccini, M. L., Monopoli, M., Salvati, A., Dawson, K. A., & Corsi, I. (2016). Nano-sized polystyrene affects feeding, behavior and physiology of brine shrimp Artemia franciscana larvae. *Ecotoxicology and Environmental Safety, 123*, 18–25.

Bergin, I. L., & Witzmann, F. A. (2013). Nanoparticle toxicity by the gastrointestinal route: Evidence and knowledge gaps. *International Journal of Biomedical Nanoscience and Nanotechnology, 3*(1–2), 163–210.

Blaise, C., Gagné, F., Ferard, J. F., & Eullaffroy, P. (2008). Ecotoxicity of selected nano-materials to aquatic organisms. *Environmental Toxicology: An International Journal, 23*(5), 591–598.

Bondarenko, O., Juganson, K., Ivask, A., Kasemets, K., Mortimer, M., & Kahru, A. (2013). Toxicity of Ag, CuO and ZnO nanoparticles to selected environmentally relevant test organisms and mammalian cells in vitro: A critical review. *Archives of Toxicology, 87*, 1181–1200.

Boraschi, D., Italiani, P., Palomba, R., Decuzzi, P., Duschl, A., Fadeel, B., & Moghimi, S. M. (2017). Nanoparticles and innate immunity: New perspectives on host defence. In *Seminars in immunology* (Vol. 34, pp. 33–51). Academic Press.

Brandelli, A. (2020). The interaction of nanostructured antimicrobials with biological systems: Cellular uptake, trafficking and potential toxicity. *Food Science and Human Wellness, 9*(1), 8–20.

Brun, N. R., Koch, B. E., Varela, M., Peijnenburg, W. J., Spaink, H. P., & Vijver, M. G. (2018). Nanoparticles induce dermal and intestinal innate immune system responses in zebrafish embryos. *Environmental Science: Nano, 5*(4), 904–916.

Bundschuh, M., Filser, J., Lüderwald, S., McKee, M. S., Metreveli, G., Schaumann, G. E., Schulz, R., & Wagner, S. (2018). Nanoparticles in the environment: Where do we come from, where do we go to? *Environmental Sciences Europe, 30*(1), 1–17.

Buzea, C., & Pacheco, I. (2017). Nanomaterial and nanoparticle: Origin and activity. In *Nanoscience and plant–soil systems* (pp. 71–112). Springer.

Canesi, L., Ciacci, C., Vallotto, D., Gallo, G., Marcomini, A., & Pojana, G. (2010). In vitro effects of suspensions of selected nanoparticles (C60 fullerene, TiO_2, SiO_2) on Mytilus hemocytes. *Aquatic Toxicology, 96*(2), 151–158.

Čapek, J., & Roušar, T. (2021). Detection of oxidative stress induced by nanomaterials in cells—the roles of reactive oxygen species and glutathione. *Molecules, 26*(16), 4710.

Carbery, M., O'Connor, W., & Palanisami, T. (2018). Trophic transfer of microplastics and mixed contaminants in the marine food web and implications for human health. *Environment International, 115*, 400–409.

Carmo, T. L., Siqueira, P. R., Azevedo, V. C., Tavares, D., Pesenti, E. C., Cestari, M. M., Martinez, C. B., & Fernandes, M. N. (2019). Overview of the toxic effects of titanium dioxide nanoparticles in blood, liver, muscles, and brain of a neotropical detritivorous fish. *Environmental Toxicology, 34*(4), 457–468.

Chen, M., Sun, Y., Liang, J., Zeng, G., Li, Z., Tang, L., Zhu, Y., Jiang, D., & Song, B. (2019). Understanding the influence of carbon nanomaterials on microbial communities. *Environment International, 126*, 690–698.

Choudhuri, S., Kaur, T., Jain, S., Sharma, C., & Asthana, S. (2021). A review on genotoxicity in connection to infertility and cancer. *Chemico-Biological Interactions, 345*, 109531.

Ciacci, C., Grimmelpont, M. V., Corsi, I., Bergami, E., Curzi, D., Burini, D., Bouchet, V. M., Ambrogini, P., Gobbi, P., Ujiié, Y., & Ishitani, Y. (2019). Nanoparticle-biological interactions in a marine benthic foraminifer. *Scientific Reports, 9*(1), 19441.

Connolly, M., Hernández-Moreno, D., Conde, E., Garnica, A., Navas, J. M., Torrent, F., Rucandio, I., & Fernandez-Cruz, M. L. (2022). Influence of citrate and PEG coatings on the bioaccumulation of TiO_2 and CeO_2 nanoparticles following dietary exposure in rainbow trout. *Environmental Sciences Europe, 34*(1), 1–17.

Côté, I. M., & Darling, E. S. (2010). Rethinking ecosystem resilience in the face of climate change. *PLoS Biology, 8*(7), e1000438.

Dabrowski, B., Zuchowska, A., & Brzozka, Z. (2022). Graphene oxide internalization into mammalian cells–a review. *Colloids and Surfaces B: Biointerfaces, 221*, 112998.

Dagar, G., & Bagchi, G. (2020). Nanoparticles as potential endocrine disruptive chemicals. In *NanoBioMedicine* (pp. 411–429). Springer.

Das, P., Metcalfe, C. D., & Xenopoulos, M. A. (2014). Interactive effects of silver nanoparticles and phosphorus on phytoplankton growth in natural waters. *Environmental Science and Technology, 48*(8), 4573–4580.

Davarpanah, E., & Guilhermino, L. (2019). Are gold nanoparticles and microplastics mixtures more toxic to the marine microalgae *Tetraselmis chuii* than the substances individually? *Ecotoxicology and Environmental Safety, 181*, 60–68.

do Carmo, T. L. L., Azevedo, V. C., de Siqueira, P. R., Galvão, T. D., Dos Santos, F. A., dos Reis Martinez, C. B., Appoloni, C. R., & Fernandes, M. N. (2018). Reactive oxygen species and other biochemical and morphological biomarkers in the gills and kidneys of the Neotropical freshwater fish, *Prochilodus lineatus*, exposed to titanium dioxide (TiO_2) nanoparticles. *Environmental Science and Pollution Research, 25*, 22963–22976.

Dong, S., Wu, Z., Wang, M., Sun, X., & Mao, L. (2023). Assessing comparable bioconcentration potentials for nanoparticles in aquatic organisms via combined utilization of machine learning and toxicokinetic models. *SmartMat, 4*(3), e1155.

Duan, L., Song, J., Li, X., Yuan, H., & Zhuang, W. (2023). Potential risks of CO2 removal project based on carbonate pump to marine ecosystem. *Science of the Total Environment, 862*, 160728.

Dube, E., & Okuthe, G. E. (2023). Engineered nanoparticles in aquatic systems: Toxicity and mechanism of toxicity in fish. *Emerging Contaminants, 9*, 100212.

Fard, J. K., Jafari, S., & Eghbal, M. A. (2015). A review of molecular mechanisms involved in toxicity of nanoparticles. *Advanced Pharmaceutical Bulletin, 5*(4), 447.

Gambardella, C., Aluigi, M. G., Ferrando, S., Gallus, L., Ramoino, P., Gatti, A. M., Rottigni, M., & Falugi, C. (2013). Developmental abnormalities and changes in cholinesterase activity in sea urchin embryos and larvae from sperm exposed to engineered nanoparticles. *Aquatic Toxicology, 130*, 77–85.

Ghosh, M., Chakraborty, A., & Mukherjee, A. (2013). Cytotoxic, genotoxic and the hemolytic effect of titanium dioxide (TiO_2) nanoparticles on human erythrocyte and lymphocyte cells in vitro. *Journal of Applied Toxicology, 33*(10), 1097–1110.

Gondikas, A., von der Kammer, F., Kaegi, R., Borovinskaya, O., Neubauer, E., Navratilova, J., Praetorius, A., Cornelis, G., & Hofmann, T. (2018). Where is the nano? Analytical approaches for the detection and quantification of TiO_2 engineered nanoparticles in surface waters. *Environmental Science: Nano, 5*(2), 313–326.

Gupta, R., & Xie, H. (2018). Nanoparticles in daily life: Applications, toxicity and regulations. *Journal of Environmental Pathology, Toxicology and Oncology, 37*(3).

Handy, R. D., Von der Kammer, F., Lead, J. R., Hassellöv, M., Owen, R., & Crane, M. (2008). The ecotoxicology and chemistry of manufactured nanoparticles. *Ecotoxicology, 17*, 287–314.

Handy, R. D., Clark, N. J., Boyle, D., Vassallo, J., Green, C., Nasser, F., Botha, T. L., Wepener, V., Van Den Brink, N. W., & Svendsen, C. (2022). The bioaccumulation testing strategy for

nanomaterials: Correlations with particle properties and a meta-analysis of in vitro fish alternatives to in vivo fish tests. *Environmental Science: Nano, 9*(2), 684–701.
Hochella, M. F., Aruguete, D., Kim, B., & Madden, A. S. (2012). Naturally occurring inorganic nanoparticles: General assessment and a global budget for one of earth's last unexplored major geochemical components. In *Nature's nanostructures* (pp. 1–31). Jenny Stanford Publishing.
Hu, P., Zhang, X., Li, Y., Pichan, C., & Chen, Z. (2018). Molecular interactions between silver nanoparticles and model cell membranes. *Topics in Catalysis, 61*, 1148–1162.
Ismanto, A., Hadibarata, T., Kristanti, R. A., Maslukah, L., Safinatunnajah, N., & Kusumastuti, W. (2022). Endocrine disrupting chemicals (EDCs) in environmental matrices: Occurrence, fate, health impact, physio-chemical and bioremediation technology. *Environmental Pollution, 302*, 119061.
Iswarya, V., Manivannan, J., De, A., Paul, S., Roy, R., Johnson, J. B., Kundu, R., Chandrasekaran, N., Mukherjee, A., & Mukherjee, A. (2016). Surface capping and size-dependent toxicity of gold nanoparticles on different trophic levels. *Environmental Science and Pollution Research, 23*, 4844–4858.
Jackson, P., Jacobsen, N. R., Baun, A., Birkedal, R., Kühnel, D., Jensen, K. A., Vogel, U., & Wallin, H. (2013). Bioaccumulation and ecotoxicity of carbon nanotubes. *Chemistry Central Journal, 7*, 1–21.
Jovanović, B., & Palić, D. (2012). Immunotoxicology of non-functionalized engineered nanoparticles in aquatic organisms with special emphasis on fish—Review of current knowledge, gap identification, and call for further research. *Aquatic Toxicology, 118*, 141–151.
Justice, J. R., & Bernot, R. J. (2014). Nanosilver inhibits freshwater gastropod (*Physa acuta*) ability to assess predation risk. *The American Midland Naturalist, 171*(2), 340–349.
Kar, S., Sangem, P., Anusha, N., & Senthilkumaran, B. (2021). Endocrine disruptors in teleosts: Evaluating environmental risks and biomarkers. *Aquaculture and Fisheries, 6*(1), 1–26.
Karlsson, H. L., Gliga, A. R., Calléja, F. M., Gonçalves, C. S., Wallinder, I. O., Vrieling, H., Fadeel, B., & Hendriks, G. (2014). Mechanism-based genotoxicity screening of metal oxide nanoparticles using the ToxTracker panel of reporter cell lines. *Particle and Fibre Toxicology, 11*, 1–14.
Kaya, H., Aydın, F., Gürkan, M., Yılmaz, S., Ates, M., Demir, V., & Arslan, Z. (2016). A comparative toxicity study between small and large size zinc oxide nanoparticles in tilapia (*Oreochromis niloticus*): Organ pathologies, osmoregulatory responses and immunological parameters. *Chemosphere, 144*, 571–582.
Khan, F. A. (2020). Nanomaterials: Types, classifications, and sources. In *Applications of nanomaterials in human health* (pp. 1–13). Springer.
Khan, I., Saeed, K., & Khan, I. (2019). Nanoparticles: Properties, applications and toxicities. *Arabian Journal of Chemistry, 12*(7), 908–931.
Khan, S., Naushad, M., Al-Gheethi, A., & Iqbal, J. (2021). Engineered nanoparticles for removal of pollutants from wastewater: Current status and future prospects of nanotechnology for remediation strategies. *Journal of Environmental Chemical Engineering, 9*(5), 106160.
Khanna, P., Ong, C., Bay, B. H., & Baeg, G. H. (2015). Nanotoxicity: An interplay of oxidative stress, inflammation and cell death. *Nanomaterials, 5*(3), 1163–1180.
Khoshnamvand, M., Hao, Z., Fadare, O. O., Hanachi, P., Chen, Y., & Liu, J. (2020). Toxicity of biosynthesized silver nanoparticles to aquatic organisms of different trophic levels. *Chemosphere, 258*, 127346.
Knight, J. (2021). Endocrine system 1: Overview of the endocrine system and hormones. *Nursing Times [Online], 117*(5), 38–42.
Kraśniewska, K., Galus, S., & Gniewosz, M. (2020). Biopolymers-based materials containing silver nanoparticles as active packaging for food applications–a review. *International Journal of Molecular Sciences, 21*(3), 698.
Kumar, N., Thorat, S. T., & Reddy, K. S. (2023). Multi biomarker approach to assess manganese and manganese nanoparticles toxicity in *Pangasianodon hypophthalmus*. *Scientific Reports, 13*(1), 8505.

Kurwadkar, S., Pugh, K., Gupta, A., & Ingole, S. (2015). Nanoparticles in the environment: Occurrence, distribution, and risks. *Journal of Hazardous, Toxic, and Radioactive Waste, 19*(3), 04014039.

Land, M. (2015). *Effects of nano-and microplastic particles on plankton and marine ecosystem functioning. An evidence overview.* EviEM.

Li, L., Hartmann, G., Döblinger, M., & Schuster, M. (2013). Quantification of nanoscale silver particles removal and release from municipal wastewater treatment plants in Germany. *Environmental Science and Technology, 47*(13), 7317–7323.

Li, F., Li, R., Lu, F., Xu, L., Gan, L., Chu, W., Yan, M., & Gong, H. (2023). Adverse effects of silver nanoparticles on aquatic plants and zooplankton: A review. *Chemosphere*, 139459.

Luo, D., Wang, X., Burda, C., & Basilion, J. P. (2021). Recent development of gold nanoparticles as contrast agents for cancer diagnosis. *Cancers, 13*(8), 1825.

Ma, S., & Lin, D. (2013). The biophysicochemical interactions at the interfaces between nanoparticles and aquatic organisms: Adsorption and internalization. *Environmental Science: Processes & Impacts, 15*(1), 145–160.

Ma, Y., Cai, F., Li, Y., Chen, J., Han, F., & Lin, W. (2020). A review of the application of nanoparticles in the diagnosis and treatment of chronic kidney disease. *Bioactive Materials, 5*(3), 732–743.

Magdolenova, Z., Collins, A., Kumar, A., Dhawan, A., Stone, V., & Dusinska, M. (2014). Mechanisms of genotoxicity. A review of in vitro and in vivo studies with engineered nanoparticles. *Nanotoxicology, 8*(3), 233–278.

Manke, A., Wang, L., & Rojanasakul, Y. (2013). Mechanisms of nanoparticle-induced oxidative stress and toxicity. *BioMed Research International, 2013.*

Marana, M. H., Poulsen, R., Thormar, E. A., Clausen, C. G., Thit, A., Mathiessen, H., Jaafar, R., Korbut, R., Hansen, A. M. B., Hansen, M., & Limborg, M. T. (2022). Plastic nanoparticles cause mild inflammation, disrupt metabolic pathways, change the gut microbiota and affect reproduction in zebrafish: A full generation multi-omics study. *Journal of Hazardous Materials, 424*, 127705.

Markus, A. A., Parsons, J. R., Roex, E. W. M., De Voogt, P., & Laane, R. W. P. M. (2015). Modeling aggregation and sedimentation of nanoparticles in the aquatic environment. *Science of the Total Environment, 506*, 323–329.

Mattsson, K., Johnson, E. V., Malmendal, A., Linse, S., Hansson, L. A., & Cedervall, T. (2017). Brain damage and behavioural disorders in fish induced by plastic nanoparticles delivered through the food chain. *Scientific Reports, 7*(1), 11452.

Maurer-Jones, M. A., Gunsolus, I. L., Murphy, C. J., & Haynes, C. L. (2013). Toxicity of engineered nanoparticles in the environment. *Analytical Chemistry, 85*(6), 3036–3049.

McGee, C. F. (2020). The effects of silver nanoparticles on the microbial nitrogen cycle: A review of the known risks. *Environmental Science and Pollution Research, 27*(25), 31061–31073.

Moreno-Garrido, I., Pérez, S., & Blasco, J. (2015). Toxicity of silver and gold nanoparticles on marine microalgae. *Marine Environmental Research, 111*, 60–73.

Muhammad, A., He, J., Yu, T., Sun, C., Shi, D., Jiang, Y., Xianyu, Y., & Shao, Y. (2022). Dietary exposure of copper and zinc oxides nanoparticles affect the fitness, enzyme activity, and microbial community of the model insect, silkworm Bombyx mori. *Science of the Total Environment, 813*, 152608.

Nam, S. H., Shin, Y. J., Lee, W. M., Kim, S. W., Kwak, J. I., Yoon, S. J., & An, Y. J. (2015). Conducting a battery of bioassays for gold nanoparticles to derive guideline value for the protection of aquatic ecosystems. *Nanotoxicology, 9*(3), 326–335.

Navarro, E., Baun, A., Behra, R., Hartmann, N. B., Filser, J., Miao, A. J., Quigg, A., Santschi, P. H., & Sigg, L. (2008). Environmental behavior and ecotoxicity of engineered nanoparticles to algae, plants, and fungi. *Ecotoxicology, 17*, 372–386.

Nel, A. E., Mädler, L., Velegol, D., Xia, T., Hoek, E. M., Somasundaran, P., Klaessig, F., Castranova, V., & Thompson, M. (2009). Understanding biophysicochemical interactions at the nano–bio interface. *Nature Materials, 8*(7), 543–557.

Nirmal, D. D. (2019). High performance flexible nanoparticles based organic electronics. *Journal of Electronics and Informatics, 1*(2), 99–106.

Pang, C., Selck, H., Misra, S. K., Berhanu, D., Dybowska, A., Valsami-Jones, E., & Forbes, V. E. (2012). Effects of sediment-associated copper to the deposit-feeding snail, *Potamopyrgus antipodarum*: A comparison of Cu added in aqueous form or as nano-and micro-CuO particles. *Aquatic Toxicology, 106*, 114–122.

Paul, W., & Sharma, C. P. (2020). Inorganic nanoparticles for targeted drug delivery. In *Biointegration of medical implant materials* (pp. 333–373).

Pentsak, E. O., Galushko, A. S., Cherepanova, V. A., & Ananikov, V. P. (2021). How to make a cocktail of palladium catalysts with cola and alcohol: Heteroatom doping vs. nanoscale morphology of carbon supports. *Nanomaterials, 11*(10), 2599.

Perreault, F., Melegari, S. P., Fuzinatto, C. F., Bogdan, N., Morin, M., Popovic, R., & Matias, W. G. (2014). Toxicity of pamam-coated gold nanoparticles in different unicellular models. *Environmental Toxicology, 29*(3), 328–336.

Pitt, J. A., Kozal, J. S., Jayasundara, N., Massarsky, A., Trevisan, R., Geitner, N., Wiesner, M., Levin, E. D., & Di Giulio, R. T. (2018). Uptake, tissue distribution, and toxicity of polystyrene nanoparticles in developing zebrafish (*Danio rerio*). *Aquatic Toxicology, 194*, 185–194.

Prajitha, N., Athira, S. S., & Mohanan, P. V. (2019). Bio-interactions and risks of engineered nanoparticles. *Environmental Research, 172*, 98–108.

Prüst, M., Meijer, J., & Westerink, R. H. (2020). The plastic brain: Neurotoxicity of micro-and nanoplastics. *Particle and Fibre Toxicology, 17*(1), 1–16.

Purushothaman, S., Raghunath, A., Dhakshinamoorthy, V., Panneerselvam, L., & Perumal, E. (2014). Acute exposure to titanium dioxide (TiO_2) induces oxidative stress in zebrafish gill tissues. *Toxicological and Environmental Chemistry, 96*(6), 890–905.

Rakib, M. R. J., Sarker, A., Ram, K., Uddin, M. G., Walker, T. R., Chowdhury, T., Uddin, J., Khandaker, M. U., Rahman, M. M., & Idris, A. M. (2023). Microplastic toxicity in aquatic organisms and aquatic ecosystems: A review. *Water, Air, and Soil Pollution, 234*(1), 52.

Rana, S., & Kalaichelvan, P. (2013). *Ecotoxicity of nanoparticles*. International Scholarly Research Notices.

Rawat, N., Singla-Pareek, S. L., & Pareek, A. (2021). Membrane dynamics during individual and combined abiotic stresses in plants and tools to study the same. *Physiologia Plantarum, 171*(4), 653–676.

Ren, E., Zhang, C., Li, D., Pang, X., & Liu, G. (2020). Leveraging metal oxide nanoparticles for bacteria tracing and eradicating. *Viewpoints, 1*(3), 20200052.

Renault, S., Baudrimont, M., Mesmer-Dudons, N., Gonzalez, P. M., Mornet, S., & Brisson, A. (2008). Impacts of gold nanoparticle exposure on two freshwater species: A phytoplanktonic alga (*Scenedesmus subspicatus*) and a benthic bivalve (*Corbicula fluminea*). *Gold Bulletin, 41*, 116–126.

Rhazouani, A., Gamrani, H., El Achaby, M., Aziz, K., Gebrati, L., Uddin, M. S., & Aziz, F. (2021). Synthesis and toxicity of graphene oxide nanoparticles: A literature review of in vitro and in vivo studies. *BioMed Research International, 2021*.

Rocha, T. L., Mestre, N. C., Sabóia-Morais, S. M. T., & Bebianno, M. J. (2017). Environmental behaviour and ecotoxicity of quantum dots at various trophic levels: A review. *Environment International, 98*, 1–17.

Sajid, M., Ilyas, M., Basheer, C., Tariq, M., Daud, M., Baig, N., & Shehzad, F. (2015). Impact of nanoparticles on human and environment: Review of toxicity factors, exposures, control strategies, and future prospects. *Environmental Science and Pollution Research, 22*, 4122–4143.

Sarkar, N., Chaudhary, S., & Kaushik, M. (2021). Nano-fertilizers and nano-pesticides as promoters of plant growth in agriculture. In *Plant-Microbes-Engineered Nano-particles (PM-ENPs) nexus in agro-ecosystems: Understanding the interaction of plant, microbes and engineered nano-particles (ENPS)* (pp. 153–163). Springer.

Scown, T. M., Van Aerle, R., & Tyler, C. R. (2010). Do engineered nanoparticles pose a significant threat to the aquatic environment? *Critical Reviews in Toxicology, 40*(7), 653–670.

Sharma, A., Mollier, J., Brocklesby, R. W., Caves, C., Jayasena, C. N., & Minhas, S. (2020). Endocrine-disrupting chemicals and male reproductive health. *Reproductive Medicine and Biology, 19*(3), 243–253.

Shi, Q., Wang, C. L., Zhang, H., Chen, C., Zhang, X., & Chang, X. L. (2020). Trophic transfer and biomagnification of fullerenol nanoparticles in an aquatic food chain. *Environmental Science: Nano, 7*(4), 1240–1251.

Shukla, R. K., Badiye, A., Vajpayee, K., & Kapoor, N. (2021). Genotoxic potential of nanoparticles: Structural and functional modifications in DNA. *Frontiers in Genetics, 12*, 728250.

Sibiya, A., Gopi, N., Jeyavani, J., Mahboob, S., Al-Ghanim, K. A., Sultana, S., Mustafa, A., Govindarajan, M., & Vaseeharan, B. (2022). Comparative toxicity of silver nanoparticles and silver nitrate in freshwater fish *Oreochromis mossambicus*: A multi-biomarker approach. *Comparative Biochemistry and Physiology Part C: Toxicology and Pharmacology, 259*, 109391.

Singer, A., Barakat, Z., Mohapatra, S., & Mohapatra, S. S. (2019). Nanoscale drug-delivery systems: In vitro and in vivo characterization. In *Nanocarriers for drug delivery* (pp. 395–419). Elsevier.

Singh, D., & Gurjar, B. R. (2022). Nanotechnology for agricultural applications: Facts, issues, knowledge gaps, and challenges in environmental risk assessment. *Journal of Environmental Management, 322*, 116033.

Spurgeon, D. J., Lahive, E., & Schultz, C. L. (2020). Nanomaterial transformations in the environment: effects of changing exposure forms on bioaccumulation and toxicity. *Small, 16*(36), 2000618.

Srinivasan, M., Venkatesan, M., Arumugam, V., Natesan, G., Saravanan, N., Murugesan, S., Ramachandran, S., Ayyasamy, R., & Pugazhendhi, A. (2019). Green synthesis and characterization of titanium dioxide nanoparticles (TiO_2 NPs) using *Sesbania grandiflora* and evaluation of toxicity in zebrafish embryos. *Process Biochemistry, 80*, 197–202.

Stara, A., Pagano, M., Albano, M., Savoca, S., Di Bella, G., Albergamo, A., Koutkova, Z., Sandova, M., Velisek, J., Fabrello, J., & Matozzo, V. (2021). Effects of long-term exposure of *Mytilus galloprovincialis* to thiacloprid: A multibiomarker approach. *Environmental Pollution, 289*, 117892.

Suman, T. Y., Rajasree, S. R., & Kirubagaran, R. (2015). Evaluation of zinc oxide nanoparticles toxicity on marine algae *Chlorella vulgaris* through flow cytometric, cytotoxicity and oxidative stress analysis. *Ecotoxicology and Environmental Safety, 113*, 23–30.

Tee, J. K., Ong, C. N., Bay, B. H., Ho, H. K., & Leong, D. T. (2016). Oxidative stress by inorganic nanoparticles. *Wiley Interdisciplinary Reviews: Nanomedicine and Nanobiotechnology, 8*(3), 414–438.

Tian, Y., Tian, Z., Dong, Y., Wang, X., & Zhan, L. (2021). Current advances in nanomaterials affecting morphology, structure, and function of erythrocytes. *RSC Advances, 11*(12), 6958–6971.

Topuz, E., & van Gestel, C. A. (2016). An approach for environmental risk assessment of engineered nanomaterials using Analytical Hierarchy Process (AHP) and fuzzy inference rules. *Environment International, 92*, 334–347.

Uddin, M. N., Desai, F., & Asmatulu, E. (2020). Engineered nanomaterials in the environment: Bioaccumulation, biomagnification and biotransformation. *Environmental Chemistry Letters, 18*, 1073–1083.

Uddin, M. N., Desai, F., & Asmatulu, E. (2021). Review of bioaccumulation, biomagnification, and biotransformation of engineered nanomaterials. In *Nanotoxicology and nanoecotoxicology* (Vol. 2, pp. 133–164). Springer International Publishing.

Vlachogianni, T., Fiotakis, K., Loridas, S., Perdicaris, S., & Valavanidis, A. (2013). Potential toxicity and safety evaluation of nanomaterials for the respiratory system and lung cancer. *Lung Cancer: Targets and Therapy, 4*, 71–82.

Vogt, R., Hartmann, S., Kunze, J., Jupke, J. F., Steinhoff, B., Schönherr, H., Kuhnert, K. D., Witte, K., Lamatsch, D. K., & Wanzenböck, J. (2022). Silver nanoparticles adversely affect the swimming behavior of European Whitefish (*Coregonus lavaretus*) larvae within the low μg/L range. *Journal of Toxicology and Environmental Health, Part A, 85*(21), 867–880.

Walke, G., Gaurkar, S. S., Prasad, R., Lohakare, T., & Wanjari, M. (2023). The impact of oxidative stress on male reproductive function: Exploring the role of antioxidant supplementation. *Cureus, 15*(7).
Wang, J., & Wang, W. X. (2014). Salinity influences on the uptake of silver nanoparticles and silver nitrate by marine medaka (*Oryzias melastigma*). *Environmental Toxicology and Chemistry, 33*(3), 632–640.
Wang, W., Zhang, J., Qiu, Z., Cui, Z., Li, N., Li, X., Wang, Y., Zhang, H., & Zhao, C. (2022). Effects of polyethylene microplastics on cell membranes: A combined study of experiments and molecular dynamics simulations. *Journal of Hazardous Materials, 429*, 128323.
Werlin, R., Priester, J. H., Mielke, R. E., Krämer, S., Jackson, S., Stoimenov, P. K., Stucky, G. D., Cherr, G. N., Orias, E., & Holden, P. A. (2011). Biomagnification of cadmium selenide quantum dots in a simple experimental microbial food chain. *Nature Nanotechnology, 6*(1), 65–71.
Williams, D. F. (2017). Biocompatibility pathways: Biomaterials-induced sterile inflammation, mechanotransduction, and principles of biocompatibility control. *ACS Biomaterials Science and Engineering, 3*(1), 2–35.
Williamson, P., & Guinder, V. A. (2021). Effect of climate change on marine ecosystems. In *The impacts of climate change* (pp. 115–176). https://doi.org/10.1016/B978-0-12-822373-4.00024-0
Wright, S. L., Thompson, R. C., & Galloway, T. S. (2013). The physical impacts of microplastics on marine organisms: A review. *Environmental Pollution, 178*, 483–492.
Wu, Y., & Zhou, Q. (2013). Silver nanoparticles cause oxidative damage and histological changes in medaka (*Oryzias latipes*) after 14 days of exposure. *Environmental Toxicology and Chemistry, 32*(1), 165–173.
Xiang, L., Fang, J., & Cheng, H. (2018). Toxicity of silver nanoparticles to green algae *M. aeruginosa* and alleviation by organic matter. *Environmental Monitoring and Assessment, 190*, 1–9.
Xiong, S., George, S., Ji, Z., Lin, S., Yu, H., Damoiseaux, R., France, B., Ng, K. W., & Loo, S. C. J. (2013). Size of TiO_2 nanoparticles influences their phototoxicity: An in vitro investigation. *Archives of Toxicology, 87*, 99–109.
Yamini, V., Shanmugam, V., Rameshpathy, M., Venkataraman, G., Ramanathan, G., Garalleh, H. A., Hashmi, A., Brindhadevi, K., & Rajeswari, V. D. (2023). Environmental effects and interaction of nanoparticles on beneficial soil and aquatic microorganisms. *Environmental Research*, 236(1):116776. https://doi.org/10.1016/j.envres.2023.116776
Yang, S., Li, M., Kong, R. Y. C., Li, L., Li, R., Chen, J., & Lai, K. P. (2023). Reproductive toxicity of micro-and nanoplastics. *Environment International*, 177:108002. https://doi.org/10.1016/j.envint.2023.108002
Yao, Y., Zhang, T., & Tang, M. (2022). The DNA damage potential of quantum dots: Toxicity, mechanism and challenge. *Environmental Pollution, 317*, 120676.
Yu, S., Liu, J., Yin, Y., & Shen, M. (2018). Interactions between engineered nanoparticles and dissolved organic matter: A review on mechanisms and environmental effects. *Journal of Environmental Sciences, 63*, 198–217.
Yu, F., Yang, C., Zhu, Z., Bai, X., & Ma, J. (2019). Adsorption behavior of organic pollutants and metals on micro/nanoplastics in the aquatic environment. *Science of the Total Environment, 694*, 133643.
Yu, S. P. Y., Cole, M. C., & Chan, B. K. (2020). Effects of microplastic on zooplankton survival and sublethal responses. In *Oceanography and marine biology*. https://doi.org/10.1201/9780429351495-7
Zhang, C., Hu, Z., Li, P., & Gajaraj, S. (2016a). Governing factors affecting the impacts of silver nanoparticles on wastewater treatment. *Science of the Total Environment, 572*, 852–873.
Zhang, Y. Q., Dringen, R., Petters, C., Rastedt, W., Köser, J., Filser, J., & Stolte, S. (2016b). Toxicity of dimercaptosuccinate-coated and un-functionalized magnetic iron oxide nanoparticles towards aquatic organisms. *Environmental Science: Nano, 3*(4), 754–767.
Zhang, J., Guo, W., Li, Q., Wang, Z., & Liu, S. (2018). The effects and the potential mechanism of environmental transformation of metal nanoparticles on their toxicity in organisms. *Environmental Science: Nano, 5*(11), 2482–2499.

Zhou, J., Cai, Z. H., & Zhu, X. S. (2010). Are endocrine disruptors among the causes of the deterioration of aquatic biodiversity? *Integrated Environmental Assessment and Management, 6*(3), 492–498.

Zhou, S., Li, H., Wang, H., Wang, R., Song, W., Li, D., Wei, C., Guo, Y., He, X., & Deng, Y. (2023). Nickel nanoparticles induced hepatotoxicity in mice via lipid-metabolism-dysfunction-regulated inflammatory injury. *Molecules, 28*(15), 5757.

Zhu, B., He, W., Hu, S., Kong, R., & Yang, L. (2019). The fate and oxidative stress of different sized SiO2 nanoparticles in zebrafish (*Danio rerio*) larvae. *Chemosphere, 225*, 705–712.

Chapter 9
Nanoparticle-Based Remediation and Environmental Cleanup

Mordecai Gana, Patrick Omoregie Isibor, Josephine Iyosomi Damisa, Frances Iseghohi, Innocent Ojeba Musa, and Oluwafemi Adebayo Oyewole

9.1 Introduction

Pollution of the environment is a global issue with far-reaching consequences for the planet and its inhabitants. Numerous pollutants have been discharged into the environment as a result of industrialization, urbanization, and agricultural practices, resulting in deteriorating air quality, contaminated water bodies, and soil degradation. Consequently, there is an urgent need for innovative and effective remediation methods that can effectively address these environmental problems while mitigating their impact on ecosystems and human health.

Nanoparticle-based remediation represents a cutting-edge approach to address environmental pollution by employing specially engineered nanoparticles as agents for targeting and removing contaminants from various environmental matrices (Chaudhary et al., 2023). These nanoparticles, typically ranging in size from 1 to 100 nanometers, offer a novel and promising solution to mitigate the adverse effects of pollution and restore the natural balance of ecosystems (Boregowda et al., 2022). Nanoparticle-based remediation utilizes the distinctive physicochemical characteristics of nanoparticles to engage with and counteract various forms of contaminants.

M. Gana (✉) · J. I. Damisa · O. A. Oyewole
Department of Microbiology, Federal University of Technology, Minna, Nigeria
e-mail: mordecaigana@gmail.com

P. O. Isibor
Department of Biological Sciences, Covenant University, Ota, Ogun State, Nigeria
e-mail: patrick.isibor@covenantuniversity.edu.ng

F. Iseghohi
Medical Biotechnology Department, National Biotechnology Development Agency, Abuja, Nigeria

I. O. Musa
Department of Microbiology, Skyline University Nigeria, Kano, Nigeria

P. O. Isibor et al. (eds.), *Environmental Nanotoxicology*,
https://doi.org/10.1007/978-3-031-54154-4_9

According to Chen and Riviere (2017), nanoparticles possess the capability to be customized and adapted to address specific environmental cleanup objectives, rendering them highly adaptable instruments in the battle against pollution. Nanoparticle-based remediation encompasses several crucial mechanisms, namely, adsorption, absorption, and catalytic degradation, which present unique processes for the elimination of pollutants.

One essential technique of nanoparticle-mediated remediation entails the adsorption of contaminants onto the nanoparticles' surface (Guerra et al., 2018). The nanoparticles exhibit a high ratio of surface area to volume, which results in a significant number of binding sites, rendering them highly effective as adsorbents. The adherence of contaminants, including heavy metals, organic pollutants, and pathogens, to the surface of nanoparticles is facilitated by electrostatic, chemical, or physical interactions (Huang et al., 2019). The aforementioned procedure efficiently isolates the contaminants, hence inhibiting their subsequent dispersion or detrimental impact on the environment. Furthermore, nanoparticles have the capability to not only adsorb pollutants on their surface but also absorb them within their structure or matrix. The process of absorption can take place through a range of processes, including encapsulation or trapping within the core or porous structure of the nanoparticle. Nanoparticles have the ability to absorb contaminants, thereby immobilizing them and perhaps enabling their controlled release over a period of time, hence facilitating progressive breakdown or removal (Roy et al., 2021).

Nanoparticles can additionally function as catalysts to facilitate the destruction or conversion of contaminants. According to Ettel et al. (2021), these catalysts have the ability to begin and expedite chemical processes that result in the decomposition of detrimental chemicals into substances that are either less hazardous or nontoxic. The catalytic characteristic of interest is particularly advantageous when it comes to dealing with persistent organic pollutants and recalcitrant molecules that pose difficulties in being eliminated using conventional methods. The utilization of nanoparticles for remediation purposes is a highly effective approach in environmental cleanup due to its several benefits (Roy et al., 2021). These characteristics encompass high levels of effectiveness, selectivity, and the capacity to specifically target contaminants, hence reducing any unintended harm to non-targeted creatures or chemicals. Furthermore, this approach frequently necessitates a lesser allocation of resources and exhibits a diminished environmental impact in contrast to conventional, more intrusive techniques of rehabilitation.

This chapter aims to provide a comprehensive examination of nanoparticle-based remediation, encompassing an in-depth analysis of the diverse range of nanoparticles employed, their inherent characteristics, modes of operation, utilization in various environmental contexts, safety implications, and the forthcoming obstacles and prospects. This novel methodology exhibits significant potential in tackling environmental degradation while simultaneously promoting sustainability and ecological restoration.

9.2 Significance of Nanoparticle-Based Remediation in Environmental Cleanup

The utilization of nanoparticles for remediation purposes signifies a significant shift in the realm of environmental science and remediation. The significance of this phenomenon cannot be exaggerated, as it presents novel resolutions to the urgent issues of pollution and environmental degradation that our planet currently confronts. The following points explicate the significant significance of this approach:

1. Precision and targeted remediation: The precision of nanoparticle-based cleanup is considered to be one of its most notable advantages. The manipulation of nanoparticles can be customized to selectively target particular pollutants, encompassing heavy metals, volatile chemical compounds, pesticides, and various other contaminants (Kuhn et al., 2022). The process of selectivity guarantees that solely the desired pollutants are caught or destroyed, while non-targeted chemicals remain unaffected. The implementation of this level of precision serves to mitigate the disturbance caused to natural ecosystems and mitigate the incidental harm frequently observed in conventional restoration approaches.
2. Enhanced removal efficiency: Nanoparticles exhibit a significantly elevated surface area-to-volume ratio as a consequence of their diminutive dimensions, hence affording a multitude of active sites for engaging with pollutants. According to Roy et al. (2021), this particular feature results in a greater removal efficiency when compared to bulk materials or conventional procedures. This implies that the utilization of lesser quantities of nanoparticles can provide comparable or superior outcomes in the context of pollution removal, resource utilization reduction, and cost reduction.
3. Versatility in environmental matrices: The application of nanoparticle-based remediation has demonstrated its effectiveness in several environmental matrices, encompassing soil, groundwater, surface water, air, and sediments (Martin et al., 2022). The adaptability of this solution renders it appropriate for mitigating a wide range of pollution scenarios, encompassing various contexts such as polluted industrial locations, oil spills in aquatic environments, and the dispersion of airborne particulate matter.
4. Reduced environmental footprint: Conventional approaches to environmental remediation, such as excavation and incineration, have been observed to require significant resources and have the potential to generate additional pollution. In contrast, nanoparticle-mediated cleanup frequently necessitates less resource allocation and generates little waste (Kuhn et al., 2022). This method demonstrates a commitment to sustainability and responsible environmental management by mitigating the environmental impact of cleanup activities.
5. Rapid response to emerging contaminants: The utilization of nanoparticles in remediation processes exhibits a notable capacity to promptly respond to newly identified environmental pollutants. According to Ahmad et al. (2023), nanoparticles have the potential to be tailored and enhanced in order to effectively

combat newly identified contaminants. The ability to adapt is of utmost importance in a global context where the ongoing emergence of toxins poses persistent challenges to the resilience of ecosystems.

6. Minimization of human exposure: The utilization of nanoparticle-based remediation techniques has the potential to substantially decrease human exposure to harmful compounds due to their enhanced precision and efficiency. The mitigation of toxins at their origin significantly reduces the potential harm to human health, whether through direct exposure or the ingestion of polluted resources (Foong et al., 2020).
7. Restoration of ecosystem health: The ultimate importance of nanoparticle-based remediation rests in its capacity to effectively restore the ecological health and functionality of ecosystems. The approach described in the study by Rajput et al. (2022) plays a significant role in the restoration of natural habitats, the enhancement of aquatic ecosystems, and the conservation of biodiversity by effectively reducing or alleviating the adverse impacts of pollutants.
8. Innovation and scientific advancement: The utilization of nanoparticle-based remediation has been crucial in driving innovation and promoting scientific progress across several fields such as materials science, nanotechnology, chemistry, and environmental engineering (Kumar, 2014). The utilization of an interdisciplinary approach not only effectively tackles contemporary environmental concerns but also fosters advancements in interconnected disciplines, hence facilitating the emergence of novel materials and methodologies.

The utilization of nanoparticles for remediation purposes is a promising approach in addressing the issue of environmental contamination. The relevance of this technology extends beyond its effectiveness in eliminating contaminants, as it has the potential to fundamentally transform our approach to environmental remediation. With the ongoing advancement of technology and the deepening comprehension of nanoparticles, this methodology exhibits the potential to usher in a cleaner and more sustainable future for our planet and forthcoming generations.

9.3 The Need for Innovative and Efficient Environmental Remediation Methods

The harsh reality of our modern surroundings is one of pollution, degraded habitats, and the impending threat of climate change. The demand for creative and effective environmental restoration techniques grows as we struggle with the effects of human activity. Numerous interrelated reasons are driving this demand, all of which emphasize how urgent it is to discover better solutions to deal with environmental concerns.

(i) Growing pollution levels: Pollution, in its many forms, continues to rise across the globe. Industrial emissions, agricultural runoff, urban development, and waste disposal have collectively led to the contamination of air, water, soil, and ecosystems (Singh & Singh, 2017). The ubiquity and diversity of pollutants demand versatile remediation approaches that can effectively target various contaminants.

(ii) Complex pollutants: Modern pollutants are often more complex and persistent than those of the past. For instance, industrial chemicals, pharmaceuticals, and microplastics pose unique challenges due to their intricate molecular structures and resistance to degradation (Campanale et al., 2020). Traditional remediation methods struggle to cope with these emerging pollutants, necessitating advanced techniques such as nanoparticle-based remediation.

(iii) Preservation of ecosystems: Ecosystems have a crucial role in providing important services, such as the maintenance of air quality, the purification of water, and the cycling of nutrients. According to Chu and Karr (2017), pollution has a detrimental impact on these services, resulting in disturbances within ecosystems and reducing their capacity to sustain life. The implementation of innovative remediation techniques plays a critical role in the restoration of ecosystems and the preservation of their functionality.

(iv) Human health concerns: Environmental pollution is not only an ecological issue but also a profound human health concern. Pollutants can contaminate food, water sources, and the air we breathe, leading to a wide range of health problems, including respiratory diseases, cancer, and developmental issues in children. Efficient remediation methods directly contribute to reducing human exposure to harmful substances (Naidu et al., 2021).

(v) Resource scarcity: The finite nature of the planet's natural resources is exacerbated by the acceleration of their depletion in the presence of pollution. The presence of soil contamination has the potential to render agricultural land unsuitable for use, while water pollution poses a significant threat to the supply of safe drinking water. Additionally, air pollution has adverse effects on the overall quality of life in metropolitan areas (Lu et al., 2015). The implementation of remediation techniques that facilitate the restoration and repurposing of resources is vital for the promotion of sustainable development (Ugwu et al., 2021).

(vi) Climate change mitigation: Climate change, driven by greenhouse gas emissions, is a global crisis that exacerbates environmental challenges. Innovative remediation methods that can sequester carbon dioxide or mitigate the effects of climate change-induced disasters are urgently needed to combat this existential threat (Kumar et al., 2021; Taylor, 2023).

(vii) Economic costs: The presence of pollution and environmental degradation is accompanied by significant economic implications. The remediation of contaminated areas, the purification of dirty water, and the mitigation of health concerns associated with pollution place significant financial strains on governmental entities and local populations. The implementation of efficient and cost-effective remediation solutions has the potential to mitigate the economic burdens associated with this issue.

(viii) Regulatory compliance: Stringent environmental regulations and international agreements compel industries and governments to take action against pollution. Meeting these regulatory requirements demands effective and efficient remediation solutions that can ensure compliance while minimizing disruption to operations (Liu & Xie, 2020).

(ix) Public awareness and expectations: Increasing public awareness and concern about environmental issues have led to higher expectations for pollution control and remediation efforts. Society is demanding innovative solutions that not only clean up existing pollution but also prevent future contamination (Hoang et al., 2019).

(x) Technological advancements: Advances in technology, including nanotechnology and materials science, have opened new avenues for environmental remediation. These innovations provide opportunities to develop more efficient and targeted remediation. The demand for novel and effective approaches to environmental remediation is motivated by the pressing necessity to tackle pollution, protect human well-being, conserve ecosystems, and alleviate the consequences of climate change (Palit, 2018). In the face of the intricate issues presented by modern contaminants, it is imperative to prioritize the advancement and use of state-of-the-art remediation methodologies, such as those centered around nanoparticles. These approaches play a pivotal role in safeguarding the welfare of present and future generations, as well as ensuring the long-term viability of our planet.

9.4 Nanoparticles in Environmental Cleanup

The field of environmental remediation is currently experiencing a significant paradigm shift, mostly influenced by the adoption of nanoparticles as very effective agents for the elimination and reduction of pollutants. Due to their distinctive characteristics, nanoparticles have considerable potential in augmenting the efficacy of environmental restoration endeavors. This section will examine many categories of nanoparticles utilized in these attempts and their distinct functions in environmental remediation (Shukla et al., 2023).

9.4.1 Types of Nanoparticles Used in Remediation

Nanoparticles that have exhivited notable remediation potentials include the following:

(i) Metallic nanoparticles: Metallic nanoparticles, which are frequently comprised of elements such as gold, silver, or iron, have garnered much acclaim due to their remarkable catalytic characteristics and extensive surface area (Salem et al., 2022). These characteristics render them indispensable in the context of environmental remediation. The utilization of metallic nanoparticles has demonstrated efficacy in the mitigation of diverse contaminants. One notable application of zerovalent iron nanoparticles (Fe^0 NPs) is their capacity to facilitate the degradation of chlorinated solvents and the remediation of heavy metals in polluted areas (Jiang et al., 2018). The antibacterial capabilities of silver nanoparticles have been observed, rendering them valuable in the context of water treatment for the purpose of managing bacterial contamination. Metallic nanoparticles possess the ability to expedite the degradation of organic pollutants by means of oxidation or reduction reactions, owing to their inherent catalytic properties. The catalytic nature of this activity facilitates the conversion of noxious chemicals into ones that are either less hazardous or nontoxic (Ningthoujam et al., 2022; Kumari & Pulimi, 2023).

(ii) Metal oxide nanoparticles: Metal oxide nanoparticles, such as titanium dioxide (TiO_2) and iron oxide (Fe2O3), are renowned for their robust adsorption capacities and photocatalytic characteristics. Due to their stability and reactivity, they are extensively utilized in environmental cleanup (Mondal & Sharma, 2016). Titanium dioxide (TiO_2) nanoparticles are widely employed in the process of photocatalytic destruction of organic contaminants present in both water and air environments. Iron oxide nanoparticles, specifically magnetite (Fe_3O_4), are utilized in the process of extracting heavy metals and nutrients from aqueous environments (Armaković et al., 2022). The catalytic properties of metal oxide nanoparticles can be harnessed to facilitate the degradation of organic contaminants upon exposure to light. The photocatalytic activity exhibits notable efficacy in the degradation of persistent organic compounds and microorganisms.

(iii) Carbon-based nanoparticles: Carbon-based nanoparticles, including carbon nanotubes (CNTs) and graphene, have garnered attention for their exceptional mechanical strength and high surface area. They exhibit remarkable versatility in environmental remediation applications (Noamani et al., 2019; Biswas et al., 2021). CNTs, due to their nanoscale dimensions and unique structure, are employed in groundwater and soil remediation to adsorb organic pollutants, heavy metals, and even radioactive materials. Graphene, with its two-dimensional structure, is effective in adsorbing a wide range of contaminants from both air and water (Baby et al., 2019). Carbon-based nanoparticles excel in adsorption, thanks to their large surface area and the presence of active sites on their surfaces. This adsorption capacity allows them to effectively capture and immobilize contaminants (Azari et al., 2020).

(iv) Polymer nanoparticles: Polymer nanoparticles, which are derived from a range of synthetic or natural polymers, present customized remedies for addressing environmental remediation concerns owing to their adjustable characteristics and compatibility with living organisms (Mamidi et al., 2022). Polymeric nanoparticles find application in the field of water treatment, particularly for the purpose of eliminating organic pollutants and heavy metals. In the field of soil remediation, these substances are utilized for the purpose of encapsulating and immobilizing pollutants, so effectively impeding their mobility (Sajid et al., 2018). One of the primary benefits associated with polymer nanoparticles is in their capacity to effectively encapsulate pollutants, so effectively mitigating the potential for their release into the surrounding environment. The utilization of this controlled release technique can offer distinct benefits in certain remediation settings (Zielińska et al., 2020).

The utilization of various types of nanoparticles in environmental remediation plays a crucial role in mitigating pollution and facilitating the restoration of ecosystems. The distinct characteristics exhibited by these substances, whether in terms of catalytic, adsorptive, or encapsulative capabilities, render them highly significant assets within the realm of remediation strategies. With the ongoing progress in the field of nanotechnology, it is anticipated that novel solutions utilizing nanoparticles will arise, thereby augmenting our ability to address environmental issues and facilitate the development of a more environmentally friendly and sustainable future (Del et al., 2021; Roy et al., 2021).

9.4.2 Mechanisms of Action

Environmental cleanup efforts involving nanoparticles rely on a spectrum of mechanisms to target and remove pollutants effectively. These mechanisms are finely tuned to address various types of contaminants and environmental matrices.

9.4.2.1 Absorption and Sequestration of Pollutants

Surface Properties: The efficiency of nanoparticle-based adsorption largely hinges on the surface chemistry of the nanoparticles. Functional groups on the nanoparticle surface, such as hydroxyl (–OH), carboxyl (–COOH), or amine (–NH2) groups, can facilitate chemical interactions with specific contaminants. For instance, nanoparticles with negatively charged surfaces may effectively adsorb positively charged ions or organic compounds through electrostatic attractions (Espíndola, 2023).

Binding Sites: Nanoparticles provide a high density of binding sites due to their exceptionally large surface area-to-volume ratio. Contaminants can adsorb onto these binding sites through a range of mechanisms, including physical adsorption (van der Waals forces) and chemical adsorption (covalent or ionic bonds) (Karadaghi et al., 2022). Smaller nanoparticles generally possess larger surface areas, increasing the number of available binding sites and, consequently, enhancing adsorption efficiency (Zhang et al., 2022). Tailoring nanoparticle surfaces with specific functional groups can improve their selectivity and affinity for particular contaminants. Adsorption efficiency may vary with contaminant concentration. At low concentrations, there may be an excess of binding sites on nanoparticles, whereas at high concentrations, the binding sites can become saturated (Tee et al., 2022).

The pH and ionic strength of the environment can significantly affect the adsorption process. Changes in these conditions may influence the charge and reactivity of nanoparticles, altering their adsorption capacity (Lin et al., 2017). Some nanoparticles, particularly polymer nanoparticles and certain metal oxide nanoparticles, can create a matrix or porous structure within which contaminants are trapped or encapsulated. This encapsulation effectively sequesters the pollutants, preventing their release into the environment (García-Ivars et al., 2019). The encapsulation of contaminants within nanoparticles can also serve as a controlled release mechanism. Depending on environmental conditions, such as pH or temperature, nanoparticles may release the encapsulated pollutants gradually, allowing for timed degradation or removal (Roy et al., 2014; Westlake et al., 2022).

Contaminant release from nanoparticles can be triggered by environmental cues. For example, changes in pH levels or the presence of specific ions in the environment may lead to the release of encapsulated contaminants (Zhang et al., 2019). In some cases, polymer nanoparticles may be biodegradable, allowing them to break down over time and release encapsulated contaminants into benign byproducts (De et al., 2014).

9.4.2.2 Catalytic Degradation of Contaminants

Certain nanoparticles, such as metal and metal oxide nanoparticles, exhibit catalytic activity. They can serve as catalysts in various chemical reactions, including oxidation and reduction reactions (Nair et al., 2022). Nanoparticles can activate contaminants by providing a surface where chemical reactions can occur more readily. This activation can lead to the degradation or transformation of contaminants into less toxic forms (Waring et al., 2018). Nanoparticles can catalyze the oxidation of organic pollutants, breaking them down into simpler and less harmful compounds. This is particularly useful for addressing persistent organic pollutants (Tufail et al., 2022).

Reduction: Nanoparticles can facilitate reduction reactions, which are crucial for removing heavy metals and metal ions from the environment. These reactions convert toxic metal ions into less harmful forms or precipitate them for easy removal (Liang et al., 2022).

The utilization of nanoparticles in remediation processes frequently yields advantageous outcomes due to the synergistic effects that arise from the combination of diverse nanoparticle kinds. As an illustration, a specific category of nanoparticles has the ability to adsorb contaminants onto their surface, whereas another distinct category of nanoparticles possesses catalytic capabilities that enable them to promote the destruction of the adsorbed pollutants (Mourdikoudis et al., 2021). The integration of these synergistic combinations has the potential to significantly improve the overall efficiency of remediation processes.

Nanoparticle-based remediation employs many mechanisms of action, such as adsorption, absorption, catalytic degradation, and controlled release, to efficiently target and mitigate environmental toxins (Marimuthu et al., 2020). The selection of the mechanism is contingent upon several factors, including the characteristics of the contaminants, prevailing environmental circumstances, and the desired objectives of the remediation endeavors. Furthermore, the collaborative effect of different types of nanoparticles can augment the effectiveness of remediation approaches, offering adaptable and powerful tools for the purpose of environmental restoration and pollution management (Rajput et al., 2022).

9.5 Applications in Environmental Cleanup

Nanoparticle-based remediation technologies have found diverse applications across different environmental matrices, offering innovative solutions to some of the most challenging pollution problems. Some of the recent novel applications include the following:

(i) Iron nanoparticles for groundwater remediation: Iron nanoparticles have been used to treat chlorinated organic compounds like trichloroethylene (TCE) in groundwater through reductive dechlorination. Lab tests and field trials show effective degradation of TCE to nontoxic breakdown products using iron nanoparticles (Karn et al., 2009; O'Carroll et al., 2013).
(ii) Nano-adsorbents for removal of heavy metals: Nanoparticles composed of iron oxide, titanium dioxide, aluminum oxide, and carbon nanotubes have very high specific surface area and adsorption capacities for heavy metal contaminants like lead, copper, zinc, and cadmium. They are more effective than conventional adsorbents (Fu & Wang, 2011; Gupta et al., 2012).
(iii) Nanoscale zerovalent iron for degradation of chlorinated solvents: Nanoscale zerovalent iron (nZVI) abiotically degrades chlorinated solvents like trichloroethene and carbon tetrachloride in groundwater via reductive dehalogenation. Field trials of nZVI injections have shown over 90% destruction of target contaminants (Yan et al., 2013).
(iv) Nanoemulsions for oil spill remediation: Oil-in-water nanoemulsions disperse and break down crude oil through increased surface area and interaction with oil droplets. Studies show nanoemulsion treatments increase oil biodegradation rates compared to traditional dispersants after marine spills (Kumar & Mandal, 2018).

9.6 Environmental Safety and Risk Assessment

As the use of nanoparticles in various applications, including environmental remediation, continues to expand, it is essential to comprehensively assess the environmental safety and potential risks associated with these nanomaterials (Corsi et al., 2023). This chapter explores the complex landscape of environmental safety and risk assessment concerning nanoparticles, including the identification of potential environmental risks, the ecotoxicity and biodegradability of nanoparticles, and strategies to mitigate these risks.

Nanoparticles, due to their unique physicochemical properties, can exhibit different behaviors and interactions with the environment compared to bulk materials. While nanoparticles offer tremendous potential for environmental applications, such as pollution cleanup and water treatment, they also pose certain risks that need to be carefully evaluated (Khan, 2020). These potential risks can be categorized into several key areas:

(i) Chemical toxicity: Some nanoparticles may release toxic ions or compounds when they interact with environmental conditions or other substances. These released materials can have adverse effects on aquatic life, terrestrial ecosystems, and human health if not managed properly (Reidy et al., 2013).
(ii) Physical hazards: The small size and high surface area of nanoparticles can lead to increased reactivity and potential hazards. They may pose inhalation risks to humans when airborne or cause harm to aquatic organisms upon ingestion (Taran et al., 2021).
(iii) Bioavailability and mobility: Nanoparticles can alter the bioavailability and mobility of contaminants in the environment. While this can be beneficial in some remediation applications, it may also lead to unintended consequences, such as the release of previously immobilized pollutants (Abbas et al., 2020).
(iv) Long-term environmental persistence: The long-term persistence of nanoparticles in the environment is a concern, especially if they accumulate over time and potentially disrupt ecosystems.
(v) Bioaccumulation: Nanoparticles may be ingested and accumulate in organisms at various trophic levels, potentially leading to biomagnification in food chains and affecting higher-level predators, including humans (Kumar et al., 2022).
(vi) Interaction with natural systems: Nanoparticles can interact with natural systems in unexpected ways, altering nutrient cycles, microbial communities, and other ecosystem processes.

Examples of potential risks include the following:

Silver Nanoparticles: Silver nanoparticles, commonly used for their antimicrobial properties, have raised concerns about the release of silver ions into aquatic environments. These ions can be toxic to aquatic organisms, leading to potential ecological disruption (Radwan et al., 2019).

Carbon Nanotubes: Carbon nanotubes, due to their fibrous nature, can exhibit asbestos-like behavior when inhaled, posing respiratory risks to humans and other animals. Their potential for persistence in the environment also raises concerns (Ottoor, 2021).

Metal Oxide Nanoparticles: Metal oxide nanoparticles, such as titanium dioxide (TiO2), used in sunscreens and various consumer products, may pose risks when released into aquatic ecosystems. They can accumulate in sediments and affect benthic organisms (Heilgeist et al., 2021).

9.6.1 Ecotoxicity and Biodegradability of Nanoparticles

Aquatic Ecosystems: The impact of nanoparticles on aquatic ecosystems depends on factors such as size, surface charge, and composition. For instance, positively charged nanoparticles may have a stronger affinity for negatively charged cell membranes of aquatic organisms, potentially affecting their viability (Turan et al., 2019).

Terrestrial Ecosystems: Terrestrial ecosystems may also be impacted by nanoparticles when they are released into soils. Effects on soil microorganisms, plants, and other biota can vary depending on nanoparticle characteristics (Ameen et al., 2021).

Biodegradation of Nanoparticles: The biodegradability of nanoparticles depends on their composition and surface properties. Some nanoparticles, particularly those with organic or biodegradable coatings, may undergo degradation over time. However, others, such as certain metal or metal oxide nanoparticles, may persist in the environment for extended periods (Su & Kang, 2020).

Environmental Factors: Environmental conditions, including temperature, pH, and microbial activity, play a crucial role in determining the biodegradability of nanoparticles. In some cases, nanoparticles may become less toxic as they degrade, while in others, degradation products may be more toxic than the original nanoparticles (Singh et al., 2018).

9.6.2 Risk Mitigation Strategies

Effective risk management and mitigation strategies are essential to ensure the responsible use of nanoparticles in environmental applications while minimizing potential environmental and human health risks. Establishing guidelines for the safe disposal of nanoparticles is critical. This includes regulations for the disposal of nanoparticle-containing waste and ensuring that nanoparticles do not enter landfill leachate or wastewater treatment plants untreated (Schulte et al., 2014). Implementing containment measures during the production, use, and disposal of nanoparticles can prevent their unintended release into the environment. This may involve the use of protective barriers, such as nanofiltration membranes, to capture nanoparticles before they enter water bodies or wastewater streams (Amoabediny et al., 2009). Secure storage facilities for nanoparticles and nanoparticle-containing products should be designed to prevent spills and minimize the risk of nanoparticles escaping into the environment (Vogel et al., 2013).

Regular environmental monitoring programs should be established to assess the presence and behavior of nanoparticles in natural systems. Monitoring can help identify potential hotspots of nanoparticle contamination and inform risk management strategies (Klaine et al., 2012). Governments and regulatory agencies should develop and enforce regulations specific to nanoparticle use and release. These regulations may include labeling requirements, emission limits, and reporting obligations for industries that handle nanoparticles (Besha et al., 2020). Encouraging transparency and data sharing among researchers, industries, and regulatory bodies is crucial for understanding the environmental risks associated with nanoparticles. This collaborative approach can facilitate the development of evidence-based regulations (Mech et al., 2022).

Comprehensive toxicological studies should be conducted to assess the potential risks of specific nanoparticles. These studies can help identify safe exposure limits and inform risk management strategies (Medici et al., 2021). Life cycle assessments (LCAs) should be conducted to evaluate the environmental impacts of nanoparticles at every stage, from production to disposal. This holistic approach can identify potential areas for risk mitigation and environmental improvement (Abu-Qdais et al., 2021). Research efforts should focus on the development of alternative nanoparticle materials that possess similar functionalities but lower environmental risks. This approach aligns with the principles of green chemistry and sustainable nanotechnology (Harish et al., 2023).

In order to ensure the responsible utilization of nanoparticles in environmental applications, it is imperative to adopt a proactive stance toward risk assessment and mitigation. By comprehending the potential environmental hazards, evaluating the ecotoxicological properties and biodegradability, and applying effective risk management measures, we may effectively utilize the advantages of nanoparticles while mitigating their detrimental effects on ecosystems and human well-being (Lee et al., 2010). The establishment of partnerships among scientists, companies, and regulatory authorities is necessary in order to guarantee the secure and sustainable application of nanotechnology in the realm of environmental remediation and its associated domains.

9.7 Future Perspectives and Challenges

The future of nanoparticle-based environmental cleanup holds great promise, but it also presents several emerging technologies, regulatory considerations, and ethical challenges. This chapter explores the evolving landscape of nanoparticle applications in environmental remediation and the path forward.

Nanoscale materials engineering continues to advance, enabling the design and fabrication of nanoparticles with precise properties for specific environmental cleanup tasks. Tailored nanoparticles can exhibit enhanced adsorption capacity, reactivity, and selectivity, making them more effective in targeting and removing pollutants. Future research may focus on developing multifunctional nanoparticles that can simultaneously adsorb, degrade, and encapsulate contaminants. These nanoparticles could offer versatile solutions for complex pollution scenarios (Ayub et al., 2023). The integration of nanorobotics into environmental cleanup is on the horizon. Nanorobots, guided by autonomous systems and artificial intelligence, could navigate complex environments, identify pollutant sources, and perform precise remediation tasks at the nanoscale. Autonomous nanoscale sensors and monitoring systems may provide real-time data on pollutant concentrations, allowing for rapid response and adaptive remediation strategies.

As nanoparticle technologies advance, regulatory agencies must keep pace with developments to ensure environmental and human safety. The establishment of standardized testing protocols, exposure limits, and reporting requirements for nanoparticles in environmental applications is essential (Allan et al., 2021). Ethical considerations surrounding nanoparticle use in environmental cleanup include questions about the responsible disposal of contaminated nanoparticles, transparency in research and reporting, and equitable access to nanoparticle-based remediation technologies (Babatunde et al., 2019). Ethical frameworks should guide decision-making and governance in this evolving field.

Hybrid remediation approaches entails the integration of nanoparticle-based technologies with traditional remediation methods holds promise. Hybrid approaches can leverage the strengths of both, potentially enhancing the efficiency and cost-effectiveness of environmental cleanup efforts (Hidangmayum et al., 2022). For example, nanoparticles could be used to target specific contaminants, while traditional methods handle larger-scale remediation. Public awareness and understanding of nanoparticle-based environmental cleanup are essential. Effective communication about the benefits, risks, and ethical considerations associated with nanoparticle technologies can foster informed decision-making and public support for research and regulatory measures. Engaging with local communities and stakeholders is crucial in the deployment of nanoparticle-based remediation projects (Babatunde et al., 2019). Public input and concerns should be considered in project planning, risk assessments, and monitoring efforts.

The future of nanoparticle-based environmental cleanup is marked by innovation, challenges, and a need for responsible governance. Emerging technologies, such as tailored nanoparticles and nanorobotics, offer exciting possibilities for more effective and efficient pollution control. However, these innovations must be accompanied by stringent regulations, ethical considerations, and engagement with the public and communities affected by environmental cleanup efforts (Zhang et al., 2021). By addressing these challenges and embracing responsible practices, nanoparticle-based environmental remediation can play a pivotal role in addressing pollution and restoring the health of ecosystems and our planet.

Throughout this chapter, we have delved into the multifaceted world of nanoparticle-based remediation. Nanoparticles, with their unique properties and capabilities, offer versatile solutions to address a wide range of environmental challenges. We have seen how they can adsorb, absorb, and catalyze the degradation of contaminants, enhancing the efficiency of environmental cleanup. From groundwater and soil remediation to air pollution control, water purification, and marine ecosystem restoration, nanoparticles have demonstrated their potential to mitigate pollution and contribute to environmental restoration.

The potential for the widespread adoption of nanoparticle-based remediation is substantial. These technologies have the capacity to revolutionize the way we approach environmental cleanup, making it more efficient, targeted, and sustainable. The deployment of tailored nanoparticles and the integration of nanorobotics into remediation efforts promise to unlock new frontiers in pollution control (Roco, 2023). Moreover, the synergy between nanoparticle technologies and traditional remediation methods offers a holistic approach to address environmental challenges on multiple fronts.

The impact of nanoparticle-based remediation on environmental sustainability cannot be overstated. By efficiently removing contaminants, preventing further pollution, and promoting the recovery of ecosystems, nanoparticle technologies contribute to the preservation of natural resources and the protection of human and ecosystem health (Corsi et al., 2018). They align with the goals of sustainable development, offering a pathway to cleaner, healthier, and more resilient environments.

9.8 Call for Continued Research and Responsible Implementation

While we have explored the achievements and potential of nanoparticle-based remediation, it is essential to recognize that this field is continually evolving. Continued research is paramount to address emerging challenges, refine technologies, and assess the long-term environmental and health impacts of nanoparticle use. Robust risk assessment and mitigation strategies must accompany the development and deployment of these technologies to ensure responsible implementation.

Ethical considerations, regulatory frameworks, and community engagement are integral to guiding the responsible use of nanoparticles in environmental cleanup. Transparency in research and reporting, equitable access to remediation technologies, and careful monitoring of potential risks are essential elements of responsible implementation. The utilization of nanoparticle-based remediation is a promising avenue in our endeavor to address environmental restoration. This phenomenon serves as evidence of human intellectual creativity and our capacity to utilize scientific knowledge and technological advancements to enhance the state of the natural world. It is imperative that to maintain unwavering dedication to the preservation and responsible management of our environment. The severe challenges presented by pollution and deterioration of ecosystems can be effectively addressed by the utilization of nanoparticle-based remediation, which has great promise in tackling these issues directly. By engaging in collaborative efforts, fostering innovation, and demonstrating a collective commitment to the preservation of the environment, we have the potential to pave the way toward a planet that is characterized by enhanced cleanliness, improved health, and increased resilience.

9.9 Conclusion

The innovative approach of nanoparticle-based environmental remediation holds immense significance in efforts to combat pollution, restore ecosystems, and safeguard the environment. The application of this technique however must be adopted with conscious environmental and health safety precaustions, hence the need for continued research and responsible implementation on the future of environmental cleanup with nanoparticles.

The prospects for utilizing nanoparticles in environmental remediation are promising. However, the realization of these benefits is heavily reliant on our shared obligation to protect the environment for both current and future cohorts. It is imperative to acknowledge and harness the capabilities of nanoscale technologies, while also adhering to the fundamental tenets of responsible research, governance, and environmental ethics. Collectively, it is within our capacity to collaboratively strive toward a global paradigm where the rehabilitation of ecosystems and the safeguarding of inherent natural aesthetics are not merely ideals, but rather concrete accomplishments in our continuous pursuit of a viable and enduring trajectory.

References

Abbas, Q., Yousaf, B., Ali, M. U., Munir, M. A. M., El-Naggar, A., Rinklebe, J., & Naushad, M. (2020). Transformation pathways and fate of engineered nanoparticles (ENPs) in distinct interactive environmental compartments: A review. *Environment International, 138*, 105646.

Abu-Qdais, H. A., Abu-Dalo, M. A., & Hajeer, Y. Y. (2021). Impacts of nanosilver-based textile products using a life cycle assessment. *Sustainability, 13*(6), 3436.

Ahmad, M. A., Adeel, M., Shakoor, N., Javed, R., Ishfaq, M., Peng, Y., et al. (2023). Modifying engineered nanomaterials to produce next generation agents for environmental remediation. *Science of the Total Environment, 164861*, 164861.

Allan, J., Belz, S., Hoeveler, A., Hugas, M., Okuda, H., Patri, A., et al. (2021). Regulatory landscape of nanotechnology and nanoplastics from a global perspective. *Regulatory Toxicology and Pharmacology, 122*, 104885.

Ameen, F., Alsamhary, K., Alabdullatif, J. A., & ALNadhari, S. (2021). A review on metal-based nanoparticles and their toxicity to beneficial soil bacteria and fungi. *Ecotoxicology and Environmental Safety, 213*, 112027.

Amoabediny, G. H., Naderi, A., Malakootikhah, J., Koohi, M. K., Mortazavi, S. A., Naderi, M., & Rashedi, H. (2009). Guidelines for safe handling, use and disposal of nanoparticles. In *Journal of Physics: Conference Series* (Vol. 170, p. 012037). IOP Publishing.

Armaković, S. J., Savanović, M. M., & Armaković, S. (2022). Titanium dioxide as the most used photocatalyst for water purification: An overview. *Catalysts, 13*(1), 26.

Ayub, J., Saeed, M. U., Hussain, N., Zulfiqar, I., Mehmood, T., Iqbal, H. M., & Bilal, M. (2023). Designing robust nano-biocatalysts using nanomaterials as multifunctional carriers-expanding the application scope of bio-enzymes. *Topics in Catalysis, 66*(9–12), 625–648.

Azari, A., Nabizadeh, R., Nasseri, S., Mahvi, A. H., & Mesdaghinia, A. R. (2020). Comprehensive systematic review and meta-analysis of dyes adsorption by carbon-based adsorbent materials: Classification and analysis of last decade studies. *Chemosphere, 250*, 126238.

Babatunde, D. E., Denwigwe, I. H., Babatunde, O. M., Gbadamosi, S. L., Babalola, I. P., & Agboola, O. (2019). Environmental and societal impact of nanotechnology. *IEEE Access, 8*, 4640–4667.

Baby, R., Saifullah, B., & Hussein, M. Z. (2019). Carbon nanomaterials for the treatment of heavy metal-contaminated water and environmental remediation. *Nanoscale Research Letters, 14*, 1–17.

Besha, A. T., Liu, Y., Bekele, D. N., Dong, Z., Naidu, R., & Gebremariam, G. N. (2020). Sustainability and environmental ethics for the application of engineered nanoparticles. *Environmental Science & Policy, 103*, 85–98.

Biswas, M. C., Lubna, M. M., Mohammed, Z., Ul Iqbal, M. H., & Hoque, M. E. (2021). *Graphene and carbon nanotube-based hybrid nanocomposites: Preparation to applications* (pp. 71–112). Springer.

Boregowda, N., Jogigowda, S. C., Bhavya, G., Sunilkumar, C. R., Geetha, N., Udikeri, S. S., et al. (2022). Recent advances in nanoremediation: Carving sustainable solution to clean-up polluted agriculture soils. *Environmental Pollution, 297*, 118728.

Campanale, C., Massarelli, C., Savino, I., Locaputo, V., & Uricchio, V. F. (2020). A detailed review study on potential effects of microplastics and additives of concern on human health. *International Journal of Environmental Research and Public Health, 17*(4), 1212.

Chaudhary, P., Ahamad, L., Chaudhary, A., Kumar, G., Chen, W. J., & Chen, S. (2023). Nanoparticle-mediated bioremediation as a powerful weapon in the removal of environmental pollutants. *Journal of Environmental Chemical Engineering, 109591*, 109591.

Chen, R., & Riviere, J. E. (2017). Biological surface adsorption index of nanomaterials: Modelling surface interactions of nanomaterials with biomolecules. *Modelling the Toxicity of Nanoparticles, 947*, 207–253.

Chu, E. W., & Karr, J. R. (2017). Environmental impact: Concept, consequences, measurement. In *Reference module in life sciences*. Elsevier.

Corsi, I., Winther-Nielsen, M., Sethi, R., Punta, C., Della Torre, C., Libralato, G., et al. (2018). Ecofriendly nanotechnologies and nanomaterials for environmental applications: Key issue and consensus recommendations for sustainable and ecosafe nanoremediation. *Ecotoxicology and Environmental Safety, 154*, 237–244.

Corsi, I., Venditti, I., Trotta, F., & Punta, C. (2023). Environmental safety of nanotechnologies: The eco-design of manufactured nanomaterials for environmental remediation. *Science of the Total Environment, 864*, 161181.

De, A., Bose, R., Kumar, A., & Mozumdar, S. (2014). *Targeted delivery of pesticides using biodegradable polymeric nanoparticles* (Vol. 10, pp. 978–981). Springer.

Del Prado-Audelo, M. L., García Kerdan, I., Escutia-Guadarrama, L., Reyna-González, J. M., Magaña, J. J., & Leyva-Gómez, G. (2021). Nanoremediation: Nanomaterials and nanotechnologies for environmental cleanup. *Frontiers in Environmental Science, 9*, 793765.

Espíndola, C. (2023). Some nanocarrier's properties and chemical interaction mechanisms with flavones. *Molecules, 28*(6), 2864.

Ettel, D., Havelka, O., Isik, S., Silvestri, D., Wacławek, S., Urbánek, M., et al. (2021). Laser-synthesized Ag/TiO nanoparticles to integrate catalytic pollutant degradation and antifouling enhancement in nanofibrous membranes for oil–water separation. *Applied Surface Science, 564*, 150471.

Foong, S. Y., Ma, N. L., Lam, S. S., Peng, W., Low, F., Lee, B. H., et al. (2020). A recent global review of hazardous chlorpyrifos pesticide in fruit and vegetables: Prevalence, remediation and actions needed. *Journal of Hazardous Materials, 400*, 123006.

Fu, F., & Wang, Q. (2011). Removal of heavy metal ions from wastewaters: A review. *Journal of Environmental Management, 92*(3), 407–418.

García-Ivars, J., Corbatón-Báguena, M. J., & Iborra-Clar, M. I. (2019). Development of mixed matrix membranes: Incorporation of metal nanoparticles in polymeric membranes. In *Nanoscale materials in water purification* (pp. 153–178). Elsevier.

Guerra, F. D., Attia, M. F., Whitehead, D. C., & Alexis, F. (2018). Nanotechnology for environmental remediation: Materials and applications. *Molecules, 23*(7), 1760.

Gupta, V. K., Ali, I., Saleh, T. A., Nayak, A., & Agarwal, S. (2012). Chemical treatment technologies for waste-water recycling—An overview. *RSC Advances, 2*(16), 6380–6388.

Harish, V., Ansari, M. M., Tewari, D., Yadav, A. B., Sharma, N., Bawarig, S., et al. (2023). Cutting-edge advances in tailoring size, shape, and functionality of nanoparticles and nanostructures: A review. *Journal of the Taiwan Institute of Chemical Engineers, 149*, 105010.

Heilgeist, S., Sekine, R., Sahin, O., & Stewart, R. A. (2021). Finding nano: Challenges involved in monitoring the presence and fate of engineered titanium dioxide nanoparticles in aquatic environments. *Water, 13*(5), 734.

Hidangmayum, A., Debnath, A., Guru, A., Singh, B. N., Upadhyay, S. K., & Dwivedi, P. (2022). Mechanistic and recent updates in nano-bioremediation for developing green technology to alleviate agricultural contaminants. *International Journal of Environmental Science and Technology, 20*, 1–26.

Hoang, T. C., Black, M. C., Knuteson, S. L., & Roberts, A. P. (2019). Environmental pollution, management, and sustainable development: Strategies for Vietnam and other developing countries. *Environmental Management, 63*, 433–436.

Huang, D., Wu, J., Wang, L., Liu, X., Meng, J., Tang, X., et al. (2019). Novel insight into adsorption and co-adsorption of heavy metal ions and an organic pollutant by magnetic graphene nanomaterials in water. *Chemical Engineering Journal, 358*, 1399–1409.

Jiang, D., Zeng, G., Huang, D., Chen, M., Zhang, C., Huang, C., & Wan, J. (2018). Remediation of contaminated soils by enhanced nanoscale zero valent iron. *Environmental Research, 163*, 217–227.

Karadaghi, L. R., To, A. T, Habas, S. E., Baddour, F. G., Ruddy, D. A., & Brutchey, R. L. (2022). Activating molybdenum carbide nanoparticle catalysts under mild conditions using thermally labile ligands. *Chemistry of Materials, 34*(19), 8849–8857.

Karn, B., Kuiken, T., & Otto, M. (2009). Nanotechnology and in situ remediation: A review of the benefits and potential risks. *Environmental Health Perspectives, 117*(12), 1813–1831.

Khan, S. H. (2020). Green nanotechnology for the environment and sustainable development. In *Green materials for wastewater treatment* (pp. 13–46). Springer.

Klaine, S. J., Koelmans, A. A., Horne, N., Carley, S., Handy, R. D., Kapustka, L., et al. (2012). Paradigms to assess the environmental impact of manufactured nanomaterials. *Environmental Toxicology and Chemistry, 31*(1), 3–14.

Kuhn, R., Bryant, I. M., Jensch, R., & Böllmann, J. (2022). Applications of environmental nanotechnologies in remediation, wastewater treatment, drinking water treatment, and agriculture. *Applied Nano, 3*(1), 54–90.

Kumar, A. (2014). *Nanotechnology development in India: An overview*. Research and Information System for Developing Countries.

Kumar, N., & Mandal, A. (2018). Oil-in-water nanoemulsion stabilized by polymeric surfactant: Characterization and properties evaluation for enhanced oil recovery. *European Polymer Journal, 109*, 265–276.

Kumar, A., Nagar, S., & Anand, S. (2021). Climate change and existential threats. In *Global climate change* (pp. 1–31). Elsevier.

Kumar, C. V., Karthick, V., Kumar, V. G., Inbakandan, D., Rene, E. R., Suganya, K. U., et al. (2022). The impact of engineered nanomaterials on the environment: Release mechanism, toxicity, transformation, and remediation. *Environmental Research, 212*, 113202.

Kumari, M., & Pulimi, M. (2023). Sulfate radical-based degradation of organic pollutants: A review on application of metal-organic frameworks as catalysts. *ACS Omega, 8*, 34262.

Lee, J., Mahendra, S., & Alvarez, P. J. (2010). Nanomaterials in the construction industry: A review of their applications and environmental health and safety considerations. *ACS Nano, 4*(7), 3580–3590.

Liang, W., Wang, G., Peng, C., Tan, J., Wan, J., Sun, P., et al. (2022). Recent advances of carbon-based nano zero valent iron for heavy metals remediation in soil and water: A critical review. *Journal of Hazardous Materials, 426*, 127993.

Lin, D., Story, S. D., Walker, S. L., Huang, Q., Liang, W., & Cai, P. (2017). Role of pH and ionic strength in the aggregation of TiO2 nanoparticles in the presence of extracellular polymeric substances from Bacillus subtilis. *Environmental Pollution, 228*, 35–42.

Liu, J., & Xie, J. (2020). Environmental regulation, technological innovation, and export competitiveness: An empirical study based on China's manufacturing industry. *International Journal of Environmental Research and Public Health, 17*(4), 1427.

Lu, Y., Song, S., Wang, R., Liu, Z., Meng, J., Sweetman, A. J., et al. (2015). Impacts of soil and water pollution on food safety and health risks in China. *Environment International, 77*, 5–15.

Mamidi, N., García, R. G., Martínez, J. D. H., Briones, C. M., Martinez Ramos, A. M., Tamez, M. F. L., et al. (2022). Recent advances in designing fibrous biomaterials for the domain of biomedical, clinical, and environmental applications. *ACS Biomaterials Science & Engineering, 8*(9), 3690–3716.

Marimuthu, S., Antonisamy, A. J., Malayandi, S., Rajendran, K., Tsai, P. C., Pugazhendhi, A., & Ponnusamy, V. K. (2020). Silver nanoparticles in dye effluent treatment: A review on synthesis, treatment methods, mechanisms, photocatalytic degradation, toxic effects and mitigation of toxicity. *Journal of Photochemistry and Photobiology B: Biology, 205*, 111823.

Martin, L. M., Sheng, J., Zimba, P. V., Zhu, L., Fadare, O. O., Haley, C., et al. (2022). Testing an iron oxide nanoparticle-based method for magnetic separation of nanoplastics and microplastics from water. *Nanomaterials, 12*(14), 2348.

Mech, A., Gottardo, S., Amenta, V., Amodio, A., Belz, S., Bøwadt, S., et al. (2022). Safe-and sustainable-by-design: The case of smart nanomaterials. A perspective based on a European workshop. *Regulatory Toxicology and Pharmacology, 128*, 105093.

Medici, S., Peana, M., Pelucelli, A., & Zoroddu, M. A. (2021, November). An updated overview on metal nanoparticles toxicity. In *Seminars in cancer biology* (Vol. 76, pp. 17–26). Academic Press.

Mondal, K., & Sharma, A. (2016). Recent advances in the synthesis and application of photocatalytic metal–metal oxide core–shell nanoparticles for environmental remediation and their recycling process. *RSC Advances, 6*(87), 83589–83612.

Mourdikoudis, S., Kostopoulou, A., & LaGrow, A. P. (2021). Magnetic nanoparticle composites: Synergistic effects and applications. *Advanced Science, 8*(12), 2004951.

Naidu, R., Biswas, B., Willett, I. R., Cribb, J., Singh, B. K., Nathanail, C. P., et al. (2021). Chemical pollution: A growing peril and potential catastrophic risk to humanity. *Environment International, 156*, 106616.

Nair, G. M., Sajini, T., & Mathew, B. (2022). Advanced green approaches for metal and metal oxide nanoparticles synthesis and their environmental applications. *Talanta Open, 5*, 100080.

Ningthoujam, R., Singh, Y. D., Babu, P. J., Tirkey, A., Pradhan, S., & Sarma, M. (2022). Nanocatalyst in remediating environmental pollutants. *Chemical Physics Impact, 4*, 100064.

Noamani, S., Niroomand, S., Rastgar, M., & Sadrzadeh, M. (2019). Carbon-based polymer nanocomposite membranes for oily wastewater treatment. *npj Clean Water, 2*(1), 20.

O'Carroll, D., Sleep, B., Krol, M., Boparai, H., & Kocur, C. (2013). Nanoscale zero valent iron and bimetallic particles for contaminated site remediation. *Advances in Water Resources, 51*, 104–122.

Ottoor, D. P. (2021). Assessment of the risks associated with carbon nanotubes. In *Handbook of carbon nanotubes* (pp. 1–26). Springer International Publishing.

Palit, S. (2018). Recent advances in the application of engineered nanomaterials in the environment industry—A critical overview and a vision for the future. In *Handbook of nanomaterials for industrial applications* (pp. 883–893). Elsevier.

Radwan, I. M., Gitipour, A., Potter, P. M., Dionysiou, D. D., & Al-Abed, S. R. (2019). Dissolution of silver nanoparticles in colloidal consumer products: Effects of particle size and capping agent. *Journal of Nanoparticle Research, 21*, 1–13.

Rajput, V. D., Minkina, T., Upadhyay, S. K., Kumari, A., Ranjan, A., Mandzhieva, S., et al. (2022). Nanotechnology in the restoration of polluted soil. *Nanomaterials, 12*(5), 769.

Reidy, B., Haase, A., Luch, A., Dawson, K. A., & Lynch, I. (2013). Mechanisms of silver nanoparticle release, transformation and toxicity: A critical review of current knowledge and recommendations for future studies and applications. *Materials, 6*(6), 2295–2350.

Roco, M. C. (2023). National Nanotechnology Initiative at 20 years: Enabling new horizons. *Journal of Nanoparticle Research, 25*(10), 197.

Roy, A., Singh, S. K., Bajpai, J., & Bajpai, A. K. (2014). Controlled pesticide release from biodegradable polymers. *Central European Journal of Chemistry, 12*, 453–469.

Roy, A., Sharma, A., Yadav, S., Jule, L. T., & Krishnaraj, R. (2021). Nanomaterials for remediation of environmental pollutants. *Bioinorganic Chemistry and Applications, 2021*, 1–16.

Sajid, M., Nazal, M. K., Baig, N., & Osman, A. M. (2018). Removal of heavy metals and organic pollutants from water using dendritic polymers based adsorbents: A critical review. *Separation and Purification Technology, 191*, 400–423.

Salem, S. S., Hammad, E. N., Mohamed, A. A., & El-Dougdoug, W. (2022). A comprehensive review of nanomaterials: Types, synthesis, characterization, and applications. *Biointerface Research in Applied Chemistry, 13*(1), 41.

Schulte, P. A., Geraci, C. L., Murashov, V., Kuempel, E. D., Zumwalde, R. D., Castranova, V., et al. (2014). Occupational safety and health criteria for responsible development of nanotechnology. *Journal of Nanoparticle Research, 16*, 1–17.

Shukla, S., Khan, R., Saxena, A., Patel, A., & Hussain, C. M. (2023). Nanoperception in remediation arena. In *Nanoremediation* (pp. 3–21). Elsevier.

Singh, R. L., & Singh, P. K. (2017). Global environmental problems. In *Principles and applications of environmental biotechnology for a sustainable future* (pp. 13–41). Springer.

Singh, J., Dutta, T., Kim, K. H., Rawat, M., Samddar, P., & Kumar, P. (2018). 'Green'synthesis of metals and their oxide nanoparticles: Applications for environmental remediation. *Journal of Nanobiotechnology, 16*(1), 1–24.

Su, S., & Kang, P. M. (2020). Systemic review of biodegradable nanomaterials in nanomedicine. *Nanomaterials, 10*(4), 656.

Taran, M., Safaei, M., Karimi, N., & Almasi, A. (2021). Benefits and application of nanotechnology in environmental science: An overview. *Biointerface Research in Applied Chemistry, 11*(1), 7860–7870.

Taylor, N. B. (2023). Climate change peace and conflict. In *Existential risks in peace and conflict studies* (pp. 109–138). Springer International Publishing.

Tee, G. T., Gok, X. Y., & Yong, W. F. (2022). Adsorption of pollutants in wastewater via biosorbents, nanoparticles and magnetic biosorbents: A review. *Environmental Research, 212*, 113248.

Tufail, M. A., Iltaf, J., Zaheer, T., Tariq, L., Amir, M. B., Fatima, R., et al. (2022). Recent advances in bioremediation of heavy metals and persistent organic pollutants: A review. *Science of the Total Environment, 850*, 157961.

Turan, N. B., Erkan, H. S., Engin, G. O., & Bilgili, M. S. (2019). Nanoparticles in the aquatic environment: Usage, properties, transformation and toxicity—A review. *Process Safety and Environmental Protection, 130*, 238–249.

Ugwu, C. O., Ozoegwu, C. G., Ozor, P. A., Agwu, N., & Mbohwa, C. (2021). Waste reduction and utilization strategies to improve municipal solid waste management on Nigerian campuses. *Fuel Communications, 9*, 100025.

Vogel, U., Savolainen, K., Wu, Q., van Tongeren, M., Brouwer, D., Berges, M., & (Eds.). (2013). *Handbook of nanosafety: Measurement, exposure and toxicology*. Elsevier.

Waring, R. H., Harris, R. M., & Mitchell, S. C. (2018). Plastic contamination of the food chain: A threat to human health? *Maturitas, 115*, 64–68.

Westlake, J. R., Tran, M. W., Jiang, Y., Zhang, X., Burrows, A. D., & Xie, M. (2022). Biodegradable active packaging with controlled release: Principles, progress, and prospects. *ACS Food Science & Technology, 2*(8), 1166–1183.

Yan, W., Lien, H. L., Koel, B. E., & Zhang, W. X. (2013). Iron nanoparticles for environmental clean-up: Recent developments and future outlook. *Environmental Science: Processes & Impacts, 15*(1), 63–77.

Zhang, T., Lowry, G. V., Capiro, N. L., Chen, J., Chen, W., Chen, Y., et al. (2019). In situ remediation of subsurface contamination: Opportunities and challenges for nanotechnology and advanced materials. *Environmental Science: Nano, 6*(5), 1283–1302.

Zhang, P., Guo, Z., Ullah, S., Melagraki, G., Afantitis, A., & Lynch, I. (2021). Nanotechnology and artificial intelligence to enable sustainable and precision agriculture. *Nature Plants, 7*(7), 864–876.

Zhang, P., He, M., Huo, S., Li, F., & Li, K. (2022). Recent progress in metal-based composites toward adsorptive removal of phosphate: Mechanisms, behaviors, and prospects. *Chemical Engineering Journal, 446*, 137081.

Zielińska, A., Carreiró, F., Oliveira, A. M., Neves, A., Pires, B., Venkatesh, D. N., et al. (2020). Polymeric nanoparticles: Production, characterization, toxicology and ecotoxicology. *Molecules, 25*(16), 3731.

Chapter 10
Nanoparticles in Air and Their Impact on Air Quality

Ummulkhair Salamah Ilyasu, Oluwadurotimi Samuel Aworunse, Clement Shina Olusanya, Patrick Omoregie Isibor, Mordecai Gana, and Oluwafemi Adebayo Oyewole

10.1 Introduction

Nanoparticles range in size from 1 to 100 nanometers and can be made of a variety of materials, including metals, organic compounds, metal oxides, and carbon. In comparison to particles of larger dimensions, nanoparticles have prominent biological, chemical, and physical properties (Awasthi et al., 2023). Many aspects, such as a significantly greater surface area-to-volume ratio, improved mechanical strength, improved chemical reactivity or stability, and other pertinent considerations, can be linked to this occurrence. Different dimensions, forms, sizes, and compositions are displayed by nanoparticles (Khan & Hossain, 2022). A wide range of synthesis techniques are being created or modified in an effort to improve properties and lower production costs. Different methods are used to improve specific nanoparticles' chemical, physical, mechanical, and optical properties (Yaqoob et al., 2020).

The nanoparticles that are most commonly found in fabrics, paints, cosmetics, water disinfectants, and food packaging are those that come in powder, suspension, or spray form, and they provide a significant danger of inadvertent inhalation (Geiser et al., 2017). The uptake of particles varies depending on their size. Smaller particles (diameters between 0.1 and 1 μm) tend to concentrate in the tracheobronchial region, whereas bigger particles (diameters between 5 and 30 μm) generally remain in the nasopharyngeal region. On the other hand, particles smaller than 0.5 μm can penetrate the blood vessels' thin epithelium (De Matteis, 2017). Furthermore, there are two ways that nanoparticles in the nasal cavity could absorb:

U. S. Ilyasu (✉) · C. S. Olusanya · M. Gana · O. A. Oyewole
Department of Microbiology, Federal University of Technology, Minna, Nigeria
e-mail: ummulkhair3536@gmail.com

O. S. Aworunse · P. O. Isibor
Department of Biological Sciences, Covenant University, Ota, Ogun State, Nigeria
e-mail: patrick.isibor@covenantuniversity.edu.ng

P. O. Isibor et al. (eds.), *Environmental Nanotoxicology*,
https://doi.org/10.1007/978-3-031-54154-4_10

either they would cross the respiratory epithelium and reach the underlying blood vessels, or they would enter the brain through the olfactory epithelium (Fröhlich & Salar-Behzadi, 2014). But most studies on the toxicity of inhaled nanoparticles concentrate on nontherapeutic versions, which often show morbidity and mortality. Therapeutic nanoparticles are distinguished from these forms by their much smaller sizes, inorganic composition, water insolubility, and ability to be administered at different doses and frequencies. As such, the toxicological research is not transferable to biological nanoparticles (Zhang et al., 2011).

Technological advancement has made it easier to characterize nanoparticles more effectively and harness them afterward. Currently, numerous sectors use nanoparticles, including renewable energy sources, food utensils, electronics, and the aerospace sector. The integration of nanotechnology is critical to realizing a sustainable and ecologically mindful future. Grounded on a 2007 report by the Intergovernmental Panel on Climate Change (IPCC), nanoparticles influence the universal urban visibility and climate by acting as predecessors of rougher particles through their accumulation all the way through the atmosphere's aging process. They also impact the universal climate and atmospheric chemistry owing to their unique chemical constitution and reactive ability (Sonwani et al., 2021a). Particles with diameters between 0.1 and 10 μm have a residence length of about 1 week. Lesser particles can be eliminated by coagulation and diffusion only, but coarser particulate matter can be eliminated by settling (Cescon & Jiang, 2020). Furthermore, the primary causes of the challenges in removing nanoparticles (NPs) from the atmosphere and the resulting health risks to humans are their fine size and longer atmospheric retention duration. In addition to being created in the bodies of insects, plants, and people, nanoparticles are also released via combustion processes, forest fires, automobile exhaust, and industrial emissions (Khan et al., 2022). As stated by Vera-Reyes et al. (2018), nanoparticles are also important in plants because they enter through the apoplastic and symplastic pathways. From there, they interact primarily with the environment and cells through stearic, electrostatic, and van der Waals interactions. One of the agents that plants are harmed by is metal oxides attached to NPs (Giorgetti, 2019). In addition to tropospheric ozone, NPs are another phytotoxic agent for plant species that compromises the world food security system. As NPs can enter the body through a variety of routes, including the skin, the respiratory system, and the digestive system, it is speculated that they may adversely affect human health (Sonwani et al., 2021a).

In Asian countries, there is a high concentration of nanoparticles in the atmosphere because of factors like rapid urbanization, industrialization, vehicle emissions, life-threatening events (like storms, volcanic eruptions, dust, and forest fires), and sporadic events (like fireworks and burning crop residue) (Sonwani et al., 2021b). Metal, carbon, or silicon oxides make up the bulk of anthropogenic nanoparticles. Excess of 110,000 tons of particulate matter, 1.2 million tons of NOx, and 4.3 million tons of SO_2 is released annually by coal-fired power stations in India. The development of atmospheric brown clouds (ABCs) is another notable occurrence (Bundschuh et al., 2018).

Increases in mortality and visitations to emergency rooms are linked with exposure to particulate matter (PM). When inhaled, these particles penetrate into the respiratory tract from which they target various anatomical sites based on their aerodynamic size and other properties (Hartland et al., 2013).

The World Health Organization (WHO) Global Air Quality Guidelines (AQGs) have been adopted in numerous countries to protect public health from air pollution from six classic pollutants, namely, sulfur dioxide (SO_2), carbon monoxide (CO), ozone (O_2), nitrogen dioxide (NO_2), and particulate matter ($PM_{2.5}$ and PM_{10}). In addition, some air pollutants are associated with both health effects and global warming; hence, all efforts to improve air quality can enhance climate change mitigation and vice versa. About 80% of the global deaths attributed to PM2.5 exposure could be avoided if countries attain the updated annual AQG level (Table 10.1).

10.2 Source of Nanoparticles in the Air

10.2.1 Natural Sources of NPs

Human activity is responsible for around 10% of the overall aerosols found in the air; the remaining 90% are produced naturally (Sonwani & Saxena, 2021). Natural sources of nanoparticles include biogenic emissions, sea spray, forest fires, volcanic eruptions, landslides, and dust storms. They are also widely distributed in the environment (Goud et al., 2022). The natural inclination of NPs to react with clouds and air particles means that organic chemicals make up a significant fraction of atmospheric nanoparticles (Lee et al., 2019). Ambient air is high in nitrogen oxides, volatile organic compounds (VOCs), and primary organic aerosols from both

Table 10.1 Recommended 2021 Air Quality Guideline (AQG) levels

Pollutant	Averaging time	2021 AQG level
$PM_{2.5}$, μg/m^3	Annual	5
	24-h[a]	15
PM_{10}, μg/m^3	Annual	15
	24-h[a]	45
O_3, μg/m^3	Peak season[b]	60
	8-h[a]	100
NO_2, μg/m^3	Annual	10
	24-h[a]	25
SO_2, μg/m^3	24-h[a]	40
CO, mg/m^3	24-h[a]	4

μg microgram

[a]99th percentile (i.e., 3–4 exceedance days per year)

[b]Average of daily maximum 8-h mean O_3 concentration in the six consecutive months with the highest 6-month running-average O_3 concentration. Note: Annual and peak season is long-term exposure, while 24 h and 8 h is short-term exposure

natural and man-made sources. These substances eventually combine to generate secondary biological aerosols (SOA) (Tiwari & Saxena, 2021). Biogenic unstable carbon-based compounds account for 10–40% of the bulk of biological aerosol in the world and produce around 90% of SOA generation (Cao et al., 2022). The chemical composition of nanoparticles differs substantially between regions due to various types of indigenous sources and their respective contributions.

In addition, incomplete incineration and geographical sources constitute the other major atmospheric sources of nanoparticle generation. The main combustion-generated sources are forest fires and volcanic eruptions, while the main geological sources are earthquakes, glaciers, dust storms, and volcanic eruptions (Leroy et al., 2022). According to Trejos et al. (2021), volcanic ash is essential to the worldwide movement of hazardous chemical species discharged together with particulates. Ash is a mixture of solid and liquid particle debris that is emitted during volcanic eruptions. It has an extremely complicated structure. Its composition varies throughout time as a result of chemical reactions and cooling. Volcanic ash nanoparticles (NPs) tend to have higher concentrations of harmful metals (Hg, Tl, Zn, Sn, Se, Te, Cd, Ag, Bi, Pb, and Ni) than comparable samples (Sonwani et al., 2021a).

10.2.2 Anthropogenic Sources of NPs

Man-made sources of NPs are more concentrated than natural sources in urban areas (Zhang et al., 2020b). There are two categories of anthropogenic sources: purposeful and unintentional. Unintentional sources include burning biomass, incinerating nonbiodegradable trash, and incomplete combustion from autos and industry. One of the intentional sources of NPs is the application of fertilizers and insecticides. NPs are further divided into main and secondary sources according to their place of origin. Industrialized emissions, mineral mining, energy generation, and transportation events are examples of primary sources. Primary sources might be mobile or fixed (Bhat et al., 2022).

Emissions from mining operations, metallurgical and chemical industries, and thermal power plants are examples of stationary sources. One of the main sources of NPs is thermal power plants, particularly in metropolises where there are more of them than elsewhere, such as semirural/urban settlements. The majority of the mobility sources include ships, automobiles, airplanes, engines, submarines, and rockets launched into extra-atmospheric space (Sonwani et al., 2021b; Strambeanu et al., 2014). One of the main sources of NPs is vehicle exhaust, which is produced when fuel is not completely burned. As a result of a dramatic rise in the number of vehicles on the road, it is one of the core origins of air pollution (Sonwani et al., 2021a). In 2002, there were more over one billion vehicles on the planet, and the number has been gradually increasing ever since (Roosa, 2020).

Hydrocarbons (HCs), carbon monoxide (CO), sulfur oxides (SOx), nitrogen oxides (NOx), volatile organic compounds (VOCs), particulate matter (PM), and their secondary derived products are the main components of exhaust (Agarwal &

Krishnamoorthi, 2023). The majority of particles found in automobile exhaust are within the sizes of 20–130 nm (diesel engines) and 20–60 nm (petrol engines). Fuel's sulfur content is a crucial factor since it sets off the PM production nucleation mode (Han et al., 2020). The discharges and ultrafine particles represent a major threat to the well-being of the environment due to their proven carcinogenic effects.

The mining and building industries are the second largest transmitters of anthropogenic nanoparticles to the atmosphere. Nanoparticles can be directly produced by surface excavation and mining through mine channels, but they can also be indirectly produced by decantation, sedimentation, and flotation (Silva et al., 2021). The amount of NPs in the atmosphere is also influenced by demolition techniques and meteorology. At demolition sites, one can find lead, wood, asbestos, fibers, glass, and other harmful particles in addition to dust. These particles have the ability to ascend high into the air and occasionally produce dust clouds that can spread widely impacting neighboring areas (Stevulova et al., 2022).

Pneumonia and, in extreme situations, cancer are brought on by inhalation of metal dust (nickel, chromium, and cobalt) and fumes (copper and zinc)—a condition that accounts for 15% of work-related risks (Nemery, 2022).

10.2.3 Trans-Boundary Movement of Nanoparticles

The migration of air contaminants is a significant driver in escalating the concentration of particulates in the atmosphere by augmenting the pollution load within a certain location, given that it carries an assortment of minerals and salts in addition to tiny particles, air mass migration from desert and ocean areas has a substantial impact on the quality of the air in secluded regions (Sonwani & Saxena, 2021). The burning of crop residues (CRB) is a major contributor to the deterioration of air quality in neighboring states due to trans-boundary movement. A significant portion of the emissions from CRB affect the air quality in the Indo-Gangetic Plains (IGP). The annual burning crop residues in IGP annually affect the air quality over the Arabian Sea coast and the south coast of India, with smoke columns reaching up to 2.5–3.5 kilometers (Sonwani et al., 2021a).

10.3 Mechanism of Exposure to Nanoparticles

10.3.1 Inhalation

The interior surface area of the human lungs ranges from 75 to 140 m^2, with approximately 3,00,106 alveoli (Wang et al., 2022a). The lungs are directly exposed to the external environment, and they serve as the primary point of entry for nanoparticles into the body. They are the main target site for studying the impacts of nanoparticles and are also regarded as significant (Garcia-Mouton et al., 2019). Breathing in air occurs through the nose and mouth, traveling down the throat, the tracheobronchial tree, and ultimately the alveoli. Particle passage is reliant on the respiratory airway's structure (Bake et al., 2019). NPs can disperse and collect in the alveolar section due to their tiny size and high retaining period. Some of them may even pass through the capillary endothelial cells and alveolar epithelium to enter the cardiovascular system as well as other internal organs. Moreover, electron imaging demonstrates that nanoparticles can penetrate the karyoplasm and cytoplasm of pulmonary mesothelial and epithelial cells in the lungs, as well as the inner and outer cellular compartments (Ferdous & Nemmar, 2020).

The respiratory system is responsible for gaseous exchange between the human body and its surroundings. It is one of the most vital systems in the human body and serves as an important entry port for nanomaterials into the human body and is the most common route of NP exposure. As humans breathe in air, NPs in the form of aerosols and powders can be easily inhaled into the respiratory tract (Yah et al., 2012). Once inhaled, particles are transported by the electrostatic force of the air from the upper respiratory tract to the lower respiratory tract. Larger NPs (>5 μm) are usually trapped by mucus in the upper respiratory tract and excreted via the body's lymphatic system (Fig. 10.1). On the other hand, smaller NPs escape from the mucociliary system and penetrate deep into the lung's alveoli where they are deposited in the alveolar epithelium or absorbed into the bloodstream (De Matteis & Rinaldi, 2018; Thu et al., 2023). In addition to the smaller-sized NPs, faster breathing rate, oral breathing, and existing lung disease can cause an increase in the penetration of NPs and deposition deep in the lungs.

10.3.2 Deposition

The aerodynamic diameter of the particle determines the location of nanoparticles dumping in the respiratory tract. The "saddle points" where the branching respiratory airways meet are where most of the bigger diameter fibers are deposited. As a

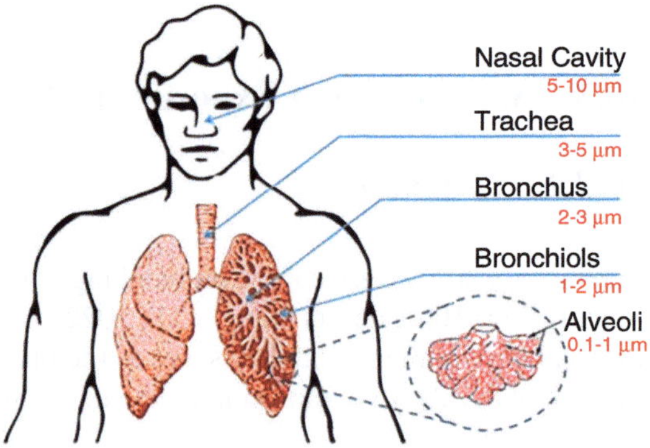

Fig. 10.1 Different size deposition of NPs in the airway after inhalation. (De Matteis & Rinaldi, 2018)

result, they can't enter the respiratory system deeply. Brownian motions primarily control deposition for smaller particles. Aerosol NPs of 20 nm size are shown to circumvent major clearance processes and deposit in the lung's alveolar region after 24 h of inhalation, according to data from energy-filtering transmission electron microscopy (EFTEM) (De Berardis et al., 2020; Si et al., 2021). The International Commission on Radiological Protection (ICRP) created a classical model that states that particulates with a diameter of less than 20 nm have a great likelihood of entering the alveolar section. Thus, by mucociliary escalation and macrophage activity, particles bigger than 20 nm in diameter are limited in the preceding regions of the respiratory tract. Prior to particles entering the pulmonary region, these responses aid in their removal from the tract (Bhat et al., 2022).

10.3.3 Clearance

Particle size, ventilator settings, and airway features are the main factors that influence deposition, but the physicochemical properties of the particulate matter ultimately decide how well it is cleared after it has been deposited. Due to impaction, bigger particles are dumped during inhalation at the intrathoracic bifurcations and extrathoracic area (nose and larynx). Mucus covers these upper respiratory airways. The primary route of clearance from the areas of the bronchus, trachea, and the upper bronchioles is the mucus layer, which traps the huge particles and transports them out by ciliary beating (Schneider-Daum et al., 2019). An effective ciliary beating cycle is facilitated by the periciliary layer, which lies beneath the outer viscoelastic mucus blanket. When the tips of the cilia come into contact with the mucus blanket, they beat in unison to speed up this layer (Hill et al., 2022). The mucus and NPs reach the pharynx, and they are eventually swallowed and continue to be processed inside the gastrointestinal system (Kaneko et al., 2020).

Tiny particles land in the bronchioles and alveoli after penetrating far into the lungs, where air velocity significantly drops (Ou et al., 2020). This is the location where macrophages engage in phagocytosis of the biogenic NPs such as bacteria, protozoa, and viruses where NPs undergo mucus transport (Joseph et al., 2023). Compared to mucociliary clearance, macrophage-associated clearance occurs far more slowly (Haque et al., 2020). The majority of these NPs move into connective tissue and then become involved in circulation of blood, which exposes them to macrophage populace in the non-alveolar area. Since surface macrophages are comparatively less effective at clearing NPs, interstitial and intravascular macrophages are primarily responsible for NP clearance. Numerous proteins and biomolecules are in contact with NPs upon their breathing in and deposition on the surfactant film with the underlying epithelium coating fluid (Garcia-Mouton et al., 2019).

10.4 Consequences of Intake of Airborne Nanoparticles

The pseudo-stratified epithelium of the human lung makes up the lung-bloodstream barrier. The mucous layer covers the thin columnar epithelium, bronchial epithelium (3–5 mm), and bronchiolar epithelium (0.5–1 mm), which make up the airways (Armstead & Li, 2016). Many cell models were created and tested in order to investigate the combined impact of nanoparticles on the lungs, as lung tissue is composed of over 40 distinct cell types. The respiratory system's epithelium is the primary target of NP exposure. A549 cell lines, which are derived from human lung adenocarcinomas, are most frequently employed in toxicity testing, and Calu-3, 16HBE14o-, and BEAS-2B cell lines are utilized as models for the bronchial-barrier system. The respiratory system is easily reached by smaller particles, which deposit at the distal portions of the respiratory tract, so the size of NPs and the harm they produce there are inversely correlated (Nho, 2020). The four major types of toxicological effects of nanoparticles (oxidative stress, inflammation, genotoxicity, and tumorigenicity) are discussed in the following sections.

10.4.1 Respiratory Injury

The responses to and lung detoxification of soluble and insoluble particulate matter differ significantly. In accordance with Khalaf et al. (2023), soluble nanoparticles enter the cardiovascular systems after dissolving in aqueous fluid. On the other hand, mucociliary escalator and macrophage phagocytosis are the methods used to eliminate the insoluble NPs (black carbon) (Osman et al., 2020). From research reports, insoluble particulates lead to increased lung tumors, inflammation, and tissue damage (Adeel et al., 2020). Lung damage results from the insoluble particulates accumulating more quickly than the ability of macrophages to clean them. As a result, the lungs' defense mechanisms are unable to function (Osman et al., 2020).

Additionally, it was found that the bronchoalveolar lavage fluid's shortened half-life of IL-1β and TGF-β1 induces acute lung inflammation. On the other hand, prolonged exposure results in the synthesis of collagen, which burdens the lungs and may lead to pulmonary fibrosis (Sun et al., 2020).

Lung burden-related health risks are closely linked to rising rates of early death, particularly in emerging countries. The rate of particle dumping, clearance, and residence time of the NPs all play a role in determining lung burden (Sun et al., 2020). Large mammals (humans and primates) have a slower rate of insoluble nanoparticle clearance and a higher propensity for reserved lung burden, which can move from the source alveolar dumping spots to the respiratory system's interstitial parts. These findings are supported by morphological observations and retrospective evidence. Long-term NP exposure results in a high rate of lung burden, which in turn supports lung tissue carcinogenesis (Sonwani et al., 2021a). Nonetheless, human exposure to this degree is uncommon, indicating a decreased risk of contracting the illness (Bouchama et al., 2022).

One of the most crucial channels for NPs to reach the body's various biological systems and for airborne contaminants to enter the body is through the respiratory system. Through the process of air particle diffusion, the reserved NPs in the lungs come into contact with the blood barrier of the alveoli, which is usually in-between the lungs' alveoli and the circulatory system's transferring vessels. This barrier aids in the exchange of stored gases in the alveoli with the whole body and vice versa. Through endocytosis in the alveolar epithelial cells, the trapped NPs permeate into these capillaries and are carried by blood to the other organs. A process known as endocytosis allows a cell to move particles into and out of the cell by creating vacuoles. The second route is the migration of these ultrafine particles via the nasal epithelial wall into the olfactory bulbs of neurons, which is how they get into the central nervous system (Kumar et al., 2022; Sonwani et al., 2021a). Nanoparticles can cause a variety of respiratory conditions when humans inhale it (Osman et al., 2020). The eight-largest hazard for the worldwide burden of disease is indoor air pollution caused by the burning of household fuels. The primary cause of indoor nanoparticle pollution in homes is smoke from biomass fuels and inefficient stoves which are mostly used in rural regions. Some of the main issues preventing rural residents from switching to modern stoves and environmentally friendly fuels include traditional practices, illiteracy, and ignorance (Altieri & Keen, 2019). Unprocessed biomass solid fuel is the most widely utilized fuel for home cooking in rural areas; it is reported to produce 50 times more emissions than gas cookers (Ravindra et al., 2019). Project Surya, which was implemented in India, recommended that upgraded cooking appliances be used by Indian homes as a way to mitigate black carbon emissions. As a result, it was advised that traditional cookstoves require more testing and better machinery (Sonwani et al., 2021a; Stoner et al., 2021). Such sustainable and environmentally friendly methods can lower the atmospheric concentration of NPs.

Another significant cause of NP exposure is cigarette smoking. It is known that one cigarette releases around 8.8×109 nanoparticles. There have been few researches done on many types of particles found in cigarette smoke, perhaps due to the large concentrations of particles and their quick dilution in air. Research indicates that those who passively smoke have a higher body mass index (BMI), a higher incidence of malignancies, and a lower lung function compared to nonsmokers (Borch et al., 2019; khan et al., 2021). Passive smoking has been shown to have negative impacts on human carcinogen levels, as demonstrated by the elevated levels of 1,3-butadiene, particulate-bound 2,5-dimethylfuran (DMF) and benzene in nonsmokers exposed smoke from cigarette. Exposure to nanoparticles (NPs) poses a significant risk to occupational health, a concern that is spreading throughout emerging and underdeveloped countries. Traffic officers have been found to have higher incidences of cardiovascular and respiratory conditions as a result of their prolonged exposure to vehicle emissions (Bajaj et al., 2017; Rodrigues et al., 2022).

10.4.2 Effect of NPs on Reproductive System

The accumulation of NPs has been identified as one of the primary causes of the current increase in cases of infertility. According to Zhang et al. (2020a), an experimental investigation conducted on rats fed a high-fat diet demonstrates that the administration of silica NPs, frequently found in workplaces, led to lower sperm motility rates and concentration and raised the rates of sperm abnormalities. Significant sperm DNA integrity loss was noted as a result of TiO_2 NPs' genotoxic effects, which are demonstrated when they penetrate the barrier of blood-testis and induce cytotoxicity and inflammation (Santonastaso et al., 2019). NPs have an impact on the reproductive system through ovarian cell cytotoxicity, which disrupts the oogenesis process and ultimately leads to an excess of reactive oxygen species (ROS), apoptosis, and an imbalance in sex hormones in the body (Ferrante et al., 2022). An animal given MoO3 in a study displayed a marked reduction in the weight of both the uterus and the right ovary, which ultimately had a negative impact on the process of reproductive (Asadi et al., 2019). Therefore, nanoparticles play a crucial role in lowering birth rates and causing defects in children. These kinds of situations are frequently observed in urban locations in females exposed to air pollutants, notably particle matter (NPs) as a result of their daily lives.

10.4.3 Effect of NPs on Endocrine System

Over the past 50 years, epidemiological data have demonstrated a significant increase in immune system disorders, including altered growth and development processes, immune system disorders, neurological system disorders, reduced fertility, and the beginning of several major diseases, such as obesity, diabetes, testicular, prostate, and ovarian cancer. The increased exposure of the general public and workers to chemicals that act as endocrine disruptors (EDCs), which can have detrimental consequences by altering the hormonal and homeostatic systems, could be one explanation for the rise in these ailments (Iavicoli et al., 2013). A study demonstrates how giving male rats 150 μg/kg of zinc oxide nanoparticles caused a net imbalance by raising the T3 thyroid hormone and dropping the hormone that stimulates thyroid (Chong et al., 2021). Palladium nanoparticles have been demonstrated to have action on hormone receptors gradually, starting with overstimulation and stopping the indicator cascades (Leso et al., 2018). According to certain research, NP exposure makes diabetic wildlife more susceptible to the hormonal imbalance than the ones that are not diabetic (Salem, 2020).

10.4.4 Effect of NPs on Neural System

The nervous tissues that comprise nervous system is composed of neurons and neuroglial cells (Rocha et al., 2020). Air pollutants enter the central nervous system (CNS) in various ways, however, the most common means is inhalation, in which substances can enter the brain through the blood-brain barrier, trigeminal nerve, olfactory bulb, the thalamus, and trigeminal nucleus, among other pathways. Nanoparticles (NPs) accumulate in many brain regions and alter gene expression that supports central nervous system growth and function, hence producing cytotoxic effects on neural cells (Vinod & Jena, 2021). Depression, inattention, and even cognitive impairment are indications of central nervous system dysfunction. Significant cytotoxicity is caused by ZnO NPs, which also cause apoptosis, changes in the cell cycle, a decrease in viability, and various forms of genetic damage, such as oxidative DNA impairment. Research indicates that inhaling manganese oxide causes inflammation in the brains of rats. In the mouse cortex and hippocampus, silver nanoparticles induce oxidative stress and upregulate genes linked to oxidative stress (Vinod & Jena, 2021).

10.4.5 Effect of NPs on Excretory System

The kidney is highly vulnerable to xenobiotic substances and the bioaccumulation of other toxins because each nephron in the kidney is equipped with a network of blood capillaries, which filter toxins from the blood. Following glomerular filtration, NPs typically gather in the proximal convoluted tubules (PCT), where endocytosis might cause tubular cells to internalize the particles. A study conducted on TH1 cells open to inorganic nanoparticles revealed that these cells had DNA damage in addition to an amplification of NP-induced nephrotoxicity (Sramkova et al., 2019). This will also cause chromosomal defects and genetic disorder. According to Naz et al. (2020), a particle's toxicity increases with its size. When exposed to SiO2 nanoparticles at dosage of 20–100 μg/ml, cells of the kidney experienced shrinking and nuclear condensation and that is an indicator of apoptosis, according to a study on HEK293 (cultured human embryonic kidney) cells. Additionally, oxidative stress induction and ROS generation in HEK293 cells suggest that nanoparticles may be nephrotoxic. Dysfunction in nephron elasticity brought on by impaired nephrotoxic potential can result in cerebral epilepsy (Sonwani et al., 2021a; Wang et al., 2022b). Additionally, the cytotoxicity of nanoparticles was assessed in the glomerular mesangial IP15 and epithelial proximal HK-2 cell lines. The tubular human renal cells and glomerular in the study were found to be cytotoxically affected by ZnO and CdS nanoparticles. Particle size, metal solubility, and metal composition were all connected to these effects (Pujalté et al., 2011; Sonwani et al., 2021a).

10.4.6 Effect of NPs on Cardiovascular System

NPs can enter human bodies by oral exposure to the gastrointestinal tract, which then travels to the circulatory system, or by inhaling (intranasal or intratracheal). These may result in notable deviations from the system's typical operation. Some of the first effects of NPs on the cardiovascular system include elevated blood pressure, lowered heart rate, changed vascular tone, and malfunction (Cao & Luo, 2019; Yu et al., 2016). NPs can affect human bodies in a variety of ways, including vasodilation/vasoconstriction, angiogenic/antiangiogenic, prooxidant/antioxidant, cytotoxic, phagocytic, and apoptotic (Gonzalez et al., 2016). These effects depend on the physical characteristics, concentration, and retention period of the NPs. People who already have cardiovascular conditions are particularly vulnerable to these alterations. Because of the incorporation of NPs in the bloodstream, these individuals are more vulnerable to heart attacks, sudden cardiac arrests, and blood clot.

The simultaneous and opposing effects of silver nanoparticles (NPs) on blood composition, angiogenesis (vessel formation), and membrane permeability are shown in pacemakers, medications, and linked antibodies (Wang et al., 2022a). TiO2 NPs can accumulate in the heart after prolonged exposure, which can result in inflammation, cellular necrosis, sparse cardiac muscle fibers, and cardiac biochemical malfunction. An investigation on SiO2 NPs in aged rats revealed elevated Fbg levels and blood viscosity in addition to atrioventricular occlusion and myocardial ischemia injury (Yu et al., 2016).

10.4.7 Effect of Nanoparticles on the Digestive System

The stomach, intestines, and esophagus make up the majority of the digestive system, sometimes referred to as the gastrointestinal tract. The most common ways that NP enters the gut are through food or water consumption (Huang & Tang, 2021). According to Polet et al. (2020), the presence of silver nanoparticles raises interleukin-8 levels, which are directly related to inflammation and increased mucus formation. Certain NPs can disrupt the mucus and epithelial cell layers in blood vessels by escaping the junctions between intestinal epithelial cells, depending on their size. Moreover, they ultimately gather in the intestinal lamina propria, where they impair goblet cell activity (Ejazi et al., 2023). The rate of NP buildup increases if an individual already has underlying medical disorders such Crohn's disease or ulcerative colitis, which causes intestinal inflammation. Cancer of the colon and other carcinomas can arise from the elevated toxicity levels (Wang et al., 2022a).

10.4.8 Oxidative Stress

The main effect of nanotoxicity is oxidative stress, which is brought forth by an imbalance between antioxidants and free radicals in the body. As a result, extra free radicals with an irregular number of electrons unintentionally react with other molecules to cause an imbalance in the respiratory system (Yahya et al., 2019). One of the key oxidative processes that lead to lung damage in humans is the generation of ROS. According to Flores-López et al. (2019), NPs' action in the cell's mitochondrial electron transport chain is what causes the overproduction of ROS. Carbon NPs are known to influence mitochondrial processes, while metallic NPs cause Fenton-type reactions that result in free radical-mediated toxicity (Canaparo et al., 2020). The surface groups' catalytic activity determines how much ROS is created by a specific NP. About 10% of the molecules in a particle with a size of 30 nm are expressed, compared to only about 20% and 50% of the molecules in particles with a size of 10 and 3 nm, respectively. The organism experiences oxidative stress when reactive oxygen species (ROS) are produced excessively or when antioxidant defense mechanisms are compromised.

10.4.9 Inflammation

There are two types of immune systems in humans: innate and adaptive. The first defense against any foreign particle that enters the body is the innate immune system. The much more sophisticated and potent adaptive immune system is triggered if the innate immune system is unable to neutralize the foreign particle (antigen) on its own. The body's dendritic cells are responsible for this activation (Keselowsky et al., 2020). Because they are foreign particles, nanoparticles also trigger the activation of dendritic cells, which release ROS, chemokines, and cytokines and stimulate naïve T-cells and different inflammasomes. The zeta potential (ξP) of a particle is a well-known indicator of its capacity to induce inflammation. The electric potential generated by the interaction of charged groups (found on a particle's surface) with the suspension medium is known as the zeta potential (ξP) (Sonwani et al., 2021a). Because the human body's medium is acidic, a particle's solubility in the medium will increase if it has more positively charged groups on its surface. This will increase the particle's contact with macrophages, which will cause inflammation (Karakashev et al., 2022). In these fluids, magnesium and zinc oxide nanoparticles exhibit great solubility. An additional indicator of inflammation in the body is the quantity of white blood cells (WBCs) in blood. When the white blood cell count rises above normal, it indicates inflammation and a decline in immunity. NPs that are inhaled cause the production of pro-inflammatory hormones upon deposition in the lungs, where they come into contact with alveolar macrophages, which are the primary immune system. The dormant macrophages are aroused to promote the delivery of several pro-inflammatory cytokines to the site of injury as a result (Sonwani et al., 2021b).

The respiratory tract's cilia are destroyed as a result of inflammation, which makes it more difficult for mucus to pass through cilia and trap dirt and infectious pathogens. Additionally, inflammation causes epithelial injury by rupturing the epithelial barrier separating the bloodstream from the lung surface. This compromises blood and oxygen circulation within the body, clotting, and reduced lung function. Pulmonary fibrosis, lung cancer, chronic obstructive pulmonary disease, asthma, and cystic fibrosis are all characterized by inflammation of the lung tissue and pneumonia (Poore & Zemanick, 2023). The primary causes of death for patients with chronic lung illnesses are different forms of pulmonary fibrosis. Additionally, inflammation causes blood clot, which triggers the thrombosis cascade at several sites. These events ultimately result in issues with the cardiovascular system, disrupt the heart's rhythm, and raise the body's risk of cardiac arrest (Stark & Massberg, 2021).

10.4.10 Genotoxicity

Direct NP-DNA contact or indirect oxidative stress or inflammatory responses are two ways that NPs might cause genotoxicity. Two categories of genotoxicity can be distinguished: primary and secondary genotoxicity, based on several research and their ensuing impacts. When minuscule particles penetrate the nucleus and alter DNA, this is known as primary genotoxicity. The cascade of DNA repair is connected to another important indirect mechanism. According to Vecchiotti et al. (2021), oxidative stress and inflammation brought on by NPs are the causes of secondary genotoxicity. Adenine and guanine undergo hydroxylation during DNA oxidation, which causes mutagenic changes and the creation of DNA adducts. DNA adducts are the product of covalent modifications to DNA caused by specific carcinogens. Particulate carcinogens can penetrate cellular membranes and the nucleus directly. Examples of such substances include asbestos and crystalline silica. They subsequently cause dysfunctionalities by interfering with many components of the mitotic spindle's operation and process. All biological functions, including mitosis, DNA replication, and DNA transcription into mRNA, can be interfered with by these foreign particles. Nanoparticles cause phagocytes (neutrophils, macrophages, etc.) to produce reactive oxygen species (ROS), which damages DNA and causes mutagenesis in nearby cells (Shukla et al., 2021). The induction of inflammation by diesel exhaust particle exposure results in epigenetic modifications, such as disruptions in DNA repair, adduct formation, gene expression impairment, and cell proliferation (Quezada-Maldonado et al., 2021).

10.4.11 Tumorigenicity

The MAPK/ERK pathway, sometimes referred to as the Ras-Raf-MEK-ERK pathway, is a network of proteins in a cell that transmits a signal between a surface receptor and the DNA located in the nucleus. Epidermal growth factor receptor (EGFR) is a cell surface receptor, and the signaling molecule that binds to it initiates the signal (Lin et al., 2020). Extracellular ligands, like epidermal growth factors, attach to this receptor in its normal state, phosphorylate it, activate it, and initiate a series of docking proteins that eventually produce mRNA, which is then translated into other proteins. Here, ROS and occasionally direct NanoPs interact with these receptors, and because of the altered DNA coding, aberrant proteins may or may not arise. One of the key steps in the development of many cancers is the possibility of a protein in the pathway becoming stuck in the "on" or "off" position due to mutation. The body's concentration of 8-hydroxy-2′-deoxyguanosine (8-OHdG) is largely increased by the introduction of NPs. Tumors can arise from G-to-T transversion mutations caused by 8-OHdG in important genes that are known to play a role in the genesis of cancer (Guo et al., 2021). As a result, there is a clear correlation between an elevated level of 8-OHdG and the development of carcinogenicity, which reduces immunity. Caveolin-1 overexpression and p53 inactivation are additional routes. NPs cause p53 inactivation by lowering phosphate levels in cells, which is necessary to activate proteins and raise the risk of tumor development (Kornberg et al., 2019).

10.5 Conclusion

A recent increase in the forms and variety of air pollutants has been attributed to urbanization, industrialization, and other human activities. A consequence of this is the increased burden of sickness brought on by pollutants that harm both the environment and humans. The benefits of nanoparticles (NPs), like high reactivity, small size, and great capacity, may turn into deadly aspects if they cause detrimental and toxic effects on cells that are not seen in micron-sized counterparts. Additionally, at that size, the optical properties also predominate, which emphasizes the significance of these materials in photocatalytic applications. It can be helpful to utilize synthetic methods to regulate the precise size, shape, and magnetic characteristics of NPs. Even though NPs have a wide range of applications, their uncontrolled use and discharge into the environment raise certain health risks. These issues should be taken into account while developing more environmentally friendly and convenient ways to employ NPs. Ongoing research and its benefits offer hope for new product breakthroughs. It is feasible to create artificial "living-like" or "nano-bot" entities that can remove heavy metals and dangerous toxins from surroundings. Therefore, the environment and industry will advance with the widespread use of these organisms and products.

References

Adeel, M., Tingting, J., Hussain, T., He, X., Ahmad, M. A., Irshad, M. K., et al. (2020). Bioaccumulation of ytterbium oxide nanoparticles insinuate oxidative stress, inflammatory, and pathological lesions in ICR mice. *Environmental Science and Pollution Research, 27*, 32944–32953.

Agarwal, A. K., & Krishnamoorthi, M. (2023). Review of morphological and chemical characteristics of particulates from compression ignition engines. *International Journal of Engine Research, 24*(7), 2807–2865.

Altieri, K. E., & Keen, S. L. (2019). Public health benefits of reducing exposure to ambient fine particulate matter in South Africa. *Science of the Total Environment, 684*, 610–620.

Armstead, A., & Li, B. (2016). Nanotoxicity: Emerging concerns regarding nanomaterial safety and occupational hard metal (WC-Co) nanoparticle exposure. *International Journal of Nanomedicine, 11*, 6421–6433. https://doi.org/10.2147/ijn.s121238

Asadi, F., Sadeghzadeh, M., Jalilvand, A., Nedaei, K., Asadi, Y., & Heidari, A. (2019). Effect of molybdenum trioxide nanoparticles on ovary function in female rats. *Journal of Advances in Medical and Biomedical Research, 27*(121), 48–53.

Awasthi, G., Maheshwari, T., Sharma, R., Kumawat, T. K., Singh, G. P., & Lodha, P. (2023). Actions and reactions of plants derived zinc-oxide nano-particles. *Materials Today: Proceedings.* https://doi.org/10.1016/j.matpr.2023.08.157. Accessed 21 Sept 2023

Bajaj, N., Sharma, T., Suneja, D., Jain, S., & Kumar, P. (2017). Determinants of respiratory and cardiovascular health effects in traffic policemen: A perception-based comparative analysis. *Journal of Transport & Health, 4*, 30–39.

Bake, B., Larsson, P., Ljungkvist, G., Ljungström, E., & Olin, A. C. (2019). Exhaled particles and small airways. *Respiratory Research, 20*, 1–14.

Bhat, S. A., Gambril, J., Azali, L., Chen, S. T., Rosen, L., Palettas, M., Wiczer, T. E., Kalathoor, S., Zhao, Q., Rogers, K. A., Kittai, A., Grever, M., Awan, F., Ruz, P., Byrd, J. C., Woyach, J., & Addison, D. (2022). Ventricular arrhythmias and sudden death events following acalabrutinib initiation. *Blood, 140*(20), 2142–2145. https://doi.org/10.1182/blood.2022016953

Borch, K. B., Weiderpass, E., Braaten, T., Hansen, M. S., & Licaj, I. (2019). Risk of lung cancer and physical activity by smoking status and body mass index, the Norwegian Women and Cancer Study. *European Journal of Epidemiology, 34*, 489–498.

Bouchama, A., Abuyassin, B., Lehe, C., Laitano, O., Jay, O., O'Connor, F. G., & Leon, L. R. (2022). Classic and exertional heatstroke. *Nature Reviews Disease Primers, 8*(1), 8.

Bundschuh, M., Filser, J., Lüderwald, S., McKee, M. S., Metreveli, G., Schaumann, G. E., Schulz, R., & Wagner, S. (2018). Nanoparticles in the environment: Where do we come from, where do we go to? *Environmental Sciences Europe, 30*(1). https://doi.org/10.1186/s12302-018-0132-6

Canaparo, R., Foglietta, F., Limongi, T., & Serpe, L. (2020). Biomedical applications of reactive oxygen species generation by metal nanoparticles. *Materials, 14*(1), 53.

Cao, Y., & Luo, Y. (2019). Pharmacological and toxicological aspects of carbon nanotubes (CNTs) to vascular system: A review. *Toxicology and Applied Pharmacology, 385*, 114801.

Cao, J., Situ, S., Hao, Y., Xie, S., & Li, L. (2022). Enhanced summertime ozone and SOA from biogenic volatile organic compound (BVOC) emissions due to vegetation biomass variability during 1981–2018 in China. *Atmospheric Chemistry and Physics, 22*(4), 2351–2364.

Cescon, A., & Jiang, J. Q. (2020). Filtration process and alternative filter media material in water treatment. *Water, 12*(12), 3377.

Chong, C. L., Fang, C. M., Pung, S. Y., Ong, C. E., Pung, Y. F., Kong, C., & Pan, Y. (2021). Current updates on the in vivo assessment of zinc oxide nanoparticles toxicity using animal models. *BioNanoScience, 11*(2), 590–620.

De Berardis, B., Marchetti, M., Risuglia, A., Ietto, F., Fanizza, C., & Superti, F. (2020). Exposure to airborne gold nanoparticles: A review of current toxicological data on the respiratory tract. *Journal of Nanoparticle Research, 22*, 1–41.

De Matteis, V. (2017). Exposure to inorganic nanoparticles: Routes of entry, immune response, biodistribution and in vitro/in vivo toxicity evaluation. *Toxics, 5*(4), 29. https://doi.org/10.3390/toxics5040029

De Matteis, V., & Rinaldi, R. (2018). Toxicity assessment in the nanoparticle era. In *Advances in experimental medicine and biology* (pp. 1–19). https://doi.org/10.1007/978-3-319-72041-8_1

Ejazi, S. A., Louisthelmy, R., & Maisel, K. (2023). Mechanisms of nanoparticle transport across intestinal tissue: An Oral delivery perspective. *ACS Nano, 17*(14), 13044–13061.

Ferdous, Z., & Nemmar, A. (2020). Health impact of silver nanoparticles: A review of the biodistribution and toxicity following various routes of exposure. *International Journal of Molecular Sciences, 21*(7), 2375.

Ferrante, M. C., Monnolo, A., Del Piano, F., Mattace Raso, G., & Meli, R. (2022). The pressing issue of micro-and nanoplastic contamination: Profiling the reproductive alterations mediated by oxidative stress. *Antioxidants, 11*(2), 193.

Flores-López, L. Z., Espinoza-Gómez, H., & Somanathan, R. (2019). Silver nanoparticles: Electron transfer, reactive oxygen species, oxidative stress, beneficial and toxicological effects. Mini review. *Journal of Applied Toxicology, 39*(1), 16–26.

Fröhlich, E., & Salar-Behzadi, S. (2014). Toxicological assessment of inhaled nanoparticles: Role of in vivo, ex vivo, in vitro, and in silico studies. *International Journal of Molecular Sciences, 15*(3), 4795–4822. https://doi.org/10.3390/ijms15034795

Garcia-Mouton, C., Hidalgo, A., Cruz, A., & Pérez-Gil, J. (2019). The Lord of the Lungs: The essential role of pulmonary surfactant upon inhalation of nanoparticles. *European Journal of Pharmaceutics and Biopharmaceutics, 144*, 230–243.

Geiser, M., Jeannet, N., Fierz, M., & Burtscher, H. (2017). Evaluating adverse effects of inhaled nanoparticles by realistic in vitro technology. *Nanomaterials (Basel, Switzerland), 7*(2), 49. https://doi.org/10.3390/nano7020049

Giorgetti, L. (2019). Effects of nanoparticles in plants: Phytotoxicity and genotoxicity assessment. *Nanomaterials in Plants, Algae and Microorganisms, 1*, 65–87. https://doi.org/10.1016/B978-0-12-811488-9.00004-4

Gonzalez, C., Rosas-Hernandez, H., Ramirez-Lee, M. A., Salazar-García, S., & Ali, S. F. (2016). Role of silver nanoparticles (AgNPs) on the cardiovascular system. *Archives of Toxicology, 90*, 493–511.

Goud, B. R., Raghavendra, M., Prasad, P. S., Hatti, V., Halli, H. M., Nayaka, G. V., et al. (2022). *Sustainable management and restoration of the fertility of damaged soils* (p. 113). Agriculture Issues and Policies.

Guo, C., Liu, Y., & Li, Y. (2021). Adverse effects of amorphous silica nanoparticles: Focus on human cardiovascular health. *Journal of Hazardous Materials, 406*, 124626.

Han, W., Lu, Y., Jin, C., Tian, X., Peng, Y., Pan, S., et al. (2020). Study on influencing factors of particle emissions from a RCCI engine with variation of premixing ratio and total cycle energy. *Energy, 202*, 117707.

Haque, S., Pouton, C. W., McIntosh, M. P., Ascher, D. B., Keizer, D. W., Whittaker, M. R., & Kaminskas, L. M. (2020). The impact of size and charge on the pulmonary pharmacokinetics and immunological response of the lungs to PLGA nanoparticles after intratracheal administration to rats. *Nanomedicine: Nanotechnology, Biology and Medicine, 30*, 102291.

Hartland, A., Lead, J. R., Slaveykova, V. I., O'Carroll, D., & Valsami-Jones, E. (2013). The environmental significance of natural nanoparticles. *Nature Education Knowledge, 4*(8), 7.

Hill, D. B., Button, B., Rubinstein, M., & Boucher, R. C. (2022). Physiology and pathophysiology of human airway mucus. *Physiological Reviews, 102*(4), 1757–1836.

Huang, X., & Tang, M. (2021). Review of gut nanotoxicology in mammals: Exposure, transformation, distribution and toxicity. *Science of the Total Environment, 773*, 145078.

Iavicoli, I., Fontana, L., Leso, V., & Bergamaschi, A. (2013). The effects of nanomaterials as endocrine disruptors. *International Journal of Molecular Sciences, 14*(8), 16732–16801.

Joseph, T. M., Kar Mahapatra, D., Esmaeili, A., Piszczyk, Ł., Hasanin, M. S., Kattali, M., et al. (2023). Nanoparticles: Taking a unique position in medicine. *Nanomaterials, 13*(3), 574.

Kaneko, K., Osman, N., Carini, V., Scagnetti, G., & Saleem, I. (2020). Overview of the advantages and disadvantages of different mucosal sites for the delivery of nanoparticles. In *Mucosal delivery of drugs and biologics in nanoparticles* (pp. 61–82). Springer Nature.

Karakashev, S. I., Grozev, N. A., Ozdemir, O., Guven, O., Ata, S., Bournival, G., et al. (2022). Physical restrictions of the flotation of fine particles and ways to overcome them. *Physicochemical Problems of Mineral Processing, 58*(5), 153944.

Keselowsky, B. G., Acharya, A., & Lewis, J. S. (2020). Innate and adaptive immunity: The immune response to foreign materials. In *Biomaterials science* (pp. 747–775). Academic.

Khalaf, E. M., Abood, N. A., Atta, R. Z., Ramírez-Coronel, A. A., Alazragi, R., Parra, R. M. R., et al. (2023). Recent progressions in biomedical and pharmaceutical applications of chitosan nanoparticles: A comprehensive review. *International Journal of Biological Macromolecules, 231*, 123354.

Khan, S., & Hossain, M. K. (2022). Classification and properties of nanoparticles. In *Nanoparticle-based polymer composites* (pp. 15–54). Woodhead Publishing.

Khan, H. S., Sheikh, N. S., Ayub, M. M., Tariq, M., Zahid, S., & Hafeez, F. (2021). Self-reported exposure to second-hand tobacco smoke and lung function in undergraduate male medical students in Pakistan. *Pakistan Journal of Physiology, 17*(3), 32–35.

Khan, S., Dewang, Y., Raghuwanshi, J., Shrivastava, A., & Sharma, V. (2022). Nanoparticles as fuel additive for improving performance and reducing exhaust emissions of internal combustion engines. *International Journal of Environmental Analytical Chemistry, 102*(2), 319–341.

Kornberg, T. G., Stueckle, T. A., Coyle, J., Derk, R., Demokritou, P., Rojanasakul, Y., & Rojanasakul, L. W. (2019). Iron oxide nanoparticle-induced neoplastic-like cell transformation in vitro is reduced with a protective amorphous silica coating. *Chemical Research in Toxicology, 32*(12), 2382–2397.

Kumar, R., Manna, C., Padha, S., Verma, A., Sharma, P., Dhar, A., et al. (2022). Micro (nano) plastics pollution and human health: How plastics can induce carcinogenesis to humans? *Chemosphere, 298*, 134267.

Lee, S. H., Gordon, H., Yu, H., Lehtipalo, K., Haley, R., Li, Y., & Zhang, R. (2019). New particle formation in the atmosphere: From molecular clusters to global climate. *Journal of Geophysical Research: Atmospheres, 124*(13), 7098–7146.

Leroy, S. A., Gracheva, R., & Medvedev, A. (2022). Natural hazards and disasters around the Caspian Sea. *Natural Hazards, 114*(3), 2435–2478.

Leso, V., Fontana, L., Marinaccio, A., Leopold, K., Fanali, C., Lucchetti, D., et al. (2018). Palladium nanoparticle effects on endocrine reproductive system of female rats. *Human & Experimental Toxicology, 37*(10), 1069–1079.

Lin, Y., Hu, C., Chen, A., Feng, X., Liang, H., Yin, S., et al. (2020). Neurotoxicity of nanoparticles entering the brain via sensory nerve-to-brain pathways: Injuries and mechanisms. *Archives of Toxicology, 94*, 1479–1495.

Naz, S., Gul, A., & Zia, M. (2020). Toxicity of copper oxide nanoparticles: A review study. *IET Nanobiotechnology, 14*(1), 1–13.

Nemery, B. (2022). Metals and the respiratory tract. In *Handbook on the toxicology of metals* (pp. 421–443). Academic.

Nho, R. (2020). Pathological effects of nano-sized particles on the respiratory system. *Nanomedicine: Nanotechnology, Biology and Medicine, 29*, 102242.

Osman, N. M., Sexton, D. W., & Saleem, I. Y. (2020). Toxicological assessment of nanoparticle interactions with the pulmonary system. *Nanotoxicology, 14*(1), 21–58.

Ou, C., Hang, J., & Deng, Q. (2020). Particle deposition in human lung airways: Effects of airflow, particle size, and mechanisms. *Aerosol and Air Quality Research, 20*(12), 2846–2858.

Polet, M., Laloux, L., Cambier, S., Ziebel, J., Gutleb, A. C., & Schneider, Y. J. (2020). Soluble silver ions from silver nanoparticles induce a polarised secretion of interleukin-8 in differentiated Caco-2 cells. *Toxicology Letters, 325*, 14–24.

Poore, T. S., & Zemanick, E. T. (2023). Infection, allergy, and inflammation: The role of aspergillus fumigatus in cystic fibrosis. *Microorganisms, 11*(8), 2013.

Pujalté, I., Passagne, I., Brouillaud B., Tréguer, M., Durand, E., Ohayon-Courtès, C., & L'Azou, B. (2011). Cytotoxicity and oxidative stress induced by different metallic nanoparticles on human kidney cells. *Particle and Fibre Toxicology, 8*, 10. https://doi.org/10.1186/1743-8977-8-10

Quezada-Maldonado, E. M., Sánchez-Pérez, Y., Chirino, Y. I., & García-Cuellar, C. M. (2021). Airborne particulate matter induces oxidative damage, DNA adduct formation and alterations in DNA repair pathways. *Environmental Pollution, 287*, 117313.

Ravindra, K., Singh, T., Mor, S., Singh, V., Mandal, T. K., Bhatti, M. S., et al. (2019). Real-time monitoring of air pollutants in seven cities of North India during crop residue burning and their relationship with meteorology and transboundary movement of air. *Science of the Total Environment, 690*, 717–729.

Rocha, L. A., Silva, D., Barata-Antunes, S., Cavaleiro, H., Gomes, E. D., Silva, N. A., & Salgado, A. J. (2020). Cell and tissue instructive materials for central nervous system repair. *Advanced Functional Materials, 30*(44), 1909083.

Rodrigues, A. C. B., de Jesus, G. P., Waked, D., Gomes, G. L., Silva, T. M., Yariwake, V. Y., et al. (2022). Scientific evidence about the risks of micro and nanoplastics (MNPLs) to human health and their exposure routes through the environment. *Toxics, 10*(6), 308.

Roosa, S. A. (2020). What is sustainable development? In *Sustainable development handbook* (2nd ed., pp. 1–33). River Publishers.

Salem, E. S. B. (2020). *The functional role of hepatic Argonaute (ago)-2 slicer activity in Metformin's action and glucose metabolism in obese mice*. Doctoral dissertation, University of Cincinnati).

Santonastaso, M., Mottola, F., Colacurci, N., Iovine, C., Pacifico, S., Cammarota, M., et al. (2019). In vitro genotoxic effects of titanium dioxide nanoparticles (n-TiO2) in human sperm cells. *Molecular Reproduction and Development, 86*(10), 1369–1377.

Schneider-Daum, N., Hittinger, M., Murgia, X., & Lehr, C. M. (2019). Cellular and non-cellular barriers to particle transport across the lungs. In *Biological responses to nanoscale particles: Molecular and cellular aspects and methodological approaches* (pp. 171–189).

Shukla, R. K., Badiye, A., Vajpayee, K., & Kapoor, N. (2021). Genotoxic potential of nanoparticles: Structural and functional modifications in DNA. *Frontiers in Genetics, 12*, 728250.

Si, X. A., Talaat, M., Su, W. C., & Xi, J. (2021). Inhalation dosimetry of nasally inhaled respiratory aerosols in the human respiratory tract with locally remodeled conducting lungs. *Inhalation Toxicology, 33*(4), 143–159.

Silva, L. F., Santosh, M., Schindler, M., Gasparotto, J., Dotto, G. L., Oliveira, M. L., & Hochella, M. F., Jr. (2021). Nanoparticles in fossil and mineral fuel sectors and their impact on environment and human health: A review and perspective. *Gondwana Research, 92*, 184–201.

Sonwani, S., & Saxena, P. (2021). Water-insoluble carbonaceous components in rainwater over an urban background location in Northern India during pre-monsoon and monsoon seasons. *Environmental Science and Pollution Research, 28*(38), 53058–53073.

Sonwani, S., Madaan, S., Arora, J., Suryanarayan, S., Rangra, D., Mongia, N., et al. (2021a). Inhalation exposure to atmospheric nanoparticles and its associated impacts on human health: A review. *Frontiers in Sustainable Cities, 3*, 690444.

Sonwani, S., Saxena, P., & Shukla, A. (2021b). Carbonaceous aerosol characterization and their relationship with meteorological parameters during summer monsoon and winter monsoon at an industrial region in Delhi, India. *Earth Space Sciience, 8*, 1–16.

Sramkova, M., Kozics, K., Masanova, V., Uhnakova, I., Razga, F., Nemethova, V., et al. (2019). Kidney nanotoxicity studied in human renal proximal tubule epithelial cell line TH1. *Mutation Research/Genetic Toxicology and Environmental Mutagenesis, 845*, 403017.

Stark, K., & Massberg, S. (2021). Interplay between inflammation and thrombosis in cardiovascular pathology. *Nature Reviews Cardiology, 18*(9), 666–682.

Stevulova, N., Estokova, A., Holub, M., & Singovszka, E. (2022). Demolition waste contaminated with asbestos. In *Advances in the toxicity of construction and building materials* (pp. 261–283). Woodhead Publishing.

Stoner, O., Lewis, J., Martínez, I. L., Gumy, S., Economou, T., & Adair-Rohani, H. (2021). Household cooking fuel estimates at global and country level for 1990 to 2030. *Nature Communications, 12*(1), 5793.

Strambeanu, N., Demetrovici, L., Dragoş, D., & Lungu, M. (2014). Nanoparticles: Definition, classification and general physical properties. In *Springer eBooks* (pp. 3–8). https://doi.org/10.1007/978-3-319-11728-7_1

Sun, B., Shi, Y., Li, Y., Jiang, J., Liang, S., Duan, J., & Sun, Z. (2020). Short-term PM2. 5 exposure induces sustained pulmonary fibrosis development during post-exposure period in rats. *Journal of Hazardous Materials, 385*, 121566.

Thu, H. E., Haider, M., Khan, S., Sohail, M., & Hussain, Z. (2023). Nanotoxicity induced by nanomaterials: A review of factors affecting nanotoxicity and possible adaptations. *OpenNano*, 100190. https://doi.org/10.1016/j.onano.2023.100190

Tiwari, S., & Saxena, P. (2021). *Air pollution and its complications* (pp. 1–178). Springer-Nature.

Trejos, E. M., Silva, L. F., Hower, J. C., Flores, E. M., González, C. M., Pachón, J. E., & Aristizábal, B. H. (2021). Volcanic emissions and atmospheric pollution: A study of nanoparticles. *Geoscience Frontiers, 12*(2), 746–755.

Vecchiotti, G., Colafarina, S., Aloisi, M., Zarivi, O., Di Carlo, P., & Poma, A. (2021). Genotoxicity and oxidative stress induction by polystyrene nanoparticles in the colorectal cancer cell line HCT116. *Public Library of Science One, 16*(7), e0255120.

Vera-Reyes, I., Vázquez-Núñez, E., Lira-Saldivar, R. H., & Méndez-Argüello, B. (2018). Effects of nanoparticles on germination, growth, and plant crop development. In *Agricultural nanobiotechnology* (pp. 77–110). Springer.

Vinod, C., & Jena, S. (2021). Nano-neurotheranostics: Impact of nanoparticles on neural dysfunctions and strategies to reduce toxicity for improved efficacy. *Frontiers in Pharmacology, 12*, 612692.

Wang, C. P. J., Byun, M. J., Kim, S. N., Park, W., Park, H. H., Kim, T. H., et al. (2022a). Biomaterials as therapeutic drug carriers for inflammatory bowel disease treatment. *Journal of Controlled Release, 345*, 1–19.

Wang, L., Shi, Y., Jiang, J., Li, C., Zhang, H., Zhang, X., et al. (2022b). Micro-nanocarriers based drug delivery technology for blood-brain barrier crossing and brain tumor targeting therapy. *Small, 18*(45), 2203678.

Yah, C., Iyuke, S., & Simate, G. (2012). A review of nanoparticles toxicity and their routes of exposures. *Iranian Journal of Pharmaceutical Sciences, 8*(1), 299–314.

Yahya, R. A., Azab, A. E., & Shkal, K. E. M. (2019). Effects of copper oxide and/or zinc oxide nanoparticles on oxidative damage and antioxidant defense system in male albino rats. *EASJ Pharm. Pharmacol, 1*, 135–144.

Yaqoob, A. A., Umar, K., & Ibrahim, M. N. M. (2020). Silver nanoparticles: Various methods of synthesis, size affecting factors and their potential applications–a review. *Applied Nanoscience, 10*, 1369–1378.

Yu, X., Hong, F., & Zhang, Y. Q. (2016). Bio-effect of nanoparticles in the cardiovascular system. *Journal of Biomedical Materials Research Part A, 104*(11), 2881–2897.

Zhang, J., Wu, L., Chan, H.-K., & Watanabe, W. (2011). Formation, characterization, and fate of inhaled drug nanoparticles. *Advanced Drug Delivery Reviews, 63*(6), 441–455. https://doi.org/10.1016/j.addr.2010.11.002

Zhang, K., Yang, W., Liu, Y., Zhang, K., Chen, Y., & Yin, X. (2020a). Laccase immobilized on chitosan-coated Fe3O4 nanoparticles as reusable biocatalyst for degradation of chlorophenol. *Journal of Molecular Structure, 1220*, 128769.

Zhang, Q., Lu, D., Wang, D., Yang, X., Zuo, P., Yang, H., et al. (2020b). Separation and tracing of anthropogenic magnetite nanoparticles in the urban atmosphere. *Environmental Science & Technology, 54*(15), 9274–9284.

Chapter 11
Nanoparticles in Food Chains: Bioaccumulation and Trophic Transfer

Patrick Omoregie Isibor, Samuel Adeniyi Oyegbade, Jerry Gbotemi Oni, Wopa Wurie Ahmed, Eniola Opeyemi Abiodun, and Oluwafemi Adebayo Oyewole

11.1 Introduction

Since the turn of the century, nanotechnology has been a recognized area of study. Since Nobel laureate Richard P. Feynman introduced the term "nanotechnology" in his well-known 1959 lecture "There's Plenty of Room at the Bottom" (Khan et al., 2019), the area of nanotechnology has undergone numerous breakthrough advancements. At the nanoscale level, nanotechnology creates materials of many sorts. According to Laurent et al. (2008), particulate compounds with at least one dimension smaller than 100 nm are classified as nanoparticles (NPs), a broad class of materials can be 0D, 1D, 2D, or 3D depending on the overall shape (Tiwari et al., 2012). When scientists discovered that a substance's size might affect its physiochemical qualities, such as its optical capabilities, they understood the significance of these materials. The typical colors of 20-nm gold (Au), platinum (Pt), silver (Ag), and palladium (Pd) NPs are, respectively, wine red, yellowish gray, black, and dark black.

The original version of the chapter has been revised. A correction to this chapter can be found at https://doi.org/10.1007/978-3-031-54154-4_17

P. O. Isibor (✉) · S. A. Oyegbade
Department of Biological Sciences, College of Science and Technology, Covenant University, Ota, Nigeria
e-mail: patrick.isibor@covenantuniversity.edu.ng

J. G. Oni · W. W. Ahmed
Department of Biochemistry, Federal University of Technology, Minna, Nigeria

E. O. Abiodun
Department of Microbiology, University of Ibadan, Oyo, Nigeria

O. A. Oyewole
Department of Microbiology, Federal University of Technology, Minna, Nigeria

P. O. Isibor et al. (eds.), *Environmental Nanotoxicology*,
https://doi.org/10.1007/978-3-031-54154-4_11

The field of nanotechnology has since witnessed rapid advancements in recent years, leading to the creation of a wide array of engineered nanoparticles with various properties and applications. These emerging nanoparticles have the potential to bring transformative changes to multiple industries, including healthcare, electronics, and environmental remediation. However, their introduction into the environment raises concerns about their ecological and health impacts. The environmental consequences of nanoparticle release into air, water, and soil are of primary concern, including their persistence, mobility, and potential toxicity. In the intricate wall-hanging of our interconnected world, the health of ecosystems and human populations is inextricably linked. Long-term ecological and health studies offer a lens through which we can unravel the complexities of this relationship, tracing the nuanced interactions between environmental factors, biodiversity, and the well-being of communities over extended periods. The urgency of addressing global challenges, such as climate change, biodiversity loss, and emerging health threats, necessitates a deeper understanding gained through sustained and comprehensive research efforts. This research theme focuses on exploring the existing knowledge landscape, identifying critical gaps in our understanding, and charting the course for future investigations in the realm of long-term ecological and health studies.

According to Dreaden et al. (2012), NPs displayed distinctive colors and features along with variations in size and shape that can be used in bioimaging applications. Since NPs are not simple molecules in and of themselves, they are made up of three layers, namely, the surface layer, which can be functionalized with a wide range of small molecules, metal ions, surfactants, and polymers, the core, which is essentially the center of the NP and commonly refers to the NP itself, and the shell layer, which is chemically distinct material from the core in all respects (Shin et al., 2016). Due to these unique qualities, these materials have attracted the attention of several researchers in a variety of sectors. The use of the NPs is possible in a variety of applications, including medication administration (Lee et al., 2011), chemical and biological sensing (Barrak et al., 2019), gas sensing (Mansha et al., 2016), CO_2 capture (Ganesh et al., 2017), and other related ones (Shaalan et al., 2016). A 25-fold growth in the number of goods between 2005 and 2010 that either contain nanoparticles (NP) or need them to be produced is evidence of the tremendous advancements made by nanotechnology over the past several decades (Bundschuh et al., 2018). This process is probably made easier by their special general characteristics, which differ from those of their bulk or dissolved counterparts in terms of particle size, surface area, surface reactivity, charge, and shape. This opens up a wide range of potential uses, such as cosmetic, pharmacological, and medicinal ones (Semenzin et al., 2015). Inorganic and carbon-based types of engineered NP are present, some of which have functionalized surfaces (Nowack & Bucheli, 2007). Environmental scientists have been faced with numerous challenges due to the high diversity of NPs' elemental and structural composition in addition to their distinctive general characteristics, including NP characterization and fate in complex matrixes as well as individual and combined NP effects in aquatic and terrestrial ecosystems (Klaine et al., 2008). Based on the mass concentration, the crystalline structure of titanium dioxide nanoparticles (TiO_2 NP), for example, affects their toxicity for the water flea *Daphnia magna* in a manner that is independent of initial particle size. According to these data, other dosage measurements, such as particle number, surface area, or body load, may be required in order to accurately describe the exposure scenario (Petersen et al., 2015). In fact, according to Bundschuh et al. (2018), NPs' surface area accounted for a significant portion of the variation in the toxicity that resulted from crystalline forms.

11.2 Fate of Nanoparticles

The fate of nanoparticles refers to their behavior and destiny once introduced into biological systems or the environment. In biological systems, nanoparticles can undergo various interactions. They might be absorbed, metabolized, excreted, or accumulated within tissues. Cellular uptake mechanisms, intracellular trafficking, and potential transformation within cells affect their fate.

Nanoparticles may distribute widely throughout the body or localize in specific organs or tissues, influenced by their size, surface properties, and route of exposure. This distribution can impact their potential toxicity and effects on organs. The body's mechanisms work to clear nanoparticles from circulation. This process involves elimination through the liver, kidneys, or immune system. However, some nanoparticles might resist clearance and persist in the body, potentially leading to accumulation and long-term effects. In the environment, nanoparticles can undergo dispersion, sedimentation, or aggregation, affecting their movement and distribution. They may interact with soil, water, or air components, altering their behavior and potential ecological impact (Hochella et al., 2019).

Nanoparticles can undergo degradation or transformation processes. They may change in chemical composition, size, or surface properties due to interactions with biological fluids, environmental conditions, or other substances. Studying the fate of nanoparticles is essential to comprehend their potential impacts on biological systems and the environment. It helps in assessing risks, designing safe nanomaterials, and developing strategies for their sustainable use and management.

11.2.1 Toxicokinetics of Nanoparticles

Toxicokinetics refers to the study of how the body interacts with and processes substances. The toxicokinetics of nanoparticles are complex and involve interactions between their physicochemical properties and biological systems. Studying these processes is crucial for assessing the risks associated with nanoparticle exposure and ensuring the safe development and application of nanotechnology.

Absorption

Nanoparticles demonstrate the capability to ingress into the body via diverse routes, encompassing inhalation, ingestion, or dermal contact. Their diminutive size and distinctive physicochemical properties wield significant influence over their absorption dynamics across biological barriers, including the respiratory and gastrointestinal tracts, orchestrating their subsequent distribution and disposition within the body.

Inhalation stands as a primary route of nanoparticle entry, facilitated by their small aerodynamic diameters that enable their deposition deep within the respiratory system. Nanoparticles, owing to their minute sizes, traverse the upper airways and reach the alveolar regions, where they may undergo retention and subsequent absorption into the systemic circulation. Similarly, ingestion of nanoparticles, either intentionally through oral intake or inadvertently via food or water, permits their entry into the gastrointestinal tract. The small sizes of nanoparticles enable their passage across the intestinal epithelium, influencing their absorption into the systemic circulation or their uptake by specialized cells within the gut-associated lymphoid tissue (Wu et al., 2016; Loui et al., 2016).

Moreover, dermal contact with nanoparticle-containing substances can facilitate their penetration through the skin barrier. Nanoparticles may translocate through the epidermal layers, exploiting skin appendages or disrupted skin barriers, thereby gaining access into the systemic circulation or local tissues. The size-dependent characteristics of nanoparticles, coupled with their surface properties, influence their interactions with biological barriers and cellular constituents. These properties dictate the extent and efficiency of nanoparticle absorption across these barriers, guiding their subsequent distribution and disposition within various tissues or organs. Comprehending the mechanisms underlying nanoparticle entry routes and their interactions with biological barriers is pivotal for evaluating their potential health risks, understanding their biodistribution patterns, and guiding the design of safer nanomaterials for biomedical and technological applications (Ivask et al., 2014).

Upon absorption, nanoparticles exhibit systemic distribution throughout the body via the circulation, guided by an interplay of several influential factors, including their size, shape, surface charge, and coating characteristics. These determinants collectively dictate the intricate distribution patterns of nanoparticles within biological systems, potentially leading to their accumulation in specific tissues or organs and subsequently impacting their toxicity profiles.

Distribution

The size of nanoparticles significantly influences their distribution kinetics, with smaller nanoparticles typically displaying enhanced tissue penetration and wider systemic dissemination due to their propensity to traverse biological barriers and access deeper tissue compartments. Furthermore, the shape of nanoparticles modulates their circulation behavior and tissue deposition, guiding their interactions with cells and extracellular components, thereby shaping their biodistribution patterns.

Surface charge and coating properties play pivotal roles in determining nanoparticle interactions within the biological milieu and influencing their distribution. Surface charge dictates interactions with biological entities, potentially affecting the cellular uptake, clearance, or localization of nanoparticles in tissues. Coating materials, including polymers or biomolecules, can impart stealth properties or facilitate specific targeting, thereby influencing nanoparticle accumulation in distinct tissues or organs. The unique physicochemical attributes of nanoparticles steer their accumulation in specific tissues or organs, which may have critical implications for their toxicity profiles. Accumulation in certain tissues can lead to localized toxicity, disrupting cellular functions, inducing inflammation, or eliciting adverse effects specific to those tissues.

Understanding the complex interplay between nanoparticle characteristics and their distribution patterns within the body is crucial for evaluating their safety profiles, discerning potential tissue-specific toxicities, and devising strategies to mitigate adverse effects. Unraveling the nuances of nanoparticle distribution is instrumental in advancing the design and development of nanomaterials with tailored properties for safer biomedical and technological applications.

Metabolism

Following their introduction into the body, nanoparticles are susceptible to undergoing biochemical alterations orchestrated by various enzymatic reactions or cellular processes. These modifications have the potential to induce changes in the structure or composition of nanoparticles, exerting profound influences on their behavior, toxicity profiles, and clearance kinetics within biological systems.

Enzymatic transformations represent a significant avenue through which nanoparticles undergo modifications within the body. Enzymes present in biological fluids or cellular compartments may catalyze reactions leading to the alteration of nanoparticle surface properties, modifications in their coating materials, or fragmentation of nanoparticle structures. These enzymatic modifications can result in alterations to the physicochemical characteristics of nanoparticles, potentially impacting their stability, biocompatibility, or interactions with biological entities.

Furthermore, cellular processes within specific tissues or organs can instigate modifications in nanoparticle structure or composition. Cellular internalization mechanisms, such as endocytosis, may trigger nanoparticle degradation or intracellular processing, resulting in alterations to their morphology, size distribution, or surface functionalities. Subsequent cellular trafficking or interactions with intracellular components may lead to modifications influencing their biological fate or toxicological behavior. The biochemical alterations experienced by nanoparticles within the biological milieu can significantly impact their behavior, including changes in their interaction profiles with biomolecules, altered biokinetics, or variations in their toxicity. Moreover, modifications to nanoparticle structure or composition may influence their recognition by clearance mechanisms, potentially affecting their elimination rates and residence time within the body (Gigault et al., 2012).

Elucidating the pathways and consequences of biochemical alterations undergone by nanoparticles within biological systems is pivotal for comprehending their fate, assessing potential toxicological effects, and guiding the design of safer nanomaterials for biomedical applications. Understanding the dynamic interplay between nanoparticles and biochemical processes within the body is instrumental in advancing the development of nanomaterials with optimized biocompatibility and reduced adverse effects.

Excretion

The elimination pathways for nanoparticles primarily encompass excretion via urine, feces, bile, or exhalation, each serving as avenues for their clearance from the body. The efficiency and kinetics of nanoparticle clearance are contingent upon a confluence of factors, including nanoparticle characteristics and the intricate mechanisms orchestrating their recognition and elimination within biological systems.

Renal excretion stands as a notable pathway for the elimination of smaller nanoparticles, facilitated by their filtration through the glomerular barrier in the kidneys. Nanoparticles below certain size thresholds can traverse the renal filtration barrier and subsequently be excreted through urine. Conversely, larger nanoparticles evade renal clearance and may be eliminated via alternative routes, such as hepatobiliary excretion or translocation to the feces via the gastrointestinal tract. Hepatobiliary excretion represents a significant elimination pathway, particularly for nanoparticles absorbed from the gastrointestinal tract or those metabolized within hepatocytes. Nanoparticles can undergo hepatic processing, leading to their incorporation into bile and subsequent excretion via the biliary system into the gastrointestinal tract, ultimately being eliminated in feces.

Additionally, some nanoparticles may find their way to the lungs via inhalation and subsequently be eliminated through exhalation, contributing to their clearance from the body. The exhalation route primarily applies to certain types of nanoparticles that can be excreted from the respiratory system in the form of exhaled air. The efficiency of nanoparticle clearance is intricately tied to their physicochemical properties, including size, shape, surface characteristics, and coating materials. These properties govern their recognition by clearance mechanisms and their susceptibility to renal filtration or recognition by hepatic or pulmonary clearance processes (von Moos et al., 2014).

Moreover, the body's intricate recognition and elimination mechanisms play a pivotal role in nanoparticle clearance kinetics. The reticuloendothelial system, comprising macrophages and mononuclear phagocytes, participates in recognizing and eliminating nanoparticles from circulation, contributing significantly to their clearance. Understanding the interplay between nanoparticle characteristics and clearance mechanisms is crucial for evaluating their biocompatibility, elucidating their fate within biological systems, and designing nanomaterials with enhanced clearance properties for safer biomedical and technological applications.

Biological Fate and Persistence

Nanoparticles possess inherent characteristics that may confer prolonged persistence within the body, leading to potential long-term effects. Their unique physicochemical properties, including size, surface modifications, and coatings, contribute to their extended residence time, which necessitates a comprehensive understanding of their kinetics to evaluate potential health implications and establish safe exposure thresholds.

The distinctive attributes of nanoparticles, such as their small sizes and surface properties, can influence their interactions with biological components, thereby affecting their biokinetics and tissue distribution. These interactions may lead to prolonged circulation, evasion of clearance mechanisms, or accumulation in specific tissues or organs, contributing to their extended retention within the body. The complexities of nanoparticle behavior within biological systems, including their biodegradation, cellular internalization, or entrapment within cellular compartments, can hinder their rapid elimination and contribute to their persistence over extended periods. Additionally, nanoparticles may undergo transformations or metabolic alterations within biological matrices, influencing their kinetics and contributing to their prolonged presence. The persistence of nanoparticles within the body raises concerns regarding potential long-term effects, including cumulative toxicity, chronic inflammation, or adverse health outcomes. Understanding the kinetics of nanoparticle clearance, tissue distribution, and potential mechanisms of long-term retention is indispensable for assessing their safety profiles and establishing permissible exposure limits to mitigate potential health risks (El-Naggar et al., 2021).

Evaluating the complex interplay between nanoparticle properties and their kinetic behavior within biological systems is pivotal for elucidating their fate, predicting potential long-term effects, and guiding the design of nanomaterials with optimized biocompatibility and reduced risks for various biomedical and technological applications.

11.2.2 Toxicodynamics of Nanoparticles

The toxicodynamics of nanoparticles involve the study of how these tiny particles interact within biological systems to induce toxic effects. This encompasses their behavior, distribution, metabolism, and mechanisms of action upon entering the body. When released into the environment, nanoparticles are subject to processes including agglomeration, sedimentation, biological buildup, biomagnification, dissolution, chemical and physical changes, etc. (Baalousha et al., 2011). Understanding these toxicodynamic processes is crucial for assessing the potential risks associated with nanoparticle exposure. It helps in elucidating the mechanisms underlying nanoparticle-induced toxicity and guides the development of safety assessments and regulations for these materials.

Distribution

In organisms featuring gastrointestinal systems, nanoparticles exhibit the capacity for absorption across intestinal cells, thereby gaining access into the systemic circulation and subsequently disseminating to diverse tissues.

Upon ingestion, nanoparticles traverse the gastrointestinal tract, encountering the specialized epithelial cells lining the intestinal wall. These nanoparticles may undergo various processes to cross the intestinal epithelium, including paracellular and transcellular transport mechanisms. Paracellular transport involves the movement of nanoparticles through the intercellular spaces between adjacent epithelial cells, while transcellular transport entails their internalization and subsequent translocation across these cells. Transcellular uptake mechanisms often involve nanoparticles undergoing endocytosis or absorption by intestinal epithelial cells, followed by their intracellular trafficking and eventual release into the underlying blood vessels or lymphatics. These pathways enable nanoparticles to bypass the formidable intestinal barrier, granting them access into the systemic circulation. Once nanoparticles gain entry into the circulation, they embark on a journey throughout the body, disseminating to various organs and tissues through blood circulation. The distribution of nanoparticles to specific tissues may be influenced by their physicochemical properties, surface modifications, and interactions with blood components, thereby dictating their biodistribution and targeting to specific sites (El-Naggar et al., 2022).

The ability of nanoparticles to be absorbed through intestinal cells and subsequently enter the systemic circulation contributes to their potential applications in oral drug delivery, targeted therapies, or systemic nanomedicine. The mechanisms underlying nanoparticle absorption in the gastrointestinal tract is pivotal for optimizing oral drug formulations and designing nanoparticle-based delivery systems for efficient therapeutic interventions.

Metabolism and Biotransformation

Following their entry into the body, nanoparticles may undergo metabolic transformations, potentially leading to alterations in their chemical structure or properties. These metabolic processes, orchestrated by various enzymatic systems or biochemical pathways, hold the capacity to modify the inherent characteristics of nanoparticles, thereby influencing their behavior, biological interactions, and clearance mechanisms.

Metabolism of nanoparticles within biological systems encompasses a spectrum of biotransformation processes, including enzymatic degradation, conjugation reactions, or surface modifications. These processes may lead to modifications in the nanoparticle's surface chemistry, alterations in their size distribution, or changes in their coating materials, consequently affecting their physicochemical properties and biological fate. The metabolic biotransformation of nanoparticles can significantly impact their toxicity profile by modulating their biocompatibility, bioavailability, or cellular interactions. Metabolically altered nanoparticles may exhibit modified cellular uptake mechanisms, altered distribution patterns, or varied interactions with biomolecules, potentially influencing their toxicological characteristics (El-Naggar et al., 2021).

Furthermore, metabolic transformations of nanoparticles may also impact their clearance mechanisms from the body. Metabolically modified nanoparticles might undergo alterations in their recognition by clearance mechanisms, affecting their elimination kinetics or routes, and ultimately influencing their persistence or residence time within biological systems. The metabolic fate of nanoparticles is crucial for comprehending their biological behavior, potential toxicity, and clearance pathways within living organisms. Elucidating the interplay between nanoparticle metabolism, toxicity, and clearance mechanisms is crucial for assessing their safety profiles, optimizing their biomedical applications, and advancing strategies for the development of safer nanomaterials.

Cellular Uptake and Intracellular Effects

Nanoparticles exhibit the capability to be internalized by cells through a spectrum of mechanisms, each imparting distinct implications for cellular functions and structures. Their diminutive size endows nanoparticles with the ability to infiltrate cells, potentially exerting influences on intracellular organelles, intricate cellular signaling pathways, and even genetic material, thereby culminating in adverse effects.

The diverse mechanisms facilitating nanoparticle internalization encompass processes such as endocytosis (including clathrin-mediated, caveolae-mediated, or macropinocytosis), passive diffusion, or receptor-mediated uptake. These pathways dictate the cellular entry of nanoparticles and vary in their intracellular trafficking, directing nanoparticles to specific cellular compartments and influencing their interactions with intracellular components. Once internalized, the small size of nanoparticles enables their interaction with cellular organelles, including mitochondria, endoplasmic reticulum, or lysosomes. These interactions may disrupt organelle functions, perturb cellular homeostasis, or induce oxidative stress, thereby contributing to cellular dysfunction or damage.

Furthermore, nanoparticles can influence intricate cellular signaling pathways, impacting processes crucial for cell survival, proliferation, or differentiation. Their interactions with signaling molecules or cellular receptors may trigger aberrant signaling cascades, disrupting normal cellular functions and potentially leading to adverse cellular responses. Moreover, the ability of nanoparticles to access genetic material within cells holds implications for potential genotoxic effects. Nanoparticles may interact with DNA or RNA molecules, potentially inducing DNA damage, mutations, or alterations in gene expression patterns, which can contribute to cellular malfunction or pathological conditions (El-Naggar et al., 2022).

The multifaceted impacts of nanoparticle internalization on cellular functions and structures underscore the importance of comprehensively assessing their intracellular interactions and biological effects. Understanding the intricate interplay between nanoparticles and cellular components is crucial for evaluating their safety profiles, elucidating potential adverse effects, and guiding the development of safer nanomaterials for various biomedical and technological applications.

Inflammatory Responses and Oxidative Stress

Nanoparticles possess the inherent capacity to incite inflammatory responses and instigate the generation of reactive oxygen species (ROS) within cells, thereby precipitating oxidative stress. This oxidative milieu engendered by nanoparticles can engender cellular damage, perturb essential cellular functions, and disrupt the intricate balance of normal physiological processes.

Upon interaction with cells, nanoparticles may provoke immune cells or cellular machinery to initiate inflammatory cascades. This inflammatory response manifests through the activation of signaling pathways and the release of pro-inflammatory mediators, fostering an inflammatory microenvironment within tissues. Concomitantly, nanoparticles may instigate the production of ROS, including superoxide anions, hydrogen peroxide, or hydroxyl radicals, within cells. These ROS are generated as a consequence of nanoparticle-cell interactions, disruptions in cellular redox balance, or activation of oxidative stress pathways. The accumulation of ROS surpasses the cellular antioxidant defense mechanisms, leading to oxidative stress. The resultant oxidative stress induces cellular damage by oxidizing biomolecules such as lipids, proteins, or DNA, compromising their structural integrity and functionality. Lipid peroxidation, protein oxidation, and DNA damage are among the deleterious effects instigated by ROS, disrupting cellular membranes, impairing protein functions, and inducing genetic alterations.

Furthermore, oxidative stress triggered by nanoparticles can perturb cellular signaling pathways, disrupt enzymatic activities, and perturb cellular homeostasis. These disruptions to normal cellular processes can manifest as impaired cell proliferation, altered cell differentiation, or aberrant cell death mechanisms, leading to adverse cellular outcomes. The propensity of nanoparticles to incite inflammatory responses and induce oxidative stress, culminating in cellular damage and dysregulation of physiological processes, underscores the importance of elucidating their potential adverse effects. Understanding the mechanisms underlying nanoparticle-induced oxidative stress is crucial for assessing their safety profiles and devising strategies to mitigate their detrimental impacts in various biomedical and technological applications.

Immune System Interactions

Nanoparticles have the capacity to interact with the immune system, eliciting intricate modulations in immune responses. These interactions may stimulate immune reactions, provoke the activation of immune cells, and thereby engender immunotoxicity or hypersensitivity reactions, posing significant implications for biological systems.

Upon encountering nanoparticles, the immune system may perceive them as foreign entities, initiating immune recognition processes. Nanoparticles can act as immunostimulatory agents, eliciting immune responses by activating immune cells, including macrophages, dendritic cells, or lymphocytes. This activation can manifest through the release of inflammatory mediators, cytokines, or chemokines, thereby establishing an inflammatory microenvironment. Conversely, nanoparticles may also induce immunosuppressive effects, dampening immune responses and impairing immune cell functions. Such immunosuppression can compromise immune surveillance and regulatory mechanisms, potentially rendering the host more susceptible to infections or impairing immune-mediated processes (El-Naggar et al., 2021).

Moreover, the interactions between nanoparticles and immune cells can provoke hypersensitivity reactions, triggering exaggerated immune responses upon subsequent exposures to nanoparticles. These hypersensitivity reactions may manifest as allergic responses or immune-mediated adverse effects, leading to heightened inflammatory responses or tissue damage. The intricate interplay between nanoparticles and the immune system can orchestrate a range of immune reactions, from immune activation to immune suppression or hypersensitivity reactions. Understanding the immunomodulatory effects of nanoparticles is crucial for assessing their safety profiles, unraveling potential immunotoxic effects, and guiding the development of safer nanomaterials for biomedical, therapeutic, and technological applications.

11.3 Nanoparticle Uptake and Bioaccumulation

Bioaccumulation and biomagnification represent complex processes by which nanoparticles, along with other environmental contaminants, possess the capacity to accumulate and amplify in concentration within ecosystems, thus presenting potential risks to organisms and the equilibrium of ecological systems. Bioaccumulation unfolds as organisms assimilate nanoparticles from their surroundings, typically via ingestion or uptake. Nanoparticles tend to amass within the tissues and organs of organisms since they are not easily metabolized or excreted, eventually culminating in an augmented nanoparticle concentration within the organism. It is noteworthy that the potential for bioaccumulation elevates with an organism's trophic level, particularly in higher-level consumers. On the other hand, biomagnification pertains to the progressive escalation of nanoparticle concentrations as they traverse up the trophic hierarchy. Predatory organisms consuming prey laden with nanoparticles accumulate more substantial amounts of these particles. Consequently, this process may yield perilously elevated nanoparticle levels in apex species and top-level predators (Xingchen et al., 2017).

These occurrences can impart deleterious impacts on ecosystems and organisms. For instance, bioaccumulated or biomagnified nanoparticles have the potential to impede cellular processes, influence reproductive and immune systems, and induce toxicity in organisms, carrying the potential for repercussions on both wildlife and human health. Consequently, an appreciation of and vigilant oversight over the dynamics of bioaccumulation and biomagnification are of critical import. This is pivotal for the assessment of risks associated with nanoparticle exposure within ecosystems and is indispensable for upholding the sustainability of natural and human environments alike.

11.3.1 Mechanisms of Nanoparticle Uptake by Organisms

Nanoparticles can be taken up by organisms through various mechanisms, primarily determined by the size, surface properties, and characteristics of the nanoparticles. Several uptake pathways facilitate the entry of nanoparticles into biological systems.

11.3.1.1 Endocytosis

Nanoparticles exhibit the remarkable ability to infiltrate cell membranes through a process termed endocytosis, which orchestrates their internalization into cells by enclosing these particles within specialized vesicles. This intricate mechanism encompasses various pathways, each with distinct characteristics. Phagocytosis, predominantly employed by specialized cells like macrophages, involves the engulfment of nanoparticles within large vesicles called phagosomes. Pinocytosis, akin to cell drinking, engulfs nanoparticles in smaller vesicles formed by invagination of the cell membrane. Additionally, clathrin-mediated endocytosis entails the formation of coated pits on the cell surface, leading to the internalization of nanoparticles within clathrin-coated vesicles (El-Naggar et al., 2021).

The choice of a specific endocytic route is contingent upon a confluence of factors, including the inherent properties of the nanoparticles themselves and the particular type of cell involved in the uptake process. Nanoparticles with distinctive characteristics, such as size, surface charge, and functionalization, may favor one endocytic pathway over others. Similarly, different cell types possess varying endocytic machinery, influencing the selection of the most suitable pathway for nanoparticle internalization.

The versatility of endocytosis as a mechanism for nanoparticle uptake underscores its significance in cellular interactions. It facilitates the internalization of nanoparticles into cells, allowing them access to the intracellular environment where they may interact with cellular organelles, biomolecules, or signaling pathways. Understanding the intricacies of endocytic pathways and their relationship with nanoparticle characteristics is pivotal for elucidating cellular uptake mechanisms and assessing the potential biological effects of nanoparticles within living systems (Qi et al., 2022).

11.3.1.2 Passive Diffusion

Nanoparticles, particularly those of smaller sizes, can penetrate cellular membranes via passive diffusion, enabling direct entry into cells by exploiting their diminutive dimensions. This process hinges on the ability of nanoparticles to traverse cellular membranes due to their small size, interacting with intracellular components upon entry. It is especially pertinent for nanoparticles whose sizes are below the threshold of the cell membrane pores.

Passive diffusion allows nanoparticles to permeate cell membranes in a manner dictated by their size and physical properties. Nanoparticles smaller than the cell membrane pores can navigate across the lipid bilayer through interstitial spaces or transient disruptions in the membrane structure. This mechanism permits their ingress into the cellular interior without the requirement for active cellular processes or specialized uptake mechanisms. The ability of small nanoparticles to passively diffuse through cell membranes offers them direct access to the intracellular medium. Once inside the cells, these nanoparticles may interact with various cellular constituents, including organelles or biomolecules, potentially influencing cellular functions or eliciting specific biological responses (El-Naggar et al., 2022).

This mode of passive diffusion-mediated cellular uptake is particularly relevant for nanoparticles that possess dimensions smaller than the limiting size of the cell membrane pores. Understanding the dynamics of nanoparticle penetration through passive diffusion is essential for elucidating their interactions with cellular components and assessing their biological effects within organisms.

11.3.1.3 Receptor-Mediated Uptake

Certain nanoparticles possess surface modifications or ligands that endow them with the ability to selectively recognize and bind to cellular receptors, facilitating their uptake into cells through receptor-mediated pathways. These surface modifications, often comprised of functional groups or biomolecules, are intricately designed to interact with specific receptors present on the cell membrane.

The incorporation of specific ligands or surface modifications on nanoparticles allows for precise targeting and recognition of cellular receptors, initiating a sequence of events that leads to the internalization of these nanoparticles into cells. Upon binding to the appropriate receptors, nanoparticles initiate receptor-mediated endocytosis, a sophisticated process orchestrating their uptake into cells through specialized vesicles. This mechanism capitalizes on the cell's internalization machinery, ensuring the effective internalization of nanoparticles through receptor-ligand interactions (Agarwal et al., 2013).

The specificity conferred by these surface modifications or ligands enables nanoparticles to target particular cell types or tissues expressing the matching receptors. This targeted approach enhances the efficiency of nanoparticle uptake into desired cell populations while minimizing nonspecific interactions with other cell types. Furthermore, employing receptor-mediated pathways for nanoparticle uptake offers a controlled and regulated means for cellular internalization, contributing to precise and selective nanoparticle delivery into cells.

Incorporating surface modifications or ligands that facilitate receptor-mediated uptake broadens the scope of nanoparticle applications in targeted drug delivery, diagnostics, and therapeutics. Exploring and leveraging these specific interactions between nanoparticles and cellular receptors offer promising avenues for tailored and efficient cellular uptake, holding significant implications in biomedical and nanomedicine research.

11.3.1.4 Transcytosis

Nanoparticles exhibit the capacity to traverse epithelial or endothelial barriers through transcytosis, a sophisticated process facilitating their translocation across cell layers within tissues. This intricate mechanism enables nanoparticles to breach biological barriers, such as epithelial linings or endothelial cell layers, by engaging in vesicle-mediated transport mechanisms.

During transcytosis, nanoparticles are internalized by cells on one side of the barrier, encapsulated within vesicles, and transported across the cell to the opposing side. These vesicles, carrying nanoparticles, traverse the cell's cytoplasm and are subsequently released at the opposite cellular membrane, enabling the nanoparticles to transit the barrier intact. Epithelial or endothelial cells that form biological interfaces utilize transcytosis to regulate the passage of molecules, nutrients, or particles across these barriers. Nanoparticles exploit this transcellular transport mechanism, utilizing it to navigate these barriers and traverse tissues while circumventing the restrictive nature of cellular barriers.

Transcytosis furnishes nanoparticles with a means to navigate biological barriers with selectivity and control. This process offers a regulated and directed mode for nanoparticle movement across epithelial or endothelial barriers, presenting prospects for targeted drug delivery, tissue-specific therapeutic interventions, or the transport of nanomaterials to specific cellular compartments.

11.3.1.5 Extracellular Uptake

Nanoparticles, beyond their cellular uptake mechanisms, engage in intricate interactions with extracellular components within biological systems, notably with proteins or membranes. These interactions engender surface adsorption phenomena or molecular associations that significantly modulate the nanoparticles' bioavailability and subsequent cellular uptake dynamics. The knowledge about the intricacies of nanoparticle interactions with extracellular components is imperative in delineating their fate, behavior, and biological responses within biological systems. The dynamic interplay between nanoparticles and extracellular elements fundamentally shapes their bioavailability, cellular interactions, and ultimate biological effects, thereby delineating crucial considerations for the design and utilization of nanomaterials in biomedical and therapeutic applications.

Upon introduction into biological environments, nanoparticles encounter a diverse milieu of extracellular proteins, forming a protein corona that envelops their surface. This corona arises from the rapid adsorption of proteins onto the nanoparticle surface, forming a dynamic and complex interfacial layer. The composition and conformation of proteins within this corona are influenced by the physicochemical properties of nanoparticles, such as size, shape, surface chemistry, and charge. The formation of this protein corona can markedly alter the surface characteristics of nanoparticles, thereby modulating their recognition and interaction with cellular membranes or receptors. These altered surface properties influence the nanoparticles' bioavailability, cellular uptake kinetics, and intracellular fate. Moreover, the protein corona may act as a bridge between nanoparticles and cellular membranes, facilitating their internalization or altering their biodistribution within biological systems (El-Naggar et al., 2021). Furthermore, nanoparticles can directly interact with cellular membranes, eliciting changes in membrane integrity, fluidity, or receptor signaling. These nanoparticle-membrane interactions can trigger cellular responses or alter membrane properties, influencing subsequent cellular uptake processes or signaling cascades.

11.3.1.6 Gut Uptake

In organisms that possess gastrointestinal systems, nanoparticles exhibit the capacity for absorption across intestinal cells, thereby gaining access into the systemic circulation and subsequently disseminating to diverse tissues.

Upon ingestion, nanoparticles traverse the gastrointestinal tract, encountering the specialized epithelial cells lining the intestinal wall. These nanoparticles may undergo various processes to cross the intestinal epithelium, including paracellular and transcellular transport mechanisms. Paracellular transport involves the movement of nanoparticles through the intercellular spaces between adjacent epithelial cells, while transcellular transport entails their internalization and subsequent translocation across these cells. Transcellular uptake mechanisms often involve nanoparticles undergoing endocytosis or absorption by intestinal epithelial cells, followed by their intracellular trafficking and eventual release into the underlying blood vessels or lymphatics. These pathways enable nanoparticles to bypass the formidable intestinal barrier, granting them access into the systemic circulation (Zimmermann & Sures, 2004).

Once nanoparticles gain entry into the circulation, they embark on a journey throughout the body, disseminating to various organs and tissues through blood circulation. The distribution of nanoparticles to specific tissues may be influenced by their physicochemical properties, surface modifications, and interactions with blood components, thereby dictating their biodistribution and targeting to specific sites. The ability of nanoparticles to be absorbed through intestinal cells and subsequently enter the systemic circulation contributes to their potential applications in oral drug delivery, targeted therapies, or systemic nanomedicine. Understanding the mechanisms underlying nanoparticle absorption in the gastrointestinal tract is pivotal for optimizing oral drug formulations and designing nanoparticle-based delivery systems for efficient therapeutic interventions (El-Naggar et al., 2022).

These mechanisms of nanoparticle uptake are influenced by several factors, including nanoparticle size, shape, surface charge, surface modifications, and the physiological characteristics of the organisms. Understanding these uptake pathways is critical for assessing the potential risks and impacts of nanoparticle exposure on organisms and ecosystems.

11.3.2 Factors Influencing Nanoparticle Bioaccumulation

Movement, biological fate, and bioavailability of nanoparticles are influenced by their physicochemical characteristics, surface charge, size, and shape. Nanoparticle bioaccumulation is influenced by several key factors that dictate their behavior and accumulation within living organisms. Understanding these factors is crucial for assessing the potential risks associated with nanoparticle exposure and their subsequent accumulation in organisms, contributing to more informed regulatory measures and risk management strategies.

11.3.2.1 Particle Size and Surface Characteristics

The accumulation of nanoparticles within biological systems and their potential toxicological effects are influenced by various factors, prominently encompassing nanoparticle size and surface characteristics.

Smaller nanoparticles exhibit heightened bioavailability owing to their larger surface area-to-volume ratio. This characteristic enhances their propensity for interaction with biological entities and facilitates their uptake into cells or tissues. Their diminutive size grants them the capability to traverse biological barriers more readily compared to larger counterparts. This heightened mobility allows smaller nanoparticles to infiltrate cell membranes and navigate through intricate biological structures, enabling access to cellular compartments or targeted sites that might be less accessible to larger particles. Surface properties of nanoparticles, notably their charge and surface functionalization, exert substantial influence on their interactions within biological systems. Surface charge governs the electrostatic interactions of nanoparticles with biomolecules, cellular membranes, and tissues. Positively or negatively charged nanoparticles can modulate their cellular uptake mechanisms or adsorption onto cellular surfaces, thereby influencing their bioaccumulation and potential toxicological effects (Lähde et al., 2013).

Moreover, surface functional groups play a pivotal role in determining the nanoparticle's biocompatibility and biological interactions. The presence of specific functional groups on the nanoparticle surface can dictate their affinity toward biomolecules or cellular receptors. Functionalization facilitates tailored interactions, potentially enhancing or mitigating the nanoparticle's toxicity profile by influencing their cellular uptake, biodistribution, and subsequent biological responses. Understanding the intricate interplay between nanoparticle characteristics and biological systems is crucial in evaluating nanotoxicology, the study of potential adverse effects of nanoparticles on living organisms. The ability of nanoparticles to traverse biological barriers and interact with cellular components warrants a comprehensive assessment of their potential toxicity and long-term impacts on biological systems (Strong & Frank, 2010).

Investigations into nanoparticle bioaccumulation and nanotoxicity necessitate a multidisciplinary approach, integrating techniques from nanotechnology, biology, and toxicology. This holistic understanding is pivotal for elucidating the mechanisms underlying nanoparticle-induced biological responses and developing strategies to harness the benefits of nanotechnology while mitigating potential adverse effects on living organisms and the environment (Ferry et al., 2009). It was observed that some nanoparticles attached to algae did not undergo higher trophic level transfer. Meanwhile, the measured fish and daphnids had nanoparticles attached to them. The study revealed trophic transfer of gold nanomaterials is dependent on the physical and chemical properties of nanomaterials. The nanomaterials that are persistent in the environment have capacity to move to greater trophic levels within the aquatic food chain. By implication, particle size is a crucial factor in the trophic transfer of nanomaterials. Also, the bioavailability of bigger nanomaterials will be higher than their smaller counterparts.

11.3.2.2 Chemical Composition

The uptake and distribution of nanoparticles within organisms are intricately influenced by the diverse compositions of these nanoscale materials. Variations in nanoparticle composition, particularly among metals, metal oxides, and organic nanoparticles, engender distinct bioaccumulation potentials contingent upon their intrinsic chemical properties. Metal nanoparticles, owing to their elemental nature, can exhibit differential bioaccumulation tendencies predicated on their specific chemical compositions. Factors such as the metallic element, its oxidation state, and the surface functionalization profoundly impact their uptake and distribution within biological systems. Certain metal nanoparticles may possess higher bioaccumulation potential due to their affinity for cellular structures or biomolecules, while others might display lower uptake rates based on their physicochemical characteristics (Xingchen et al., 2017).

Metal oxide nanoparticles, characterized by their metal cation and oxygen composition, showcase varied bioaccumulation tendencies contingent upon their chemical makeup. The valence state of the metal cation and the crystalline structure of the oxide significantly influence their interaction with biological entities. Surface modifications and size variations within metal oxide nanoparticles also play a pivotal role in dictating their bioavailability and subsequent distribution within organisms. In contrast, organic nanoparticles, derived from organic compounds or polymers, present diverse bioaccumulation profiles based on their structural attributes and functional groups. The chemical composition of organic nanoparticles determines their solubility, stability, and interactions with biological matrices, thereby influencing their uptake and biodistribution. Factors such as molecular weight, hydrophobicity, and surface modifications dictate the bioavailability and propensity for accumulation within living systems (Wang et al., 2016).

Understanding the nuanced intricacies of nanoparticle compositions and their inherent chemical properties is crucial for assessing their bioaccumulation potential and potential ecological implications. Diverse nanoparticle compositions may elicit varying degrees of uptake and distribution within organisms, thereby necessitating comprehensive evaluations to discern their ecological impact and potential risks in biological systems. This knowledge forms the cornerstone for delineating strategies to optimize the beneficial applications of nanoparticles while minimizing potential adverse effects on living organisms and the environment.

11.3.2.3 Exposure Route

Nanoparticles, due to their minuscule size, possess the capability to ingress into organisms through multiple exposure routes, including inhalation, ingestion, dermal contact, or injection. The distinctive physicochemical properties of nanoparticles dictate their interaction mechanisms with biological systems, and the route of exposure significantly influences their subsequent distribution and accumulation patterns within the body.

Inhalation: When nanoparticles are inhaled, they traverse the respiratory tract and can access the deeper regions of the lungs. Their ultrafine dimensions allow them to bypass natural filtration mechanisms and reach alveolar spaces or even translocate into systemic circulation, facilitating their dissemination to various organs and tissues. The size and surface characteristics of inhaled nanoparticles play a pivotal role in determining their retention within the lungs and potential systemic distribution (Qi et al., 2022).

Ingestion: Nanoparticles encountered through ingestion enter the gastrointestinal tract, where they might undergo various interactions with the digestive system. Factors such as particle size, surface modifications, and chemical composition influence their absorption across the intestinal barrier. Once absorbed, nanoparticles may traverse through the circulatory system, reaching different organs or accumulating in specific tissues based on their physicochemical properties.

Dermal Contact: Nanoparticles can permeate the skin barrier upon direct contact. Factors including particle size, surface charge, and surface functionalization influence their ability to penetrate the skin's layers. Upon entry, nanoparticles can circulate in the bloodstream or accumulate in skin tissues, depending on their characteristics and interactions with the skin's components.

Injection: Direct injection of nanoparticles bypasses many natural barriers and introduces them directly into the bloodstream or targeted tissues. This route allows for precise delivery and immediate access to systemic circulation, influencing their distribution and accumulation patterns within the body.

Each route of exposure imparts distinct biodistribution and accumulation profiles for nanoparticles within organisms. The physicochemical properties of nanoparticles, combined with the body's biological barriers and interactions, dictate their fate within the organism. Understanding these pathways of entry and subsequent distribution patterns is crucial for assessing the potential health risks and designing safer nanoparticle-based applications in various fields, including medicine, consumer products, and environmental technologies.

11.3.2.4 Species Specificity

Knowledge of the interplay between species-specific characteristics and nanoparticle behavior is essential for comprehending the ecological implications and potential risks associated with nanoparticle exposure across diverse organisms. These considerations form the basis for assessing the environmental impact and health implications of nanoparticles and aid in the development of safer nanomaterials with reduced ecological and biological risks. The uptake, accumulation, and subsequent fate of nanoparticles within organisms are profoundly influenced by species-specific traits, encompassing metabolic processes, physiological factors, and the presence of biological barriers. These inherent characteristics delineate varying abilities among different organisms to metabolize or eliminate nanoparticles, contributing to diverse nanoparticle bioaccumulation patterns and potential effects (Ezoe et al., 2002).

Metabolic Processes: Species exhibit distinct metabolic pathways and enzymatic systems that govern the transformation and breakdown of foreign substances, including nanoparticles. Variations in enzymatic activities and metabolic rates among species influence their capacity to metabolize or biotransform nanoparticles, leading to differences in their retention or elimination.

Physiological Factors: Species-specific physiological attributes, such as body size, organ morphology, and circulation dynamics, impact nanoparticle distribution within the body. Variances in organ structure, vascularization, and tissue composition influence the extent of nanoparticle uptake and accumulation in different organs or tissues among diverse organisms.

Biological Barriers: Biological barriers, including cellular membranes, epithelial linings, and specialized barrier tissues, serve as protective shields regulating the passage of nanoparticles into vital organs or systems. Variability in the integrity and permeability of these barriers across species affects the penetration and accumulation of nanoparticles in specific biological compartments.

Elimination Pathways: Organisms possess distinctive mechanisms for nanoparticle elimination, such as renal excretion, biliary excretion, or clearance via the reticuloendothelial system. Variations in these elimination pathways among species dictate their ability to expel or retain nanoparticles within the body (Moller et al., 2010).

Furthermore, interspecies variations in nanoparticle interactions with biological matrices, such as protein binding or cellular uptake mechanisms, contribute to divergent responses and effects among different organisms.

11.3.2.5 Environmental Conditions

Understanding the influence of environmental factors on nanoparticle behavior is crucial for evaluating their fate and potential impacts within ecosystems. These multifaceted interactions underscore the importance of considering environmental parameters in assessing nanoparticle bioaccumulation and ecological risks across diverse environmental settings. Such insights facilitate the development of strategies to mitigate potential adverse effects and foster the responsible utilization of nanoparticles in various environmental applications. Environmental factors wield significant influence over the behavior and bioaccumulation potential of nanoparticles within diverse ecosystems. Parameters including temperature, pH, salinity, and the presence of other substances intricately modulate nanoparticle interactions, distribution, and subsequent bioaccumulation patterns across various environmental settings (Rana and Kalaichelvan, 2013; Guénette et al., 2014; Kubo-Irie et al., 2016; Ramzan et al., 2022; Kumah et al., 2023).

Temperature: Fluctuations in temperature profoundly impact the physicochemical properties and reactivity of nanoparticles. Alterations in temperature regimes can influence nanoparticle stability, aggregation kinetics, and interactions with biological entities, thereby influencing their bioavailability and uptake within organisms across different ecosystems.

pH: Variations in pH levels within environmental compartments affect the surface charge and chemical reactivity of nanoparticles. Shifts in pH conditions can alter nanoparticle dissolution, aggregation, or surface modifications, thereby influencing their bioaccumulation potential and behavior in aquatic systems or soils.

Salinity: Salinity levels in aquatic environments significantly impact nanoparticle behavior. Changes in salinity affect nanoparticle aggregation, dispersion, and stability, altering their mobility and bioavailability for uptake by aquatic organisms. Salinity-induced variations can modify the fate and transport of nanoparticles in marine or estuarine ecosystems (Meier et al., 2016).

Presence of Other Substances: The coexistence of additional substances, such as organic matter, ions, or pollutants, can interact with nanoparticles, influencing their fate and interactions within ecosystems. These interactions may lead to changes in nanoparticle stability, aggregation kinetics, or bioavailability, subsequently affecting their bioaccumulation potential within organisms.

11.3.2.6 Exposure Duration and Concentration

Extended exposure to elevated concentrations of nanoparticles can instigate heightened bioaccumulation rates within organisms. Prolonged and sustained exposure to high nanoparticle concentrations often amplifies their uptake and accumulation within biological systems. This heightened accumulation is attributable to increased availability and prolonged contact duration, facilitating greater absorption and retention of nanoparticles within tissues or organs (Meng et al., 2022).

Conversely, short-term exposure to low concentrations of nanoparticles typically yields minimal effects on accumulation within organisms. Lower concentrations during brief exposure durations may not elicit significant uptake or retention due to limited availability or insufficient exposure time. Consequently, the impact on bioaccumulation under such conditions tends to be relatively subdued or negligible compared to prolonged or higher concentration exposures (Dingyuan et al., 2022).

The contrasting outcomes between prolonged exposure to elevated nanoparticle concentrations and short-term exposure to lower concentrations underscore the significance of exposure duration and concentration levels in dictating the extent of bioaccumulation within organisms. These differential responses highlight the nuanced relationship between exposure regimes and bioaccumulation dynamics, elucidating the importance of comprehensive assessments when evaluating the ecological implications and potential risks associated with nanoparticle exposure in various environmental settings (Xingchen et al., 2017).

11.4 Trophic Transfer of Nanoparticles

Trophic transfer of nanoparticles refers to their movement through different levels of the food chain, where these particles are transferred from one organism to another as they are consumed as food. This transfer mechanism can lead to the accumulation of nanoparticles in higher trophic levels. Understanding the potential for trophic transfer of nanoparticles is crucial in assessing the ecological risks associated with their presence in ecosystems. It highlights the need for comprehensive studies to evaluate the impacts of nanoparticles on different organisms and ecosystems within the context of food web dynamics (Hund-Rinke & Simon, 2006).

11.4.1 Uptake at Lower Trophic Levels

Knowledge of the mechanisms and extent of nanoparticle accumulation in primary producers is crucial for comprehending their ecological implications and potential effects on ecosystem dynamics. This knowledge aids in evaluating the pathways of nanoparticle transfer within food webs and elucidates the potential risks associated with nanoparticle exposure in natural environments. Nanoparticles possess the propensity to be ingested or absorbed by primary producers, such as algae and plankton in aquatic ecosystems, or plants in terrestrial environments. These primary producers serve as initial points of contact for nanoparticle uptake from the surrounding environment. Upon exposure, nanoparticles can enter these organisms through various routes, including direct absorption by cellular structures or uptake through physiological processes.

In aquatic ecosystems, primary producers like algae and plankton may uptake nanoparticles from the water column or sediment interfaces. These microscopic organisms, due to their high surface area-to-volume ratio and continual interaction with their surrounding environment, can accumulate nanoparticles through direct contact or through their uptake mechanisms. Similarly, in terrestrial ecosystems, plants can absorb nanoparticles from soil, water, or air. Nanoparticles may adhere to the plant surfaces or penetrate through roots, subsequently translocating within the plant tissues. These particles can be distributed throughout the plant structure, accumulating in different organs or compartments (Brar et al., 2010).

The accumulation of nanoparticles in primary producers is a result of their capacity to assimilate these nanoscale materials from their ambient surroundings. Once internalized, nanoparticles may persist within these organisms, accumulating in various tissues or cellular compartments. This accumulation in primary producers can serve as a starting point for the transfer of nanoparticles through food webs, potentially impacting higher trophic levels in aquatic and terrestrial ecosystems.

11.4.2 Transfer to Herbivores

Herbivorous organisms, by feeding on primary producers or lower-level organisms, can accumulate nanoparticles that are present within these plants or organisms. Nanoparticles, initially taken up and stored within primary producers, can subsequently be transferred along trophic levels when herbivores consume these nanoparticle-containing organisms as part of their diet.

In aquatic ecosystems, herbivorous organisms such as zooplankton or certain fish species may ingest algae or other primary producers that have accumulated nanoparticles (Arts et al., 2016). These nanoparticles, initially acquired by the primary producers, become integrated into the herbivores' tissues as they feed on these nanoparticle-containing organisms. Similarly, in terrestrial ecosystems, herbivorous insects, mammals, or other organisms that primarily feed on plants can accumulate nanoparticles present within the plant tissues. Nanoparticles absorbed or taken up by plants from the soil, air, or water sources can subsequently transfer to the herbivores as they consume these nanoparticle-bearing plants. The accumulation of nanoparticles in herbivorous organisms is a consequence of their dietary intake of primary producers or lower trophic level organisms that have incorporated nanoparticles. Upon ingestion, nanoparticles can traverse through the digestive tract or be assimilated within tissues, potentially accumulating in various organs or compartments of the herbivores (Lee et al., 2019).

This transfer and accumulation of nanoparticles through herbivory contribute to their biomagnification across trophic levels within ecosystems. Understanding the dynamics of nanoparticle transfer through food chains is crucial for assessing the potential risks and ecological implications associated with nanoparticle exposure in higher trophic levels and the overall functioning of ecosystems.

11.4.3 Further Transfer to Higher Trophic Levels

Predators, through their consumption of herbivores or other prey organisms that have accumulated nanoparticles, can also accumulate these particles within their tissues. Nanoparticles present in the prey organisms, acquired from primary producers or lower trophic levels, can transfer to the predators as they feed on nanoparticle-containing prey. This phenomenon allows for the successive transfer of nanoparticles through various levels of the food chain (Werlin et al., 2011).

In aquatic ecosystems, predatory species like larger fish or certain invertebrates can accumulate nanoparticles when they consume herbivorous organisms or prey species that have previously assimilated nanoparticles. These nanoparticles, originally acquired by the primary producers and subsequently passed through intermediate trophic levels, can then be transferred to the predators upon consumption. Similarly, in terrestrial ecosystems, predatory organisms such as carnivorous mammals, birds, or insects can accumulate nanoparticles by preying on herbivores or other organisms that have previously accumulated these particles. Nanoparticles present in the tissues of the prey organisms are transferred to the predators during the consumption of nanoparticle-laden prey (Chen, 2017; Polesel, 2018).

The accumulation of nanoparticles in predators occurs through the ingestion and assimilation of prey organisms that have already accumulated these nanoscale materials. Upon consumption, nanoparticles may be absorbed, distributed, and retained within various tissues or organs of the predators. This process of nanoparticle transfer through successive levels of the food chain results in the biomagnification of nanoparticles, leading to their increased concentrations in higher trophic level organisms. Understanding the mechanisms of nanoparticle transfer across different trophic levels is crucial for assessing the potential ecological impacts and risks associated with nanoparticle exposure in top predators and the overall dynamics of ecosystems.

11.4.4 Accumulation and Biomagnification

Nanoparticles have the potential to accumulate and even undergo biomagnification as they progress through successive trophic levels in food chains, culminating in higher concentrations within top predators compared to lower trophic levels. This phenomenon leads to an escalation in nanoparticle concentrations as they ascend the food chain (Werlin et al., 2011).

At lower trophic levels, such as primary producers or herbivorous organisms, nanoparticles may be absorbed or ingested from the environment, accumulating in their tissues. Subsequently, when these nanoparticle-laden organisms are consumed by higher trophic level predators, the accumulated nanoparticles are transferred to the predators through dietary intake. Throughout this transfer process, the uptake and retention of nanoparticles tend to exceed elimination rates in the organisms, leading to an increase in nanoparticle concentrations at each successive trophic level. Predatory organisms feeding on multiple lower trophic level prey organisms containing accumulated nanoparticles experience a cumulative buildup of these nanoscale materials within their tissues. This continuous transfer and accumulation of nanoparticles through each trophic level result in biomagnification, causing higher concentrations of nanoparticles in top predators compared to those present in lower trophic level organisms. The amplified accumulation of nanoparticles in top predators can lead to disproportionately elevated levels of these materials in organisms occupying higher positions within the food chain (Werlin et al., 2011; Bajpai et al., 2018; Weiskopf et al., 2020).

This phenomenon of biomagnification of nanoparticles highlights the potential for increased ecological risks and health implications in top predators due to elevated nanoparticle concentrations. Understanding the dynamics of nanoparticle biomagnification within food chains is crucial for evaluating the ecological consequences and potential impacts on higher trophic level organisms and ecosystems (Werlin et al., 2011).

11.4.5 Effects on Ecosystem Dynamics

Evaluating the impacts of nanoparticle accumulation on species interactions, ecological processes, and organismal health across trophic levels is crucial for understanding the potential risks and mitigating adverse effects on ecosystems and their inhabitants (Jakubiak et al., 2014; Majumdar et al., 2015; Zhu et al., 2010). The trophic transfer of nanoparticles has the potential to exert significant influences on ecosystem dynamics, species interactions, and the health of organisms across diverse trophic levels within ecosystems. This transfer of nanoparticles through successive levels of the food chain can instigate multifaceted effects with ecological implications (Gilro et al., 2014; Safari & Zarnegar, 2014; Qiuyue et al., 2020).

Ecosystem Dynamics: Nanoparticle trophic transfer can alter the balance and functioning of ecosystems. Accumulation and biomagnification of nanoparticles in higher trophic levels may disrupt energy flow, nutrient cycling, and trophic relationships, leading to potential shifts in ecosystem structure and stability (Axson et al., 2015; Briffa et al., 2018; Abdolahpur Monikh et al., 2019).

Species Interactions: Nanoparticle accumulation in organisms across trophic levels can influence species interactions within ecosystems. Elevated nanoparticle concentrations in certain organisms may affect predator-prey relationships, competition dynamics, or symbiotic interactions, potentially altering community compositions or ecological niches (Kiser et al., 2012, Louie et al., 2016).

Health Implications: The presence of accumulated nanoparticles in organisms at higher trophic levels can pose health risks. Elevated concentrations of nanoparticles may induce physiological effects, disrupt metabolic processes, or lead to toxicity in organisms, impacting their fitness, reproductive success, or susceptibility to diseases (Holbrook et al., 2008, Ferry et al., 2009, Werlin et al., 2011; Riediker et al., 2019).

The transfer of nanoparticles through trophic levels raises concerns about the potential for cascading effects on ecosystem health. Elevated nanoparticle concentrations in top predators, critical in regulating ecosystem dynamics, may render them more susceptible to adverse effects, potentially disrupting entire food webs or ecosystem functioning (Levy et al., 2011; Wang et al., 2014; Kolosnjaj-Tabi et al., 2015; Nasser et al., 2020).

11.5 Risk Assessment and Management

Environmental and regulatory challenges in risk assessment and management have become increasingly crucial in our rapidly changing world. This research theme encompasses a broad array of concerns, including but not limited to climate change, pollution control, natural disasters, and the preservation of ecosystems. Effective risk assessment and management strategies are essential for mitigating environmental harm, protecting public health, and ensuring regulatory compliance. Assessment of risk is fundamental and crucial in safeguarding the environment and humans, as we confront growing environmental risks and uncertainties, ongoing research and innovative strategies will be instrumental in developing robust risk assessment and management frameworks, ultimately contributing to a more resilient and sustainable future (Chaulya & Prasad, 2016).

11.5.1 Nanotoxicology Mitigation Strategies

The use of renewable resources such as biodegradable nanoparticles such as nano-chitosan represents a paradigm shift in the development and application of nanomaterials. It emphasizes sustainability, environmental responsibility, and safety while harnessing the unique properties of nanoscale materials for a wide range of applications (Saleh & Hassan, 2023). The research theme focuses on mitigating potential environmental and health risks associated with nanotechnology while exploring innovative directions for its sustainable future (Jun et al., 2016).

Addressing the potential risks of nanomaterials to human health and the environment is a critical aspect. Green nanotechnology promotes safe design and responsible handling of nanomaterials. Assessing the environmental and health risks associated with new nanomaterials and applications is a complex challenge, requiring robust methodologies and regulatory frameworks. Sustainability has become a paramount global concern in the face of environmental degradation, resource depletion, and climate change. Sustainable practices involve adopting strategies and behaviors that promote long-term well-being for both current and future generations while minimizing negative impacts on the planet. Sustainable practices are a crucial research theme that encompasses a wide range of topics and challenges (Fox et al., 2019). As societies around the world grapple with pressing environmental and social issues, research in this area will play a pivotal role in identifying and implementing innovative strategies for a more sustainable and resilient future. It requires interdisciplinary collaboration, visionary leadership, and a commitment to balance the needs of people, the planet, and prosperity.

11.5.2 Regulatory Considerations

Increasing in nanoparticle applications in the food industry, there are growing worries about human exposure to these engineered materials through food consumption. The rapid development and commercialization of nanotechnology have outpaced regulatory frameworks. It is important to create standard regulations on the use of nanoparticles in food production and food products to ensure safer food and the safety of consumers (Biswas et al., 2022).

Environmental and regulatory challenges in monitoring and detection methods have gained significant attention in recent years due to the escalating concerns over climate change, pollution, and the depletion of natural resources. Effective monitoring and detection methods are crucial for assessing the state of the environment, ensuring regulatory compliance, and implementing sustainable practices. Environmental monitoring encompasses a wide range of factors such as air quality, water quality, soil health, and biodiversity (Danielsen et al., 2022). Developing comprehensive detection methods that can cover this diversity is a significant challenge. As technology continues to advance, we can expect increasingly innovative solutions to address these challenges, ultimately leading to more effective environmental protection and sustainable practices.

Formulation and implementation of policies to address specific issues such as climate change mitigation and promoting public health play a crucial role in shaping the behavior of individuals, businesses, and governments, impacting a wide range of areas such as environmental protection, public health, consumer safety, and technological innovation. Understanding and addressing research gaps in this field are essential for ensuring effective governance and addressing emerging challenges. The review of policy is a vital research theme that underpins the functioning of modern societies and economies. As technology advances and global challenges evolve, ongoing research will be instrumental in shaping policies and regulations that promote the public interest, protect the environment, and foster innovation.

11.6 Conclusion

The varied absorption rates of nanoparticles in food and water imply that distinct exposure paths have distinct accumulation mechanisms and may function in distinct ways, such as influencing distinct organs. Numerous elements need to be taken into account, such as how NPs' physicochemical characteristics affect how they are transferred and biomagnified in food chains.

References

Abdolahpur Monikh, F., Chupani, L., Zusková, E., Peters, R., Vancová, M., Vijver, M. G., Porcal, P., & Peijnenburg, W. J. G. M. (2019). Method for extraction and quantification of metal-based nanoparticles in biological media: Number-based biodistribution and bioconcentration. *Environmental Science & Technology, 53*(2), 946–953. https://doi.org/10.1021/acs.est.8b03715

Agarwal, M., Murugan, M. S., Sharma, A., Rai, R., Kamboj, A., Sanjeev, H. S., & Roy, K. (2013). Nanoparticles and its toxic effects: A review. *International Journal of Current Microbiology and Applied Sciences, 2*(10), 76–82.

Arts, J. H., Irfan, M. A., Keene, A. M., Kreiling, R., Lyon, D., Maier, M., Michel, K., Neubauer, N., Petry, T., Sauer, U. G., Warheit, D., Wiench, K., Wohlleben, W., & Landsiedel, R. (2016). Case studies putting the decision-making framework for the grouping and testing of nanomaterials (DF4nanoGrouping) into practice. *Regulatory Toxicology and Pharmacology: RTP, 76*, 234–261. https://doi.org/10.1016/j.yrtph.2015.11.020

Axson, J. L., Stark, D. I., Bondy, A. L., Capracotta, S. S., Maynard, A. D., Philbert, M. A., Bergin, I. L., & Ault, A. P. (2015). Rapid kinetics of size and pH-dependent dissolution and aggregation of silver nanoparticles in simulated gastric fluid. *The Journal of Physical Chemistry. C, Nanomaterials and Interfaces, 119*(35), 20632–20641. https://doi.org/10.1021/acs.jpcc.5b03634

Baalousha, M., Lead, J., Ju-Nam, Y., & Wilderer, P. (2011). Natural colloids and manufactured-nanoparticlesinaquaticandterrestrialsystems. *Treatise Water Science, 3*, 89–129. https://doi.org/10.1016/b978-0-444-53199-5.00053-1

Bajpai, V. K., Kamle, M., Shukla, S., Mahato, D. K., Chandra, P., Hwang, S. K., Kumar, P., Huh, Y. S., & Han, Y. K. (2018). Prospects of using nanotechnology for food preservation, safety, and security. *Journal of Food and Drug Analysis, 26*(4), 1201–1214. https://doi.org/10.1016/j.jfda.2018.06.011

Barrak, H., Saied, T., Chevallier, P., Laroche, G., M'nif, A., & Hamzaoui, A. H. (2019). Synthesis, characterization, and functionalization of ZnO nanoparticles by N-(trimethoxysilylpropyl) ethylenediamine triacetic acid (TMSEDTA): Investigation of the interactions between Phloroglucinol and ZnO@TMSEDTA. *Arabian Journal of Chemistry, 12*(8), 4340–4347. https://doi.org/10.1016/j.arabjc.2016.04.019

Biswas, R., Alam, M., Sarkar, A., Haque, M. I., Hasan, M. M., & Hoque, M. (2022). Application of nanotechnology in food: Processing, preservation, packaging and safety assessment. *Heliyon, 8*(11), e11795. https://doi.org/10.1016/j.heliyon.2022.e11795

Brar, S. K., Verma, M., Tyagi, R. D., & Surampalli, R. Y. (2010). Engineered nanoparticles in wastewater and wastewater sludge – Evidence and impacts. *Waste Management, 30*(3), 504–520. https://doi.org/10.1016/j.wasman.2009.10.012

Briffa, S. M., Nasser, F., Valsami-Jones, E., & Lynch, I. (2018). Uptake and impacts of polyvinyl-pyrrolidone (PVP) capped metal oxide nanoparticles on Daphnia magna: Role of core composition and acquired corona. *Environmental Science. Nano, 5*, 1745–1756.

Bundschuh, M., Filser, J., Lüderwald, S., McKee, M. S., Metreveli, G., Schaumann, G. E., et al. (2018). Nanoparticles in the environment: Where do we come from, where do we go to? *Environmental Sciences Europe, 30*(1), 1–17. https://doi.org/10.1186/s12302-018-0132-6

Chaulya, S. K., & Prasad, G. M. (2016). *Sensing and monitoring technologies for mines and hazardous areas*. Elsevier.

Chen, Y. Z. (2017). Differential bioaccumulation patterns of nanosized and dissolved silver in a land snail Achatina fulica. *Environmental Pollution, 222*, 50–57.

Danielsen, F., Eicken, H., Funder, M., Johnson, N., Lee, O., Theilade, I., Argyriou, D., & Burgess, N. D. (2022). Community monitoring of natural resource systems and the environment. *Annual Review of Environment and Resources, 47*, 637–670. https://doi.org/10.1146/annurev-environ-012220-022325

Dingyuan, L., Wenhong, F., You, W., & Ying, W. (2022). Effect of organic matter on the trophic transfer of silver nanoparticles in an aquatic food chain. *Journal of Hazardous Materials, 438*, 129521. https://doi.org/10.1016/j.jhazmat.2022.129521

Dreaden, E. C., Alkilany, A. M., Huang, X., Murphy, C. J., & El-Sayed, M. A. (2012). The golden age: Gold nanoparticles for biomedicine. *Chemical Society Reviews, 41*(7), 2740–2779. https://doi.org/10.1039/c1cs15237h

El-Naggar, M. M., Salah, S., El-Shabaka, H. A., Abd El-Rahman, F. A. A., Khalil, M. T., & Suloma, A. (2021). Efficacy of dietary chitosan and chitosan nanoparticles supplementation on health status of Nile tilapia, Oreochromis niloticus (L.). *Aquaculture Reports, 19*, 100628. https://doi.org/10.1016/j.aqrep.2021.100628

El-Naggar, M. M., Medhat, F., & Taha, A. (2022). Applications of chitosan and chitosan nanoparticles in fish aquaculture. *Egyptian Journal Of Aquatic Biology And Fisheries, 26*(1), 23–43. https://doi.org/10.21608/ejabf.2022.213365

Ezoe, Y., Lin, C.-H., Noto, M., Watanabe, Y., & Yoshimura, K. (2002). Evolution of water chemistry in natural acidic environments in Yangmingshan, Taiwan. *Journal of Environmental Monitoring: JEM, 4*(4), 533. https://doi.org/10.1039/b201499h

Ferry, J. L., et al. (2009). Transfer of gold nanoparticles from the water column to the estuarine food web. *Nature Nanotechnology, 4*, 441–444.

Fox, M., Zuidema, C., Bauman, B., Burke, T., & Sheehan, M. (2019). Integrating public health into climate change policy and planning: State of practice update. *International Journal of Environmental Research and Public Health, 16*(18), 3232. https://doi.org/10.3390/ijerph16183232

Ganesh, M., Hemalatha, P., Peng, M. M., & Jang, H. T. (2017). One pot synthesized Li, Zr doped porous silica nanoparticle for low temperature CO2 adsorption. *Arabian Journal of Chemistry, 10*, S1501–S1505. https://doi.org/10.1016/j.arabjc.2013.04.031

Gigault, J., Grassl, B., & Lespes, G. (2012). Size characterization of the associations between carbon nanotubes and humic acids in aqueous media by asymmetrical flow field-flow fractionation combined with multi-angle light scattering. *Chemosphere, 86*, 177–182. https://doi.org/10.1016/j.chemosphere.2011.10.00

Gilro, K. D., Neretina, S., & Sanders, R. W. (2014). Behavior of gold nanoparticles in an experimental algal–zooplankton food chain. *Journal of Nanoparticle Research, 16*, 2414. https://doi.org/10.1007/s11051-014-2414-2

Guénette, S., Meissa, B., & Gascuel, D. (2014). Assessing the contribution of marine protected areas to the trophic functioning of ecosystems: A model for the banc d'arguin and the mauritanian shelf. *PLoS One, 2014*, 1–16.

Hochella, M. F., Jr., Mogk, D. W., Ranville, J., Allen, I. C., Luther, G. W., Marr, L. C., McGrail, B. P., Murayama, M., Qafoku, N. P., Rosso, K. M., Sahai, N., Schroeder, P. A., Vikesland, P., Westerhoff, P., & Yang, Y. (2019). Natural, incidental, and engineered nanomaterials and their impacts on the Earth system. *Science, 363*(6434), eaau8299. https://doi.org/10.1126/science.aau8299

Holbrook, R., Murphy, K., Morrow, J., & Cole, K. (2008). Trophic transfer of nanoparticles in a simplified food web. *Nature Nanotechnology, 13*, 352–355.

Hund-Rinke, K., & Simon, M. (2006). Ecotoxic effect of photocatalytic active nanoparticles (Ti2O) on algae and daphinds. *Environmental Science and Pollution Research, 13*, 225–232.

Ivask, A., Juganson, K., Bondarenko, O., Mortimer, M., Aruoja, V., Kasemets, K., & Kahru, A. (2014). Mechanisms of toxic action of ag, ZnO and CuO nanoparticles to selected ecotoxicological test organisms and mammalian cells *in vitro*: A comparative review. *Nanotoxicology, 8*(sup1), 57–71. https://doi.org/10.3109/17435390.2013.855831

Jakubiak, M., Giska, I., Asztemborska, M., & Bystrzejewska-Piotrowska, G. (2014). Bioaccumulation and biosorption of inorganic nanoparticles: Factors affecting the efficiency of nanoparticle my co extraction by liquid-grown mycelia of Pleurotus eryngii and Trametes versicolor. *Mycol Progress, 13*, 525–532.

Jun, Y.-S., Kim, D., & Neil, C. W. (2016). Heterogeneous nucleation and growth of nanoparticles at environmental interfaces. *Accounts of Chemical Research, 49*(9), 1681–1690. https://doi.org/10.1021/acs.accounts.6b00208

Khan, I., Saeed, K., & Khan, I. (2019). Nanoparticles: Properties, applications and toxicities. *Arabian Journal of Chemistry, 12*(7), 908–931. https://doi.org/10.1016/j.arabjc.2017.05.011

Kiser, M. A., Ladner, D. A., Hristovski, K. D., & Westerhoff, P. K. (2012). Nanomaterial transformation and association with fresh and freeze-dried wastewater activated sludge: Implications for testing protocol and environmental fate. *Environmental Science & Technology, 46*, 7046–7053.

Klaine, S. J., Alvarez, P. J. J., Batley, G. E., Fernandes, T. F., Handy, R. D., Lyon, D. Y., et al. (2008). Nanomaterials in the environment: Behavior, fate, bioavailability, and effects. *Environmental Toxicology and Chemistry, 27*(9), 1825. https://doi.org/10.1897/08-090.1

Kolosnjaj-Tabi, J., et al. (2015). The one year fate of iron oxide coated gold nanoparticles in mice. *ACS Nano, 9*, 7925–7939.

Kubo-Irie, M., Yokoyama, M., Shinkai, Y., Niki, R., Takeda, K., & Irie, M. (2016). The transfer of titanium dioxide nanoparticles from the host plant to butterfly larvae through a food chain. *Scientific Reports, 6*, 23819. https://doi.org/10.1038/srep23819

Kumah, E. A., Fopa, R. D., Harati, S., Boadu, P., Zohoori, F. V., & Pak, T. (2023). Human and environmental impacts of nanoparticles: A scoping review of the current literature. *BMC Public Health, 23*(1), 1059. https://doi.org/10.1186/s12889-023-15958-4

Lähde, A., Sæunn Gudmundsdottir, S., Joutsensaari, J., Tapper, U., Ruusunen, J., Ihalainen, M., & Gizurarson, S. (2013). In vitro evaluation of pulmonary deposition of airborne volcanic ash. *Atmospheric Environment, 70*, 18–27. https://doi.org/10.1016/j.atmosenv.2012.12.048

Laurent, S., Forge, D., Port, M., Roch, A., Robic, C., Vander Elst, L., & Muller, R. N. (2008). Magnetic iron oxide nanoparticles: Synthesis, stabilization, vectorization, physicochemical characterizations, and biological applications. *Chemical Reviews, 108*(6), 2064–2110. https://doi.org/10.1021/cr068445e

Lee, J. E., Lee, N., Kim, T., Kim, J., & Hyeon, T. (2011). Multifunctional mesoporous silica nanocomposite nanoparticles for theranostic applications. *Accounts of Chemical Research, 44*(10), 893–902. https://doi.org/10.1021/ar2000259

Lee, S.-H., Gordon, H., Yu, H., Lehtipalo, K., Haley, R., Li, Y., & Zhang, R. (2019). New particle formation in the atmosphere: From molecular clusters to global climate. *Journal of Geophysical Research Atmospheres, 124*(13), 7098–7146. https://doi.org/10.1029/2018jd029356

Levy, M., et al. (2011). Long term in vivo biotransformation of iron oxide nanoparticles. *Biomaterials, 32*, 3988–3999.

Louie, S. M., Tilton, R. D., & Lowry, G. V. (2016). Critical review: Impacts of macromolecular coatings on critical physicochemical processes controlling environmental fate of nanomaterials. *Environmental Science: Nano, 3*, 283–310.

Majumdar, S., Trujillo-Reyes, J., Hernandez-Viezcas, J. A., White, J. C., Peralta-Videa, J. R., & Gardea-Torresdey, J. L. (2015). Cerium biomagnification in a terrestrial food chain: Influence of particle size and growth stage. *Environmental Science & Technology, 50*(13), 6782–6792. https://doi.org/10.1021/acs.est.5b04784

Mansha, M., Qurashi, A., Ullah, N., Bakare, F. O., Khan, I., & Yamani, Z. H. (2016). Synthesis of In2O3/graphene heterostructure and their hydrogen gas sensing properties. *Ceramics International, 42*(9), 11490–11495. https://doi.org/10.1016/j.ceramint.2016.04.035

Meier, C., Voegelin, A., Pradas del Real, A., Sarret, G., Mueller, C. R., & Kaegi, R. (2016). Transformation of silver nanoparticles in sewage sludge during incineration. *Environmental Science & Technology, 50*(7), 3503–3510. https://doi.org/10.1021/acs.est.5b04804

Meng, L., Yongzhi, Z., Sen, F., Xuxiang, Z., Yilong, X., & Xianling, X. (2022). Bioaccumulation and biomagnification effects of nano-TiO2 in the aquatic food chain. *Ecotoxicology, 31*, 1023. https://doi.org/10.1007/s10646-022-02572-0

Moller, P., Jacobsen, N. R., Folkmann, J. K., Danielsen, P. H., & Mikkelsen, L. (2010). Role of oxidative damage in toxicity of particulates. *Free Radical Research, 44*(1), 1–46. https://doi.org/10.3109/10715760903300691

Nasser, F., Constantinou, J., & Lynch, I. (2020). Nanomaterials in the environment acquire an "eco-Corona" impacting their toxicity to Daphnia magna—A call for updating toxicity testing policies. *Proteomics, 20*, e1800412.

Nowack, B., & Bucheli, T. D. (2007). Occurrence, behavior and effects of nanoparticles in the environment. *Environmental Pollution, 150*(1), 5–22. https://doi.org/10.1016/j.envpol.2007.06.006

Petersen, E. J., Diamond, S. A., Kennedy, A. J., Goss, G. G., Ho, K., Lead, J., & Sayre, P. (2015). Adapting OECD aquatic toxicity tests for use with manufactured nanomaterials: Key issues and consensus recommendations. *Environmental Science & Technology, 49*(16), 9532–9547. https://doi.org/10.1021/acs.est.5b00997

Polesel, F. (2018). Occurrence, characterisation and fate of (nano)particulate Ti and Ag in two Norwegian wastewater treatment plants. *Water Research, 141*, 19.

Qi, Y., Zhenyan, Z. B., Fazel, A. M. A. C., Juan, W. A., Zhuang, W., Martina, G. V., Thijs, B., & Willie, J. G. M. (2022). Trophic transfer of Cu nanoparticles in a simulated aquatic food chain. *Ecotoxicology and Environmental Safety, 242*, 113920.

Qiuyue, S., Cheng, L. W., Han, Z., Chunying, C., Xian, Z., & Xue-Ling, C. (2020). Trophic transfer and biomagnification of fullerenol nanoparticles in an aquatic food chain. *Environmental Science: Nano, 7*(4), 1240–1251.

Ramzan, U., Majeed, W., Hussain, A. A., Qureshi, F., Qamar, S. U. R., Naeem, M., Uddin, J., Khan, A., Al-Harrasi, A., Razak, S. I. A., & Lee, T. Y. (2022). New insights for exploring the risks of bioaccumulation, molecular mechanisms, and cellular toxicities of AgNPs in aquatic ecosystem. *Water, 14*(14), 2192. https://doi.org/10.3390/w14142192

Rana, S., & Kalaichelvan, P. T. (2013). Ecotoxicity of nanoparticles. ISRN. *Toxicology, 2013*, 1–11. https://doi.org/10.1155/2013/574648

Riediker, M., et al. (2019). Particle toxicology and health – where are we? *Particle and Fibre Toxicology, 16*, 19.

Safari, J., & Zarnegar, Z. (2014). Advanced drug delivery systems: Nanotechnology of health design a review. *Journal of Saudi Chemical Society, 18*, 85. https://doi.org/10.1016/j.jscs.2012.12.009

Saleh, H. M., & Hassan, A. I. (2023). Synthesis and characterization of nanomaterials for application in cost-effective electrochemical devices. *Sustainability, 15*, 10891. https://doi.org/10.3390/su151410891

Semenzin, E., Lanzellotto, E., Hristozov, D., Critto, A., Zabeo, A., Giubilato, E., & Marcomini, A. (2015). Species sensitivity weighted distribution for ecological risk assessment of engineered nanomaterials: The n-TiO_2case study: Ecological risk assessment of engineered nanomaterials. *Environmental Toxicology and Chemistry, 34*(11), 2644–2659. https://doi.org/10.1002/etc.3103

Shaalan, M., Saleh, M., El-Mahdy, M., & El-Matbouli, M. (2016). Recent progress in applications of nanoparticles in fish medicine: A review. *Nanomedicine: Nanotechnology, Biology, and Medicine, 12*(3), 701–710. https://doi.org/10.1016/j.nano.2015.11.005

Shin, W.-K., Cho, J., Kannan, A. G., Lee, Y.-S., & Kim, D.-W. (2016). Cross-linked composite gel polymer electrolyte using mesoporous methacrylate-functionalized SiO2 nanoparticles for lithium-ion polymer batteries. *Scientific Reports, 6*(1), 1–10. https://doi.org/10.1038/srep26332

Strong, D. R., & Frank, K. T. (2010). Human involvement in food webs. *Annual Review of Environment and Resources, 35*, 1–23. https://doi.org/10.1146/annurev-environ-031809-133103

Tiwari, J. N., Tiwari, R. N., & Kim, K. S. (2012). Zero-dimensional, one-dimensional, two-dimensional and three-dimensional nanostructured materials for advanced electrochemical energy devices. *Progress in Materials Science, 57*(4), 724–803. https://doi.org/10.1016/j.pmatsci.2011.08.003

Von Moos, N., Bowen, P., & Slaveykova, V. I. (2014). Bio availability of inorganic nanoparticles to planktonic bacteria and aquatic microalgae in freshwater. *Environmental Science: Nano, 1*, 214–232. https://doi.org/10.1039/c3en00054k

Wang, H., et al. (2014). Toxicity, bioaccumulation, and biotransformation of silver nanoparticles in marine organisms. *Environmental Science & Technology, 48*, 13711–13717.

Wang, Z., Zhang, L., Zhao, J., & Xing, B. (2016). Environmental processes and toxicity of metallic nanoparticles in aquatic systems as affected by natural organic matter. *Environmental Science. Nano, 3*(2), 240–255. https://doi.org/10.1039/c5en00230c

Weiskopf, S. R., Rubenstein, M. A., Crozier, L. G., Gaichas, S., Griffis, R., Halofsky, J. E., Hyde, K. J. W., Morelli, T. L., Morisette, J. T., Muñoz, R. C., Pershing, A. J., Peterson, D. L., Poudel, R., Staudinger, M. D., Sutton-Grier, A. E., Thompson, L., Vose, J., Weltzin, J. F., & Whyte, K. P. (2020). Climate change affects biodiversity, ecosystems, ecosystem services, and natural resource management in the United States. *Science of the Total Environment, 733*, 137782. https://doi.org/10.1016/j.scitotenv.2020.137782

Werlin, R., et al. (2011). Biomagnification of cadmium selenide quantum dots in a simple experimental microbial food chain. *Nature Nanotechnology, 6*, 65–71.

Wu, C.-Y., Martel, J., Wong, T.-Y., Young, D., Liu, C.-C., Lin, C.-W., & Young, J. D. (2016). Formation and characteristics of biomimetic mineralo-organic particles in natural surface water. *Scientific Reports, 6*(1), 28817. https://doi.org/10.1038/srep28817

Xingchen, Z., Miao, Y., Dan, X., Aifeng, L., Xingwang, H., Fang, H., Yanmin, L., Qunfang, Z., & Guibin, J. (2017). Distribution, bioaccumulation, trophic transfer, and influences of CeO2 nanoparticles in a constructed aquatic food web. *Environmental Science & Technology, 51*(9), 5205–5214. https://doi.org/10.1021/acs.est.6b05875

Zhu, X., Wang, J., Zhang, X., Chang, Y., & Chen, Y. (2010). Trophic transfer of TiO2 nanoparticles from Daphnia to zebrafish in a simplified fresh water food chain. *Chemosphere, 79*, 928–933.

Zimmermann, S., & Sures, B. (2004). Significance of platinum group metals emitted from automobile exhaust gas converters for the biosphere. *Environmental Science and Pollution Research International, 11*(3), 194–199. https://doi.org/10.1007/bf02979675

Chapter 12
Human Health Implications of Environmental Nanoparticles

Patrick Omoregie Isibor, Samuel Adeniyi Oyegbade, Jerry Gbotemi Oni, Wopa Wurie Ahmed, Eniola Opeyemi Abiodun, Aishat Rabiu Sani, and Oluwafemi Adebayo Oyewole

12.1 Introduction

Nanoparticles are applied as components of industrial products such as paints, cosmetics, and pharmaceuticals containing NPs or requiring NPs to be produced. Also, nanomaterials have a variety of novel features, and their industrial application opens up new possibilities. However, recently, the application has been linked to new dangers and uncertainties (Keller et al., 2017). The use of nanoparticles (NPs; particle sizes 100 nm) is thought to be a fast-expanding area of research with a variety of applications in numerous industries due to intrinsic properties derived from their nanosized. By 2022, the market for NPs was predicted to be worth

The original version of the chapter has been revised. A correction to this chapter can be found at https://doi.org/10.1007/978-3-031-54154-4_17

P. O. Isibor (✉) · S. A. Oyegbade
Department of Biological Sciences, College of Science and Technology, Covenant University, Ota, Nigeria
e-mail: patrick.isibor@covenantuniversity.edu.ng

J. G. Oni · W. W. Ahmed
Department of Biochemistry, Federal University of Technology, Minna, Nigeria

E. O. Abiodun
Department of Microbiology, University of Ibadan, Oyo, Nigeria

A. R. Sani
Department of Microbiology, Ibrahim Badamasi Babangida University, Lapai, Niger State, Nigeria

O. A. Oyewole
Department of Microbiology, Federal University of Technology, Gidan Kwano Campus, Minna, Nigeria

P. O. Isibor et al. (eds.), *Environmental Nanotoxicology*,
https://doi.org/10.1007/978-3-031-54154-4_12

$25.26 billion worldwide (Keller et al., 2017). The development of agrochemicals (nanoformulations) with increased efficacy and reduced pollutant levels could be very profitable for agriculture. Additionally, tools like nanosensors would support the early detection of biotic or abiotic stress before it can impair output (Giraldo et al., 2019). According to a number of studies (Giraldo et al., 2019), nanotechnology will make it easier to construct high-tech agricultural fields and a variety of intelligent nanotools. The creation of these technologies will unquestionably result in a radical shift in agricultural methods and might significantly lessen the environmental effect of modern farming (Arora, 2018). Agriculture will receive new tools through the paradigm of nanotechnology to promote sustainable food production. Although nanoparticles (NPs) have many advantages, their harmful environmental emission problems are significant. Uncontrolled exposure to soil is expected to inhibit plant growth and build up in food tissue. Human exposure becomes therefore inevitable. One of the crucial factors to be discovered and understood is the issue of NP detection in soil and plant absorption mechanisms. It is therefore crucial to determine the exact amount of NP released into the environment, how it is accumulated in plants, and how it affects human health.

Research in this area is currently increasingly concentrated on the creation of innovative nano agrochemicals, such as the so-called "nano pesticides" and "nano fertilizers" (Kah, 2015). These materials can deliver different macromolecules selectively and with a regulated release of agrochemicals. By incorporating nanoscale transporters and chemicals, it is possible to increase the effectiveness of fertilizers and insecticides while reducing their application rates without sacrificing productivity (Parisi et al., 2015). For the advancement of the modern style of farming, the employment of nano agrochemicals looks to be crucial. Plants are the fundamental trophic level in the ecosystem, acting as the primary source of supply and the food chain's main component. According to Xiong et al. (2017), bioaccumulation and prolonged exposure to NPs in plants may affect the food chain. According to a number of studies, crop plants' edible tissues can absorb and store NPs. NPs, the initial structure of these accumulations, or metal ions would both manifest. Physicochemical characteristics could, however, vary depending on the situation (Cota-Ruiz et al., 2018). To accomplish this, the accumulation of NPs has the capacity to change plant physiological processes, have an effect on the structure of cellular and subcellular organelles, and change the composition of proteins, lipids, and nucleic acids by generating hydroxyl radicals. It can be concluded, in summary, that the widespread use of these NPs raises a variety of questions from an ecological, ethical, health, and safety perspective. But up until now, there hasn't been much evidence to support the claimed harmful effects of NPs on human health. With the creation of several NPs, nanotechnology is becoming more and more well-liked, giving the agriculture industry new capabilities. We must learn more and understand more about the specific benefits and drawbacks of using NPs, and we must do it immediately. The development of nanotechnologies has resulted in notable levels of produced NPs in the agro-environment. Despite the many advantages of this technique, researchers and industry professionals are concerned about the harmful disposal of different NPs in significant quantities (a few hundred tons annually). According to contemporary normative texts, NPs can be found in a variety of controlled objects, including soils, water objects, soil microorganisms, algae, fungi, land plant and animal tissues, and atmospheric air (Gupta & Xie, 2018).

It has been observed that even inert NPs can interact with biota through physical pathways like biological surface coating (Bhat et al., 2022; Thu et al., 2023; Armstead & Li, 2016; Yah et al., 2012; De Matteis & Rinaldi, 2018; Daigle et al., 2003; De Matteis & Rinaldi, 2018). The presence of NPs in the environment and its steady increase in production raise questions about the interactions of NPs with biota and even humans in a complex environment. Notably, nanoplastics are generated by widely used plastic products in the modern day. These nanoparticles are found in the environment where they are major pollutants and a cause for global concern due to increasing human exposure and harm (Bhat et al., 2022; Thu et al., 2023; Armstead & Li, 2016; Yah et al., 2012; De Matteis & Rinaldi, 2018; Daigle et al., 2003).

Environmental nanoparticles can pose human health risks when inhaled, ingested, or absorbed through the skin. Their small size allows them to enter the bloodstream, potentially reaching various organs and tissues. This raises concerns about inflammation, oxidative stress, and potential long-term effects, emphasizing the need for thorough risk assessment and mitigation strategies to protect public health. This chapter covers the interactions of environmental particles with human beings. We provide insights into exposure pathways of NPs in the human body, mechanisms of nanotoxicity in humans, and potential health impacts. We also study notable events and epidemiological consequences of exposure to environmental nanoparticles. Furthermore, we take into consideration the risk management of nanoparticle exposure and explore how we can safeguard public health through mitigation strategies and policy (Ermolin & Fedotov, 2016).

12.2 Natural and Anthropogenic Sources of Environmental Nanoparticles

Nanoparticles can be obtained from different sources or means. It could occur or be found in nature or produced by man on a commercial basis or gotten as a by-product within any deliberate attempt or intention. In recent times, naturally occurring nanoparticles as well as nanoparticles of anthropogenic origin have been widely found in the biosphere (Hochella et al., 2019). Environmental nanoparticles gain access into and interact with the body of humans and consequently have significant health effects. Some nanoparticles considered as incidental are obtained from anthropogenic sources such as wastes (industry-based, mining, and automobile) and chemical processes (corrosion and combustion). These nanoparticles can pose threat to exposed health of humans (Baalousha et al., 2016). Natural and man-made nanomaterials usually undergo transformation before they are released into the environment. There are numerous pathways through which these interactions occur in diverse environmental conditions and climate. Urban areas or communities have greater likelihood of occurrence of nanoparticles that occur prevalently in the air space. Nanoparticle accumulation often results to nanoparticles finding their routes to the environment as diverse sources of water and wastewater and soil-based sources within landfills, natural phenomenon processes, and industrial discharges (Baalousha et al., 2016; Westmeier et al., 2018; Malakar & Snow, 2020).

Natural sources of environmental nanoparticles are abundant in nature and in the biosphere. They could evolve in the form of precipitation and get to the sources of water. They are also likely to occur to soil microbiome and suspended for prolonged period of time as dust. Furthermore, sewage emanating from landfills, discharge waste from factories or automobiles, and waste from wastewater amenities often get to the soil and natural aquatic sources by discharge point sources or precipitation, thereby extending the risk of toxicity (Baalousha et al., 2016). Nanoparticles generated for commercial purposes can find their route into the environment at various stages of production or in the form of waste product at the termination of their life cycle. These environmental nanoparticles could stay in soils waiting to be absorbed in food products of plant or animal origin or move gradually to the groundwater or environment's surface. Environmental nanoparticles are primarily distributed into the natural environment by wastewater.

The lack of appropriate tools to reliably analyze and measure natural nanomaterials in the composite natural environment is often responsible for the lack of attention or consideration associated with the fate and potential toxicity of natural nanomaterials (Malakar & Snow, 2020). The dynamic evolution of the earth is also responsible for the inability to understand the environmental impact of natural nanomaterials (Hochella et al., 2019). The development of nanogeosciences via modern tools and facilities has facilitated the understanding of the diverse roles of natural nanomaterials as well as their fate, movement, and toxicity in the ecosystem.

Natural nanomaterials can be abundantly found in the air as a result of release from volcano eruptions or lightning (Griffin et al., 2018). SiO_2 nanoparticles generated as a result of volcano eruptions get into the atmosphere, thereby leading to irritation of the eye (Lungu et al., 2015). Naturally occurring mercury nanoparticles can be found in the air and then easily get through to soil and water sources (Ghoshdastidar & Ariya, 2019). Airborne nanomaterials can also get attached to fungal spores, thereby making its exposure to humans more penetrating (Westmeier et al., 2018). The gastrointestinal tract has a relatively large surface area that enables it to facilitate nanoparticle adsorption of nanoparticles in the course of ingestion and translocation (De Matteis, 2017). Ingested nanoparticles can get dissolved in the stomach and can result in further toxic impacts (Sukhanova et al., 2018). The rate of adsorption is largely influenced by nanoparticle size. As soon as nanoparticles find their way into the bloodstream, it undergoes processes just like in the exposure pathway via inhalation to cause damage to the systems and organs through the creation of ROS (De Matteis, 2017).

12.3 Human Exposure Pathways to Nanoparticles

Industrial revolution and nanotechnology also brought about higher exposure of human beings to nano-sized particles and materials (Fig. 12.1). The routes by which humans get exposed to nanoparticles include dermal, ingestion, inhalation, and drug delivery methods (Al-Mutairi et al., 2013). Human beings could get exposed to nanoparticles via complicated pathways such as skin penetration, ingestion, injection, and inhalation.

As nanoparticles are present in all environmental media including air, soil, and water, human beings are exposed to environmental nanoparticles by merely existing. Nanometer-sized foreign particles can easily make their way into the human body through several exposure pathways causing severe nanotoxicity in humans (Yah et al., 2012; Armstead & Li, 2016; De Matteis & Rinaldi, 2018; De Matteis & Rinaldi, 2018; Thu et al., 2023). The pathway of exposure to various nanoparticles from food, water, and water is carefully considered in this review. The fate, retention time, and life cycle of nanoparticles can be totally different from person to person and are largely influenced by the physicochemical properties as well as the pathway of nanoparticle exposure (Fig. 12.2). The excessive application of environmental nanoparticles has raised a lot of concerns about their toxicity, movement, life cycle, and fate when they come into contact with humans or the environment (Bour et al., 2015; Ma et al., 2016; Goswami et al., 2017; Arvidsson et al., 2018).

Diverse pathways of exposure such as ingestion, injection, penetration of skin, and inhalation often permit access of foreign nanoparticles into the human body. The minute size of nanoparticles makes them have adverse impacts on living cells. This often results in cellular death (Type 3, Type 2, and Type 1, respectively) (Ganguly et al., 2018).

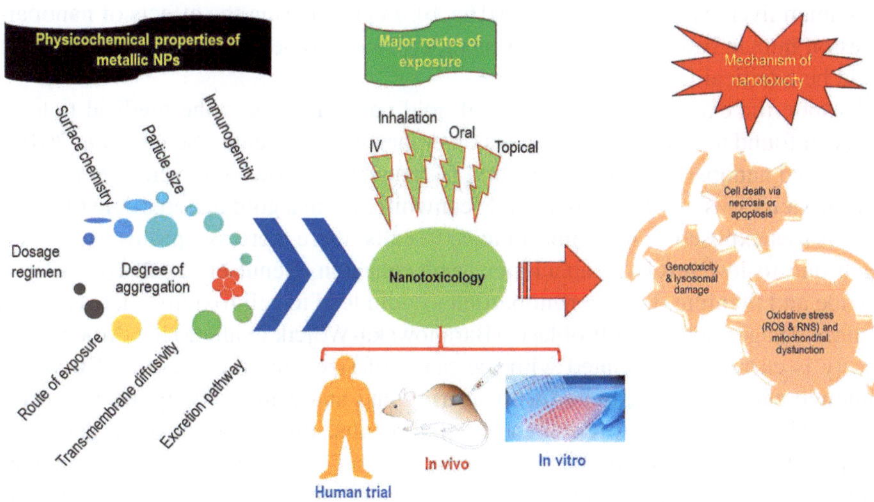

Fig. 12.1 Nanotoxicology of NPs in humans. (Thu et al., 2023)

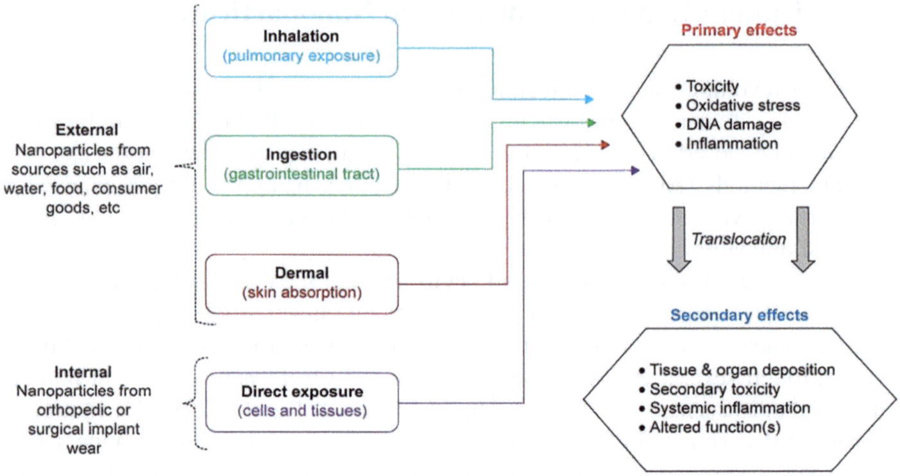

Fig. 12.2 Routes and potential detrimental effects of nanoparticle exposure. (Armstead & Li, 2016)

Chemical composition, size, and other physicochemical properties are largely responsible for nanoparticle toxicity in human beings. Cell penetration is considered easy with nanoparticles <100 nm, nanoparticles with less than 40 nm can penetrate the nucleus, and the brain-brain carrier can be bypassed with nanoparticles less than 35 nm (Ganguly et al., 2018; Malakar & Snow, 2020). Diverse research findings have affirmed that the property critically responsible for toxicity of nanoparticles in humans is size. One such study was on the cytotoxicity and genotoxicity of silver nanoparticles in the liver of humans. Observations from this study showed that the smaller the size of the nanoparticles, the more toxic the effects on the human liver (Sahu et al., 2014, 2016a, b). In considering the effects of nanoparticles on human health, the shape of nanoparticles also needs to be considered as it could be a critical factor.

Despite the extensive applications of gold nanoparticles in the medical field, it has been found to have adverse effects on embryonic stem cells (Senut et al., 2016). When gold nanoparticles between 4 and 14 nm were exposed to the embryonic stem cells, toxicity was not detected at all. Meanwhile, when a gold nanoparticle of about 1.5 nm was exposed to the embryonic stem cells, there were symptoms of cellular death due to low levels of attachment and cohesion (Senut et al., 2016). Silicon dioxide and titanium dioxide nanoparticles could lead to cell toxicity such as found in macrophages and 3T3 fibroblasts (Baranowska-Wójcik et al., 2020). Much complexity is currently associated with the range of sizes and their levels of toxicity. Although standard procedures for measuring and analyzing toxicity are not available to allow scientists to compare diverse results, there is however a generally accepted fact among researchers suggesting that a decrease in the size of nanoparticles leads to increased toxicity and that size is mainly responsible for the increase in toxicity levels. The surface charge, shape, and size of nanoparticles are considered to increase the rate of translocation through cell membranes by 60 times (Nangia & Sureshkumar, 2012).

Most nanoparticles are likely to undergo solubilization within the human cell's pH range (Sajid et al., 2015). Nanoparticles get to the lungs faster, because there is no interception within the nasopharyngeal region. This facilitates the retention of nanoparticles in the body for an extended period of time (Gupta & Xie, 2018). As soon as nanoparticles pass through the blood-air-tissue barrier in the lungs, they gain entrance in the bloodstream and inflict damage on other organs of the body. Inhalation of nanoparticle doses could induce thrombosis, vascular dysfunction, and myocardial ischemia (Miller et al., 2017). The nanoparticles exposed to the body via inhalation can be retained for a few months or eventually excreted as urine (Miller et al., 2017). After inhalation, nanoparticles are translocated to various body parts. Nanoparticles with sizes less than 10 nm are however associated with a higher level of translocation compared to relatively larger nanosizes (Miller et al., 2017). This research is crucial as soil contaminants and their sources are strongly linked to human health (Khan et al., 2010). The usage of nanotechnology in a variety of industries, such as agriculture, natural resources, and medicine, has an influence on the environment by uncontrollably releasing NPs into it. Food crops should be considered for nano-toxicity because NPs can have negative effects on crop physiology (particularly reducing photosynthesis) and human health (Xiong et al., 2017). How NPs interact with one another inside the human body is one of the crucial questions that has to be addressed. The accumulation of NPs that degrade slowly or cannot degrade in the organs is quite concerning. Phagocytes, the cells in charge of consuming and eliminating foreign matter, may get overloaded with NPs, leading to stress reactions. These reactions eventually result in inflammation and lowered defenses against subsequent infections. When exposed to tissue and body fluids, NPs can also take up a range of macromolecules onto their surfaces. The regulation mechanisms of enzymes and other proteins may be impacted by this. According to Herzog et al. (2007), NPs can also be harmful due to their chemical compositions, surface charge, surface shape, solubility, aggregation, and functional groups.

The formation of free radicals that can cause cell damage can be facilitated by the unhindered access of NPs into the body through the skin, lungs, and digestive systems. The possibility of NPs crossing the blood-brain barrier once they enter the bloodstream also raises concerns (Thomas et al., 2013). In 2013, Thomas and colleagues investigated the impact of NPs on both the environment and human health. Thomas et al. (2013) examined all potential exposure risks and analyzed the risks of being exposed to NP through multiple channels, such as dietary intake, skin absorption, inhalation, and drug delivery designed by NPs. The authors argued that it is important to consider the toxic effects of NP throughout the life cycle of a product, from manufacturing to use to disposal. This prevents harmful effects on human health that can be caused directly or through the food chain. Kirchner et al. (2005) reported some major factors that contribute to the toxicity of nanoparticles in contact with living cells: the chemical toxicity of the raw materials used for the manufacture of NPs and the inherent toxicity of NPs themselves. For example, Cd^{2+} emits cadmium selenide NP, which, due to its small size, can attach to the cell membrane and penetrate the cell. Even chemically inert NPs, which do not decompose or interact with other matrix components, can hinder cell function when they are connected to membranes and stored within cell form. The authors discussed many ways in which NPs could enter the human body, how they could be used in daily life, and how NPs manifest their toxicity. In addition, they explored the possibility of NP entering air, soil, and aquatic environments (Gupta & Xie, 2018).

Being in close proximity to people or the environment could be dangerous for NPs. The lungs of rats can be harmed by consuming Fe and TiO_2 NPs (Fraser et al., 2011). This has been shown in a variety of studies. According to research by Shaw and Handy (2011), fullerene and TiO_2 NPs can also harm the brains of fish and dogs. The cytotoxic and genotoxic effects of Ag and TiO_2 NPs are also seen in mouse and even in human cell cultures. The effects generate reactive oxygen species (ROS), which lead to DNA damage and cell proliferation (Rajput et al., 2020a).

According to Rajput et al. (2020b), ZnO and CuO NPs have toxic effects on human health when ingested by plants in the soil. These NPs may be exposed to chemical and biological interactions, have an impact on biological nitrogen fixation, and cause damage to plant cells. As a result, it may also cause serious harm to human health. Other research has shown that CuO-based NPs have fatal effects on human blood lymphocyte mitochondria and lysosomes and can cause oxidative stress in breast cancer cells in the MDA-MB 231 cell line. In order to address these problems, the authors recommend a safety evaluation and toxicological risk assessment study involving the path of exposure and the safe concentrations of ZnO and CuONPs (Rajput et al., 2020a).

12.3.1 Inhalation

Inhalation is the main pathway through which human beings are exposed to nanoparticles. Exposure of humans to airborne nanoparticles through inhalation causes severe health consequences such as diverse cancer cases prevalent in urban centers where humans are more exposed to suspended nanoparticles in the atmosphere (Malakar & Snow, 2020).

The respiratory system is responsible for gaseous exchange between the human body and its surroundings. It is one of the most vital systems in the human body, serving as an important entry point for nanomaterials into the human body and representing the most common route of NP exposure. As humans breathe in air, NPs in the form of aerosols and powders can be easily inhaled into the respiratory tract (Yah et al., 2012; Thu et al., 2023; De Matteis & Rinaldi, 2018; Daigle et al., 2003). Once inhaled, particles are transported by the electrostatic force of the air from the upper respiratory tract to the lower respiratory tract. Larger NPs (>5 μm) are usually trapped by mucus in the upper respiratory tract and excreted via the body's lymphatic system. On the other hand, smaller NPs escape from the mucociliary system and penetrate deep into the lung's alveoli where they are deposited in the alveolar epithelium or absorbed into the bloodstream (Thu et al., 2023; De Matteis & Rinaldi,

2018; Daigle et al., 2003; De Matteis & Rinaldi, 2018). In addition to the smaller-sized NPs, faster breathing rate, oral breathing, and existing lung disease can cause an increase in the penetration of NPs and deposition deep in the lungs (Daigle et al., 2003; De Matteis & Rinaldi, 2018).

Nanoparticle exposure has been connected to different respiratory conditions including fibrosis, inflammation, lung cancer, and oxidative stress, which cause serious effects on human health. The effects of nanoparticles exposure greatly depend on physicochemical properties of the exposure dose and particles. Biological models and markers have shown toxicity effects of nanoparticles such as carbon nanotubes, zinc oxide, titanium dioxide, silicon dioxide, and silver which include cell death, DNA damage, apoptosis, production of oxidative stress, and inflammatory responses (Kumah et al., 2023).

12.3.2 Skin Absorption

The human skin is the largest organ in the human body; hence, it serves as an interface between the human body and its environment. It protects the body against interference and works in temperature regulation, sensation, and more. While the skin is an effective barrier, the existence of hair follicles, sweat glands, and alterations in the skin such as wounds can serve as susceptible points and exposure routes for NP penetration (Sufian et al., 2017; Yah et al., 2012). The skin comes in contact with solid nanoparticles through direct application of lotions, creams, or other substances containing NPs. Zinc oxide and titanium dioxide NPs are commonly used in typically applied cosmetics like antiperspirants (Fu et al., 2014). Nanomaterials below 5–7 nm can be absorbed via aqueous pores or the lipidic trans-epidermal and trans-follicular routes (Baroli, 2010). In addition, during the application process of such cosmetics to the skin, there can be further exposure to NPs via inhalation (Geiser et al., 2008). When absorbed by the skin, nanoparticle distribution and toxicity are minimal and limited to the uppermost layers of skin near the hair follicles, leading to irritation and inflammation of skin cells (Yah et al., 2012).

Another route of NPs is the parenteral route or intravenous administration. NPs are often applied in drug delivery and nanotechnology-based medicines such as vaccines. This route involves complex nanoparticle movement through lymph and blood (Yah et al., 2012). Injected nanoparticles are retained for a longer time in the body and distributed to other organs (Khan et al., 2015). The rate at which the nanoparticles get into the bloodstream is largely determined by their size. Nanoparticles with small sizes can gain access in many cases and cause damage to organs of the body. Several findings have affirmed that nanoscale size facilitates easy bypass of the blood-brain barrier, which could result in neurotoxicity. It also facilitates movement of nanoparticles in the bloodstream and by extension damage the kidney, lungs, and other vital organs (Wang et al., 2013, 2018; Cai et al., 2019; Boyes & van Thriel, 2020).

Nanoparticles have found diverse applications commercially as they are now present in thousands of products in diverse sectors of the economy (Huang et al., 2017). Environmental nanoparticles such as titanium, zinc, and silver nanoparticles are prevalent in baby and adult care products in the varying proportions, and they can get into the human system via penetration through the skin (De Matteis, 2017; Ding et al., 2018).

Metal-based particles bypass the dermal layer and get through the basal strata (De Matteis, 2017). The pores of the skin are significantly small, thereby facilitating easy crossover (Larese Filon et al., 2015). Iron oxide nanoparticles find application in paint and jewelry polish production, and silver nanoparticles are applied for commercial production of antibacterial products and deodorants in the textile sector, while in the food sector, zinc oxide nanoparticles are prominently used. These applications apparently can predispose humans to exposure to these particles (Huang et al., 2017; Mantecca et al., 2017).

12.3.3 Oral Exposure Route

Nanoparticles can also enter the human body via the gastrointestinal route (McClements and Xiao, 2017; Gupta and Xie, 2018). Some NPs are naturally present in water, while other inorganic engineered NPs may be added to drinking water during the water filtration process (Simeonidis et al., 2016, 2018). Furthermore, NPs may get into food when they are added as food additives, produced during thermal food processing, migrate from food package materials or accumulate in food chains. Nanoparticle food additives may be added directly to food to improve the color or texture of food. For example, silica NPs are added as anticoagulants in instant coffee or milk powder, while titanium dioxide NPs are added to enhance the white color in candy. Furthermore, food packaging that degrades due to mechanical wear, hydrolysis, or photodegradation can release nanoparticles into the food that humans ingest. In the food chain, NPs may accumulate in plants when they absorb environmental nanoparticles in the soil and water around them (Chen et al., 2023). Human ingestion of NPs can occur from unintentional hand-to-mouth transfer of nanomaterials or by ingestion through food, drinking water, cosmetics, drugs, and drug delivery devices (Stern & McNeil, 2007; Yah et al., 2012). Once ingested, NPs are taken up across the gastrointestinal tract and absorbed by the lymph nodes, liver, and spleen. Most NPs in the GI tract are eliminated; however, small NPs not eliminated quickly are translocated into blood capillaries and move to other parts of the body (Yah et al., 2012; De Matteis & Rinaldi, 2018).

The direct injection of nanoparticles applied in drug delivery in the biomedical field has proven to be effective in combating but not without undesirable effects (Korin et al., 2013; Sykes et al., 2014; Huang et al., 2017; Wang et al., 2018; Hashem and Sallam, 2020). Increased food production as a result of the rise in the application of nano-based production such as pesticides and nanobiofertilizers has resulted in bioaccumulation of nanoparticles in crops and soil media, and by implication, humans get exposed to them at the point of application or ingestion (Iavicoli et al., 2017). The toxicological effects are still clear as regards the fate and movement of nanoparticles from the soil through various stages and processes before find its way into end food products such as milk and meat and then taken in by humans and animals (Malakar et al., 2019; Malakar & Snow, 2020).

12.4 Mechanisms of Nanoparticle Toxicity

The entrance of nanoparticles into the human body occurs through any of the aforementioned exposure pathways, and NPs may proceed from their initial site of administration to other locations in the body (Hagens et al., 2007). As a result of their large surface area, on exposure to tissue and fluids, NPs adsorb the macromolecules they encounter onto their surface. They cross biological membranes via simple diffusion or passive transport, and once in systemic circulation, NPs are free to transverse body cells by persorption and interact with local tissues (Oberdörster et al., 2005). NPs are distributed toward organs and tissues including the liver, kidney, spleen, heart, and brain. In human cells, NP toxicity may be observed as oxidative stress, inflammatory cytokine production, and cell death. They may also be eliminated through feces and urine. Inhaled NPs are cleared by the mucociliary system in the airways into the GI tract, while NPs in the blood are cleared by the liver and bile into the small intestine. If not excreted, NPs may remain in the body, causing further toxic effects (Gupta, 2023).

12.4.1 Toxicokinetics of Nanoparticles

Toxicokinetics refers to the study of how substances move through the body and their effects on biological systems. In the case of nanoparticles, their toxicokinetics involves understanding how these tiny particles are absorbed, distributed, metabolized, and excreted within the body and how these processes influence their toxicity (Bhat et al., 2022; Thu et al., 2023).

(i) Absorption: Nanoparticles can enter the body through inhalation, ingestion, or dermal contact. Once inside, their small size allows them to pass through biological barriers such as the respiratory tract, gastrointestinal tract, or skin, potentially reaching systemic circulation.
(ii) Distribution: Nanoparticles, due to their size, shape, and surface properties, can distribute throughout the body. They may accumulate in specific organs or tissues, influenced by factors like their surface chemistry, charge, and interactions with biological components.

(iii) Metabolism: Some nanoparticles might undergo metabolic changes within the body. For instance, they can be metabolized by cells or enzymes, leading to alterations in their properties or reactivity.
(iv) Excretion: The elimination of nanoparticles can occur through various routes such as exhalation, urine, feces, or through processes like shedding of skin cells. However, the elimination pathways for nanoparticles are not always well-defined and can vary based on their physicochemical properties.

Understanding the toxicokinetics of nanoparticles is crucial in assessing their potential toxicity and health implications (Hartland et al., 2013). Factors such as nanoparticle size, shape, surface charge, coating, and chemical composition influence their behavior within the body and their interaction with biological systems. Studying these factors aids in determining the potential risks associated with exposure to nanoparticles and guides the development of safety guidelines and regulations.

12.4.2 Health Effects of Nanoparticles

12.4.2.1 Oxidative Stress and Inflammation

Owing to the unique physicochemical properties of nanoparticles, they gain entrance into cells through various mechanisms. A common way through which nanoparticles enter the cell is endocytosis; during this process, the cells engulf nanoparticles as a common route of entering of particles. Factors such as particle size, shape, and surface charge influence cellular uptake. Nanoparticles can move across several cellular compartment once they are inside. The particles interact with organelles and other biological components inside the cells, which may have an impact on how effectively the cells function. They could affect the nucleus, mitochondria, and endoplasmic reticulum, altering cellular functions. Cell survival and function may be impacted by these effects, which might include modifications to cell signaling pathways and disruptions of cellular homeostasis (Strong & Frank, 2010).

After exposure, once NPs are in contact with or enter a cell, they trigger biological responses (Schirmer, 2014). The toxicity of nanoparticles has been observed in microcellular levels to whole organ systems across a variety of living organisms including humans. However, how NPs behave in the human body is still an unresolved question. The toxicity of NPs can be determined by their composition like bulk materials. However, there are additional physicochemical properties such as size, surface chemistry, shape, protein absorption gradient, and surface texture that influence the toxicity of NPs. Other factors such as pre-exposure, dose and exposure time, and environmental factors ultimately influence NP behavior and toxicity in humans (Fig. 12.3).

In humans, nanotoxicity is induced by a variety of complex mechanisms. The main processes of nanotoxicity are binding to the cell exterior, dissolution, and

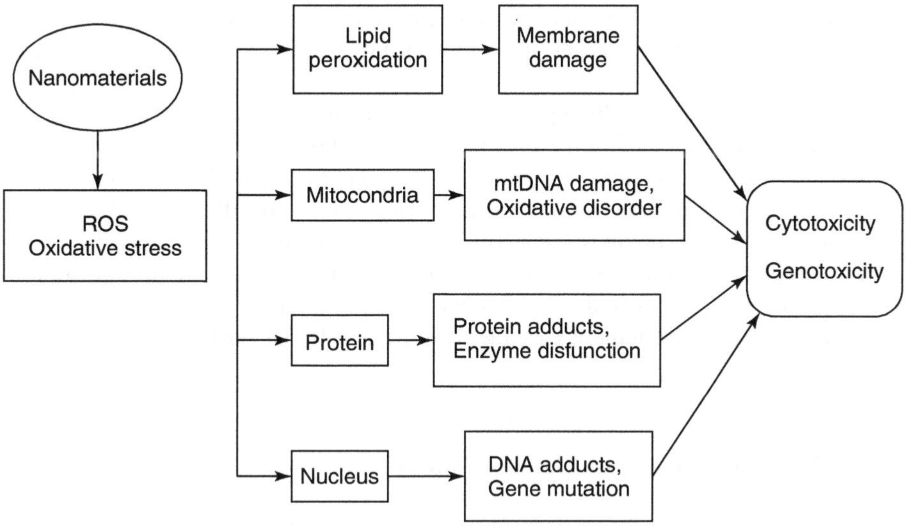

Fig. 12.3 NP-induced toxicity mediated by ROS generation in cells. (Fu et al., 2014)

generation of reactive oxygen species (ROS). When stuck to cells, NPs can directly disrupt cell membranes leading to cell death. NPs dissolved in liquid may also release toxic ions, which interact with cells causing harmful effects (Gupta, 2023). The most important mechanism of nanotoxicity is the generation of ROS and oxidative stress (Gupta, 2023; Gangwar et al., 2020; Jain & Patel, 2021).

In the mitochondria, ATP is created in a series of chemical redox reactions. In these reactions, a fraction of oxygen molecules, which are not converted, form molecules of superoxide anion radicals and other oxygen-related radicals. Collectively known as ROS, these molecules are byproducts of cellular metabolism and in controlled amounts are useful in cellular signaling. Apart from cellular metabolism, ROS are also generated in other biological reactions in the human body. Certain NPs can participate in biological reactions that generate ROS leading to excess ROS production. ROS generation in the human body serves as a precursor for other types of toxicity including inflammation, genotoxicity, and immunotoxicity. The overproduction of ROS leads to cellular oxidative stress and in turn modification of cellular proteins, lipid peroxidation, DNA damage, alterations in gene expression, signal transduction of inflammatory response, genotoxic effects, and ultimately cell death (Fu et al., 2014; Singh et al., 2020).

One significant consequence of nanoparticle exposure is the generation of reactive oxygen species. These highly reactive molecules can cause oxidative stress by disrupting the delicate balance of cellular redox systems. Nanoparticles, through direct interactions or by stimulating cellular processes, may induce the production of reactive oxygen species, triggering oxidative stress. Nanoparticles have been linked to the activation of inflammatory pathways. The inflammatory response, while essential for defending the body against harmful stimuli, can become chronic when triggered by persistent nanoparticle exposure. Chronic inflammation has been associated with various diseases, including cardiovascular disorders and respiratory ailments.

The immune system, which protects humans against disease, is an intricate system able to differentiate foreign invaders from the body's normal residents. When cells recognize a pathogen or potential NPs, signaling cytokines are secreted to attract more cells and destroy the invader. This results in inflammation, a body's defense mechanism involving several immune regulatory molecules (Schirmer, 2014). Single and multi-walled carbon nanotubes can increase the transcription of pro-inflammatory genes and trigger innate immune system responses. Other NPs may activate the complement cascade, resulting in hypersensitivity reactions and anaphylaxis (Gupta, 2023).

12.4.2.2 Genotoxicity and Carcinogenicity

Nanoparticles can also cause severe genotoxicity and DNA damage through direct interaction with genetic material or indirectly via interactions with nuclear proteins (Magdolénová et al., 2013). NPs like quantum dots can exploit the cell's active transport machinery to enter the nucleus via the nuclear pore complex. Other NPs can enter the nucleus during cell division when the nuclear envelope naturally breaks down and cytoplasmic and nuclear compartments mix. The translocation of NPs into the nucleus results in the inhibition of gene expression and the formation of protein aggregates (Magdolénová et al., 2013; Schirmer, 2014).

Genotoxicity encompasses the capacity of substances or agents to harm genetic material, primarily by causing damage to DNA. Nanoparticles, due to their small size and unique properties, have demonstrated the potential to induce genotoxic effects, initiating damage to the genetic material within cells. This capability raises significant concerns about the integrity of the genetic code, as unrepaired DNA damage can lead to mutations and chromosomal aberrations. The ability of nanoparticles to induce DNA damage is a subject of extensive research and scrutiny. Studies have shown that certain nanoparticles, based on their size, surface characteristics, composition, and reactivity, can interfere with DNA integrity, triggering various forms of damage such as strand breaks, base modifications, or oxidative stress-related DNA lesions. The consequences of DNA damage caused by nanoparticles are substantial. Unrepaired or inadequately repaired DNA lesions can lead to mutations, genomic instability, or chromosomal aberrations within cells. These genetic alterations hold implications for cellular function, potentially contributing to the initiation or progression of diseases, including cancer (Armstead & Li, 2016; Yah et al., 2012).

Indeed, research suggests a plausible link between nanoparticle exposure and carcinogenesis, the process by which normal cells transform into cancerous cells. Studies, particularly in animal models exposed to certain nanoparticles over prolonged periods, have demonstrated an increased incidence of cancer development. These findings raise concerns about the potential role of nanoparticles in promoting or exacerbating carcinogenic processes within the body. Moreover, epidemiological data is being analyzed to understand potential associations between human exposure to nanoparticles and cancer development. Although the research is ongoing, these investigations aim to evaluate the potential risks associated with human exposure to nanoparticles and their implications for health outcomes, particularly regarding cancer incidence and progression (Thu et al., 2023).

The examination of nanoparticle-induced genotoxicity and its potential contribution to carcinogenesis highlights the importance of comprehensive safety assessments and regulatory oversight. Understanding the mechanisms underlying nanoparticle-induced DNA damage is crucial for evaluating the risks associated with nanoparticle exposure and developing guidelines to mitigate potential adverse health effects. This knowledge informs the safe design, utilization, and regulation of nanoparticles in various applications to safeguard human health (Phillips & Arlt, 2009).

12.4.2.3 Immune System Response

A total comprehension of the interaction between nanoparticles and the immune system holds paramount importance in the evaluation of their biocompatibility, prospective immunotoxicity, and overarching safety profile. This profound insight serves as a guiding principle in the development of nanoparticles engineered to evoke diminished immunogenicity and heightened biocompatibility. These advancements cater to diverse applications encompassing drug delivery, diagnostic modalities, and biomedical devices.

Nanoparticles have the capacity to elicit multifaceted reactions within the immune system, which are determined by their distinct physicochemical attributes, including size, morphology, surface charge, and elemental composition. The major aspects of immune system responses to nanoparticles include innate immune response, inflammatory reactions, phagocytosis, immune cell activation, adaptive immune response, immunomodulation, and hypersensitivity reactions.

Innate Immune Response Nanoparticles possess an inherent ability to interact with vital cellular constituents intrinsic to the innate immune system, notably macrophages, dendritic cells, and natural killer cells. These interactions have the potential to initiate a plethora of immune reactions, encompassing the instigation of inflammatory signaling pathways, the release of cytokines, and the stimulation of immune cell activation. The resultant immune responses are contingent upon the unique characteristics inherent to the nanoparticles, spanning their physical and chemical attributes, surface topography, and compositional nuances. These distinctive nanoparticle features play a pivotal role in dictating the nature and magnitude of the elicited immune response, thereby influencing the overall biological outcome (De Matteis & Rinaldi, 2018). This understanding is crucial for establishing safety guidelines and regulatory frameworks to minimize potential risks associated with nanoparticle exposure, particularly concerning immune-related health outcomes.

Additionally, it guides the development of nanoparticles for therapeutic or diagnostic purposes, ensuring their safe and effective utilization without adverse immune effects. Efforts to elucidate the mechanisms underlying nanoparticle-induced immune modulation are essential for advancing our understanding of nanoparticle-immune interactions and their implications for human health (Ray et al., 2021).

12.4.2.4 Neurotoxicity and Behavioral Effects

Nanoparticles' ability to breach the blood-brain barrier, the protective membrane that shields the brain from potentially harmful substances circulating in the bloodstream, poses significant concerns regarding their impact on the central nervous system (CNS). This particular capability raises alarms due to the brain's heightened sensitivity to external influences, including foreign particles.

The potential for nanoparticles to induce neurotoxic effects stems from their ability to access and interact with neural tissues. Studies have shown that certain nanoparticles, based on their size, surface properties, and composition, can traverse the blood-brain barrier, gaining entry into the CNS. Once inside, these particles may interact with neural cells, disrupting normal cellular functions and potentially leading to neurotoxicity. Experimental models have demonstrated associations between nanoparticle exposure and alterations in behavior. Researchers have observed changes in behavioral patterns, including cognitive impairments, modifications in locomotor activity, and alterations in anxiety-like behaviors in animals exposed to nanoparticles. These behavioral changes suggest a potential impact on neurological function and raise concerns about the long-term consequences of nanoparticle exposure, especially regarding brain health (De Matteis & Rinaldi, 2018).

Understanding these behavioral consequences is crucial for a comprehensive assessment of the impact of nanoparticles on neurological function. Behavioral changes serve as observable indicators of potential neurotoxic effects and can provide insights into how nanoparticles may affect neural circuits, neurotransmitter systems, or neuronal signaling pathways within the brain. Given the intricate nature of the brain and its critical role in cognitive function and overall well-being, elucidating the effects of nanoparticles on neurological function is essential. Research endeavors aim to discern the mechanisms underlying nanoparticle-induced neurotoxicity and the specific pathways through which these particles influence brain function and behavior.

This understanding is pivotal not only for assessing the risks associated with nanoparticle exposure but also for developing strategies to mitigate potential neurotoxic effects. Establishing safety guidelines and regulatory frameworks based on a comprehensive understanding of nanoparticle-induced neurotoxicity is essential to safeguard neurological health and ensure the safe utilization of nanoparticles in various applications (Teleanu et al., 2018).

12.4.2.5 Endocrine Disruption

Nanoparticles, due to their small size and unique properties, have the potential to interact with the endocrine system, which regulates hormone synthesis, secretion, and overall physiological balance within the body. This interaction can disrupt the intricate signaling pathways of hormones, leading to what is termed as endocrine disruption.

Endocrine disruption caused by nanoparticles can have far-reaching implications for reproductive and metabolic health. Hormones play a pivotal role in controlling various bodily functions, including reproduction, growth, metabolism, and development. When nanoparticles interfere with these hormonal pathways, it can result in dysregulation of these essential processes. In terms of reproductive health, the impact of nanoparticles on fertility, pregnancy outcomes, and developmental processes is a subject of active investigation. The ability of nanoparticles to interfere with hormonal synthesis and function raises concerns about their potential effects on fertility parameters, such as sperm quality, ovarian function, and menstrual cycles. Additionally, during pregnancy, any disruption to hormonal signaling pathways could influence fetal development and potentially affect pregnancy outcomes (Rai et al., 2018).

Studies exploring the effects of nanoparticles on reproductive health aim to understand how these particles may affect the delicate balance of hormones crucial for conception, pregnancy maintenance, and healthy fetal development. Researchers are investigating the potential risks nanoparticles pose to reproductive systems and the developing fetus, seeking to identify possible adverse effects on fertility, pregnancy complications, or developmental abnormalities (Priyam et al., 2018). Given the importance of hormonal balance in reproductive and metabolic processes, the potential endocrine-disrupting effects of nanoparticles highlight the need for comprehensive assessments of their safety, particularly concerning their impact on reproductive health. Understanding the mechanisms underlying nanoparticle-induced endocrine disruption is essential for establishing safety guidelines and minimizing potential risks associated with their use or exposure, especially in settings where individuals might be at higher risk, such as occupational or environmental exposure scenarios (Ermolin & Fedotov, 2016).

12.4.2.6 Epidemiology of Nanotoxicity

Historically, research and epidemiological studies have focused on ultrafine PM_{10}, which are environmental particulate air pollutants with a diameter of 10 μm or less. Several epidemiological studies have shown a correlation between adverse effects and acute exposure to particulate air pollution (Hutchison & Malone, 2011). However, the nano-industry is young; hence, there are few controlled epidemiological studies indicating human health hazards from engineered nanoparticles (Kendall & Holgate, 2012). Donaldson et al., (2017) argue that toxic effects caused by nanomaterials are not "nano-specific," and there is no evidence to suggest a significant shift in hazardous effects below the 100 nm threshold as they have been identified to play a role in toxicity in conventional-sized particles.

In contrast, several in vitro and in vivo studies on mammalian cell lines and experimental animals show cytotoxicity, genotoxicity, reproductive toxicity, oxidative stress, and other forms of nanotoxicity on exposure to indium oxide and silver nanoparticles (Tulinská et al., 2020; Kim et al., 2019; Braakhuis et al., 2015; Paul et al., 2017). Furthermore, human studies show that the occupational exposure of workers directly involved in the production or handling of nanomaterials can lead to a range of respiratory, cardiovascular immune system adverse reactions, an increase in biological markers of disease in the body, and even epigenetic changes as seen in Fig. 12.4 (Ursini et al., 2020; Novotná et al., 2020; Liao et al., 2013). Kendall and Holgate (2012) reported that workers exposed to NPs show decreased lung function and adverse respiratory symptoms. In addition, exposure to silica nanoparticles, common with industrial workers, is associated with silicosis, lung injury, emphysema, lung cancer, and death (Chen et al., 2012; García-Mouton et al., 2019). These studies while useful are limited by a small number of participants, inconsistent exposure assessments, and short intervals between exposure and effect (Liou et al., 2015).

There is a need for more longitudinal studies to report correlations of nanoparticle exposures to human health effects for risk assessment and management purposes. Notably, the NanoExplore project is a recently developed harmonized protocol of an international multicenter prospective cohort study of nanotechnology workers in companies producing engineered nanomaterials (Canu et al., 2023). This

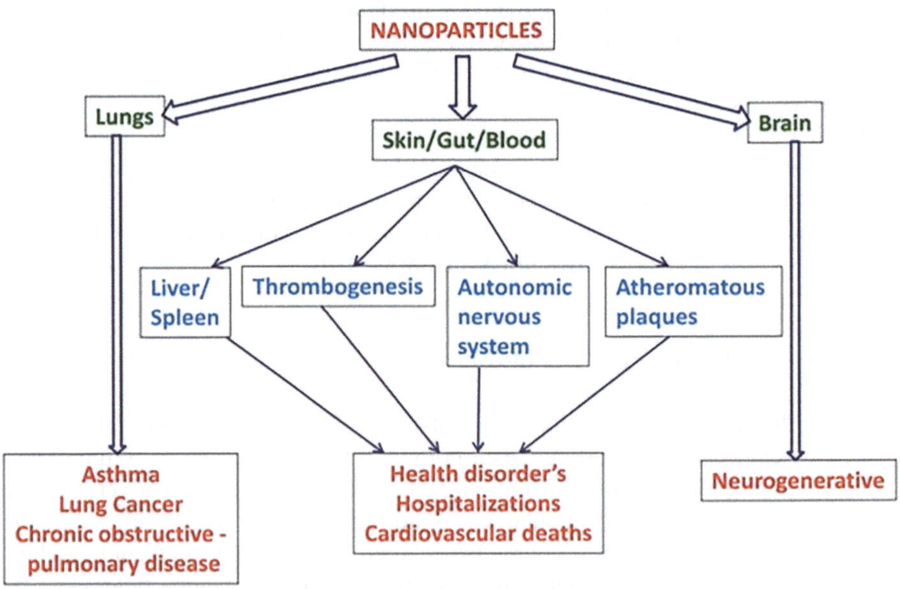

Fig. 12.4 Systemic health effects and clinical implications of nanoparticles on the human body. (Viswanath & Kim, 2016)

protocol may be useful in the future to give insights into the epidemiological discourse.

Currently, exposure to environmental particles is increasing, raising concerns for the health of workers and consumers who manufacture and use nanomaterial products (Viswanath & Kim, 2016). Extensive literature shows that nanoparticles can cause a range of adverse effects in a variety of living organisms including humans (Albrecht et al., 2006; Viswanath & Kim, 2016; Khan et al., 2015; Ferdous & Nemmar, 2020). Nanoparticles encountered through inhalation are a prominent source of exposure and cause increased adverse effects on human physiology (Oberdörster & Utell, 2002; Kendall & Holgate, 2012). In the lungs, NP exposure is associated with pneumonia, asthma, and chronic obstructive pulmonary disease, especially in individuals who are already susceptible (Kendall & Holgate, 2012). NPs deposited in the lung can affect mechanical lung functioning by inhibiting pulmonary surfactants, a complex mixture that plays an important role in reducing surface tension in the alveoli, easing breathing and preventing lung collapse (Liu et al., 2022; Arick et al., 2015; García-Mouton et al., 2019). NPs in the lungs also induce respiratory nanotoxicity by triggering ROS generation and lung inflammation. This can result in several respiratory illnesses including respiratory distress, pulmonary fibrosis, and in some cases lung cancers (Thu et al., 2023).

When not sufficiently cleared from their initial site of entrance, NPs can cross barriers to reach the bloodstream. Ultrafine blood-borne nanoparticles perfuse other vital human organs, causing more damage (Bierkandt et al., 2018). In circulation, NPs interact with epithelial and endothelial tissues altering the blood's clotting status to induce thrombogenesis and interact with coronary and brain arteries causing atherogenesis and plaque destabilization. They might also accumulate in vascular lesions. This results in a variety of cardiovascular diseases including myocardial infarction, cardiac arrhythmias, and stroke (Kendall & Holgate, 2012; Miller et al., 2017; Kan et al., 2018). Furthermore, some metallic NPs may reach sensory neurons where they produce ROS and induce neuronal damage (Yousef et al., 2022; Thu et al., 2023). For example, inhaled titanium dioxide NPs can be distributed to the brain through the olfactory bulb and accumulate in the hippocampus (Catalán-Figueroa et al., 2016). Nanotoxicity-induced oxidative stress and an increase in pro-inflammatory molecules may also cause neuroinflammation This may then manifest in neurodegenerative pathologies such as Alzheimer's and Parkinson's disease (Calderón-Garcidueñas et al., 2016).

Nanoparticles may also harm the reproductive system and offspring development in humans. Carbon nanoparticles have been shown to inhibit the production of the sex hormone, progesterone; decrease sperm mobility; and alter the maturation of oocytes (Holmannová et al., 2022; Ema et al., 2010; Zergui, 2023). Additionally, nanoparticles may breach the placental barrier, causing developmental toxicity to embryos resulting in decreased uterine weight and an increase in fetal resorption rate. While most studies have been conducted in animal models, the potential fetal toxicity in humans cannot be ignored (Ema et al., 2015; Campagnolo et al., 2017). Exposure to nanoparticles in ambient air pollution can also lead to cognitive and behavioral developmental deficits in children (Calderón-Garcidueñas et al., 2019).

Other organs of the body can also be impacted by nanotoxicity. The liver, kidneys, and spleen, which play a role in the detoxification and excretion of biological waste, may be particularly vulnerable to inflammation or metabolic disease from NPs (Hoet et al., 2004; Strojny et al., 2015). Metallic NPs may localize in the tissues of these organs causing inflammation (Thu et al., 2023). NPs may interact with a network of immune cells and act as allergens in the neonatal period, triggering the immune system to induce allergic inflammation later in life. Furthermore, NPs may interact with the gut microbiome potentially influencing microbial metabolism and nutrient absorption (Bergin & Witzmann, 2013). NPs may also have long-term effects on signaling pathways that control cellular metabolism and energy homeostasis (Chen et al., 2014).

12.5 Vulnerable Populations and Health Inequalities

Identifying vulnerable populations involves recognizing groups that are at a higher risk of experiencing adverse health outcomes due to various factors. Vulnerability can stem from socioeconomic status, age, ethnicity, disability, or other social determinants of health. Understanding and acknowledging the unique needs and challenges faced by these populations are critical for effective public health interventions. Vulnerable populations may include but are not limited to children, the elderly, low-income communities, racial and ethnic minorities, individuals with preexisting health conditions, and those lacking access to adequate healthcare resources. Each group may face distinct challenges, making it essential to tailor interventions to address their specific needs (van der Meer et al., 2022).

Health disparities related to nanoparticle exposure may arise from variations in exposure levels and individual susceptibility within different population groups. Factors such as occupation, living conditions, and access to protective measures can contribute to differential exposure, influencing health outcomes. Vulnerable populations, often already burdened with preexisting health conditions, may experience exacerbated effects of nanoparticle exposure. Understanding the interplay between environmental exposure and existing health disparities is crucial for developing targeted interventions and policies (Ray et al., 2009; Ferdous & Nemmar, 2020).

A comprehensive approach involves not only identifying at-risk groups but also addressing the institutional and structural factors causing disparities in the field of vulnerable populations and health disparities connected to nanoparticle exposure. In order to promote environmental justice and fair health outcomes, initiatives must be both proactive and reactive. Examining case studies provides concrete examples of how health disparities manifest in the context of nanoparticle exposure. These cases offer insights into the complexities of the issue, considering variables such as geographical location, socioeconomic factors, and regulatory frameworks. Case studies allow for a deeper understanding of the factors contributing to disparities in exposure and health outcomes. They offer valuable lessons for policymakers, researchers, and public health practitioners, informing strategies to address and mitigate these disparities (Gee, & Payne-Sturges, 2004; Frohlich & Potvin, 2008; Perry et al. 2021).

12.6 Mitigation and Protection Strategies

12.6.1 Environmental Justice Considerations

Environmental justice advocates for fair and equal protection from environmental and health hazards for all, regardless of race, class, or income. It emphasizes the right of every individual, irrespective of their demographic background, to live in a healthy environment. Addressing environmental justice is crucial for preventing the exacerbation of health disparities within vulnerable communities. Vulnerable populations often bear a disproportionate burden of environmental pollutants and hazards. These disparities can result from the location of hazardous facilities, unequal distribution of resources, and inadequate regulatory oversight. Examining the intersections of socioeconomic factors and environmental exposures is vital for promoting environmental justice (Kumah et al., 2023).

12.6.2 Personal Protective Measures

Personal protective measures against nanoparticles are essential, especially in occupational settings or situations where exposure to these tiny particles is likely. The use of personal protective equipment (PPE) is essential in protection from environmental nanoparticles. Depending on the nature of the exposure, appropriate PPE should be worn. This might include gloves, respirators, goggles, lab coats, or coveralls designed to prevent skin contact or inhalation of nanoparticles.

Implementation engineering controls in work environments is essential. For example, ventilation systems equipped with high-efficiency particulate air filters, local exhaust ventilation, or enclosed systems to minimize the release and spread of nanoparticles. Adhering to safe handling practices, including using containment devices, avoiding unnecessary agitation of nanoparticles, and minimizing spills or splashes, can reduce the risk of exposure. Educating individuals about the potential hazards of nanoparticles, proper handling techniques, and the use of protective equipment is crucial. Training programs should emphasize safety protocols and measures to mitigate exposure risks. Practicing good personal hygiene, such as washing hands thoroughly after handling nanoparticles or before eating or drinking, can prevent inadvertent ingestion or transfer of nanoparticles (Frohlich & Potvin, 2008; Perry et al., 2021).

It is important to minimize direct contact with products or environments containing nanoparticles unless adequate protective measures are in place. This includes avoiding unnecessary exposure in areas known to have high nanoparticle concentrations. Workers in environments with potential nanoparticle exposure, regular health checkups, and monitoring can help detect any potential health effects early on. Strict adherence to safety guidelines and protocols provided by regulatory bodies or employers regarding the handling, disposal, and cleanup of nanoparticles is paramount.

The efficacy of protective measures against nanoparticles depends on their specific properties, concentration, and the context of exposure. Therefore, it's crucial to tailor protective measures to the specific nanoparticles and work environment and to continually reassess and update safety protocols based on new information or changes in exposure conditions.

12.6.3 Environmental Management and Pollution Control

Environmental management and pollution control of nanoparticles are critical for minimizing their potential adverse effects on ecosystems and human health. Establishing comprehensive regulations and standards specific to nanoparticle use, production, and disposal is crucial. This includes setting limits on nanoparticle emissions, defining safe handling practices, and implementing guidelines for disposal to prevent environmental contamination. Conducting thorough risk assessments to evaluate the potential environmental impact of nanoparticles is essential. This involves studying their fate, transport, bioavailability, and toxicity in various environmental compartments such as air, water, soil, and biota (Brown et al., 2018).

Employing source reduction strategies by modifying manufacturing processes, implementing cleaner production technologies, or utilizing alternative materials that pose fewer risks to the environment can reduce nanoparticle release. Implementing advanced wastewater treatment methods capable of removing nanoparticles from effluents before discharge into water bodies is crucial. Conventional treatment methods may not effectively remove nanoparticles, necessitating the development of specialized filtration or purification techniques. Implementing monitoring programs to detect nanoparticle presence in environmental matrices enables early identification of potential contamination. This helps in understanding dispersion patterns and assessing the effectiveness of control measures (Frohlich & Potvin, 2008).

Developing regulations specifically tailored to nanoparticles due to their unique properties is essential. Traditional pollution control methods may not suffice, requiring specialized guidelines considering their size, reactivity, and potential for aggregation or transformation in the environment. Proper disposal or recycling of nanoparticle-containing products is crucial. Designing end-of-life strategies to handle nano-waste and prevent their release into the environment is imperative. Raising awareness among industries, researchers, policymakers, and the public about the environmental risks posed by nanoparticles is crucial. Promoting responsible use, handling, and disposal practices helps in preventing nanoparticle pollution.

Encouraging collaborative research initiatives involving multidisciplinary expertise can advance our understanding of nanoparticle behavior in the environment and facilitate the development of effective mitigation strategies. Implementing adaptive management strategies allows for flexibility in adjusting regulations and approaches based on new scientific findings and evolving knowledge about nanoparticle behavior and impacts.

Implementing a comprehensive environmental management framework for nanoparticles requires a proactive and interdisciplinary approach. By integrating research, regulation, monitoring, and education, it becomes possible to minimize the potential environmental risks associated with nanoparticle use and ensure their sustainable and responsible utilization (Perry et al., 2021).

12.6.4 Research and Monitoring Efforts

Research and monitoring efforts focused on nanoparticle toxicity are critical for understanding the potential risks these particles pose to human health and the environment.

Conducting in-depth toxicological studies to determine the adverse effects of nanoparticles on biological systems is crucial. This involves assessing their effects on cellular, molecular, and systemic levels to understand mechanisms of toxicity, including inflammation, oxidative stress, and cellular damage. Determining dose-response relationships helps identify safe exposure levels and thresholds for adverse effects. Research focuses on establishing how varying nanoparticle concentrations affect biological systems and whether certain exposure levels pose heightened risks. Detailed characterization of nanoparticles, including their size, shape, surface charge, chemical composition, and stability, is essential. Variations in these properties can significantly influence their toxicity (Miller et al., 2017; Kan et al., 2018).

Studying the bioaccumulation and biodistribution of nanoparticles within organisms and ecosystems is crucial. This involves tracing the movement of nanoparticles within biological systems to understand their persistence and potential for accumulation in different tissues or organs. Assessing the long-term effects of chronic nanoparticle exposure is vital. Understanding cumulative effects over extended periods aids in predicting health risks associated with prolonged exposure. Investigating the impact of nanoparticles on ecosystems and nonhuman organisms is important. This includes assessing effects on aquatic and terrestrial organisms, soil health, and the overall ecological balance. Furthermore, studies to assess human exposure to nanoparticles in various occupational settings, consumer products, and environmental matrices are critical. This involves measuring nanoparticle levels in air, water, soil, and biological samples to estimate exposure risks. Implementing monitoring programs to detect and quantify nanoparticle levels in different environmental compartments facilitates early identification of potential contamination. Continuous surveillance aids in understanding dispersion patterns and assessing the effectiveness of control measures, alongside developing standardized protocols and methodologies for nanoparticle toxicity assessment ensures consistency in research findings and facilitates comparisons between studies. Most importantly, integrating findings from diverse studies and disciplines contributes to a comprehensive understanding of nanoparticle toxicity. This involves collaboration between toxicologists, chemists, engineers, environmental scientists, and medical professionals.

By combining these research and monitoring efforts, it becomes possible to comprehensively evaluate the toxicity of nanoparticles, mitigate potential risks, and establish guidelines for safe handling and utilization, safeguarding both human health and the environment.

12.7 Future Perspectives

Much research efforts and focus need to be channeled toward understanding the roles and effects of nanoparticles in various ecosystems and environments, via the development of analytical tools that are easy to operate, are affordable, and can be reproduced. Robust understanding of the diverse pathways of nanoparticle interaction will fast track policy formulation, establishment, and enforcement of mitigating strategies, ground rules, and precautions on the possible effects of environmental nanoparticles.

Examining specific case studies provides valuable insights into the diverse toxicity mechanisms associated with different nanoparticles. Case studies encompassing a range of nanoparticles and exposure scenarios contribute to a comprehensive understanding of their impact on human health. Case studies offer the opportunity to identify patterns and commonalities in toxicity mechanisms. Lessons learned from these studies contribute to refining risk assessment strategies, informing regulatory frameworks, and guiding future research endeavors (Gupta & Xie, 2018).

12.8 Conclusion

There is a dire need to understand the natural and anthropogenic sources of environmental nanoparticles and routes of human exposure to nanoparticles in order to explore strategies and plans to protect humans and the environment. The widespread application and occurrence of nanoparticles in water, soil, and air media raises much concern with regard to consequences on environment and human health. Size remains a crucial factor that hinders holistic understanding of the fate of nanoparticles in the environment. More and more environmental nanoparticles are released into the environment, and this calls for proper understanding of their toxicological effects and tendencies in order to be able to take proactive actions, which may be impede the passage of nanoparticles in the environment and also reduce their impacts on human health, and the ecosystem has a clear-cut view and prediction of its consequences in the long run. Proactive measures may prevent nanomaterials entry to the environment and minimize their effects to the ecosystem and human health at large. There is still so much ambiguity about the toxicological implications of the exposure to various forms and sources of nanoparticles. The current techniques available for analysis of nanoparticle exposure and toxicity effects are not sufficient and effective in scope for the study of nanoparticles in the complicated biological systems and natural environment. Despite myriads of challenges, scientific efforts have so far brought about a more balanced understanding of nanoparticle interaction at cellular and tissue levels. There is a need to conduct more research studies to assess the effects of occurrence of environmental nanoparticles on human health.

References

Albrecht, M. A., Evans, C. W., & Raston, C. L. (2006). Green chemistry and the health implications of nanoparticles. *Green Chemistry, 8*(5), 417. https://doi.org/10.1039/b517131h

Al-Mutairi, E. M., Thomas, S. P., & Sadhan, K. D. (2013). Impact of nanomaterials on health and environment. *Arabian Journal for Science and Engineering*. https://doi.org/10.1007/s13369-012-0324-0

Arick, D. Q., Choi, Y., Kim, H. C., & Won, Y. (2015). Effects of nanoparticles on the mechanical functioning of the lung. *Advances in Colloid and Interface Science, 225*, 218–228. https://doi.org/10.1016/j.cis.2015.10.002

Armstead, A., & Li, B. (2016). Nanotoxicity: emerging concerns regarding nanomaterial safety and occupational hard metal (WC-Co) nanoparticle exposure. *International Journal of Nanomedicine, 11*: 6421–6433. https://doi.org/10.2147/ijn.s121238

Arora, K. (2018). Advances in nano based biosensors for food and agriculture. In *Environmental chemistry for a sustainable world* (pp. 1–52). Springer International Publishing.

Arvidsson, R., Baun, A., Hansen, S. F., & Molander, S. (2018). Proxy measures for simplified environmental assessment of manufactured nanomaterials. *Environmental Science & Technology, 52*, 13670–13680. https://doi.org/10.1021/acs.est.8b05405

Baalousha, M., Yang, Y., Vance, M. E., Colman, B. P., McNeal, S., Xu, J., Blaszczak, J., Steele, M., Bernhardt, E., & Hochella, M. F. (2016). Outdoor urban nanomaterials: The emergence of a new, integrated, and critical field of study. *Science of the Total Environment, 557–558*, 740–753. https://doi.org/10.1016/j.scitotenv.2016.03.132

Baranowska-Wójcik, E., Szwajgier, D., Oleszczuk, P., & Winiarska-Mieczan, A. (2020). Effects of titanium dioxide nanoparticles exposure on human health—A review. *Biological Trace Element Research, 193*, 118–129. https://doi.org/10.1007/s12011-019-01706-6

Baroli, B. M. (2010). Penetration of nanoparticles and nanomaterials in the skin: Fiction or reality? *Journal of Pharmaceutical Sciences, 99*(1), 21–50. https://doi.org/10.1002/jps.21817

Bergin, I. L., & Witzmann, F. A. (2013). Nanoparticle toxicity by the gastrointestinal route: evidence and knowledge gaps. *International Journal of Biomedical Nanoscience and Nanotechnology, 3*(1–2). https://doi.org/10.1504/IJBNN.2013.054515

Bhat, M. A., Gedik, K., & Gaga, E. O. (2022). Atmospheric micro (nano) plastics: future growing concerns for human health. Air Quality, *Atmosphere & Health, 16*(2), 233–262. https://doi.org/10.1007/s11869-022-01272-2

Bierkandt, F. S., Leibrock, L., Wagener, S., Laux, P., & Luch, A. (2018). The impact of nanomaterial characteristics on inhalation toxicity. *Toxicology Research, 7*(3), 321–346. https://doi.org/10.1039/c7tx00242d

Bour, A., Mouchet, F., Silvestre, J., Gauthier, L., & Pinelli, E. (2015). Environmentally relevant approaches to assess nanoparticles ecotoxicity: A review. *Journal of Hazardous Materials, 283*, 764–777. https://doi.org/10.1016/j.jhazmat.2014.10.021

Boyes, W. K., & van Thriel, C. (2020). Neurotoxicology of nanomaterials. *Chemical Research in Toxicology, 33*, 1121–1144. https://doi.org/10.1021/acs.chemrestox.0c00050

Braakhuis, H., Cassee, F. R., Fokkens, P., De La Fonteyne, L. J., Oomen, A. G., Krystek, P., De Jong, W. H., Van Loveren, H., & Park, M. V. D. Z. (2015). Identification of the appropriate dose metric for pulmonary inflammation of silver nanoparticles in an inhalation toxicity study. *Nanotoxicology, 10*, 1–11. https://doi.org/10.3109/17435390.2015.1012184

Brown, D., Alkilany, A. M., Omid, C., Mahmoudi, M., Hospital, M. G., Hospital, M. G., & Arabia, S. (2018). *HHS Public Access, 46*(14), 4218–4244. https://doi.org/10.1039/c6cs00636a. Cellular

Cai, J., Zang, X., Wu, Z., Liu, J., & Wang, D. (2019). Translocation of transition metal oxide nanoparticles to breast milk and offspring: The necessity of bridging mother-offspring-integration toxicological assessments. *Environment International, 133*, 105153. https://doi.org/10.1016/j.envint.2019.105153

Calderón-Garcidueñas, L., Leray, E., Heydarpour, P., Torres-Jardón, R., & Reis, J. (2016). Air pollution, a rising environmental risk factor for cognition, neuroinflammation and neurodegeneration: The clinical impact on children and beyond. *Revue Neurologique, 172*(1), 69–80. https://doi.org/10.1016/j.neurol.2015.10.008

Calderón-Garcidueñas, L., González-Maciel, A., Kulesza, R. J., González-González, L. O., Reynoso-Robles, R., Mukherjee, P. S., & Torres-Jardón, R. (2019). Air pollution, combustion and friction derived nanoparticles, and Alzheimer's disease in urban children and young adults. *Journal of Alzheimer's Disease, 70*(2), 343–360. https://doi.org/10.3233/jad-190331

Campagnolo, L., Massimiani, M., Vecchione, L., Piccirilli, D., Toschi, N., Magrini, A., Bonanno, E., Scimeca, M., Castagnozzi, L., Buonanno, G., Stabile, L., Cubadda, F., Aureli, F., Fokkens, P., Kreyling, W. G., Cassee, F. R., & Pietroiusti, A. (2017). Silver nanoparticles inhaled during pregnancy reach and affect the placenta and the foetus. *Nanotoxicology, 11*(5), 687–698. https://doi.org/10.1080/17435390.2017.1343875

Canu, I. G., Plys, E., Crézé, C., Fito, C., Hopf, N. B., Progiou, A., Riganti, C., Sauvain, J., Squillacioti, G., Suárez, G., Wild, P., & Bergamaschi, E. (2023). A harmonized protocol for an international multicenter prospective study of nanotechnology workers: The NanoExplore cohort. *Nanotoxicology, 17*(1), 1–19. https://doi.org/10.1080/17435390.2023.2180220

Catalán-Figueroa, J., Palma-Florez, S., Álvarez, G., Fritz, H., Jara, M., & Morales, J. (2016). Nanomedicine and nanotoxicology: The pros and cons for neurodegeneration and brain cancer. *Nanomedicine, 11*(2), 171–187. https://doi.org/10.2217/nnm.15.189

Chen, W., Liu, Y., Wang, H., Hnizdo, E., Sun, Y., Su, L., Zhang, X., Weng, S., Frank, B., Hearl, F. J., Chen, J., & Wu, T. (2012). Long-term exposure to silica dust and risk of Total and cause-specific mortality in Chinese workers: a cohort study. *PLoS Medicine, 9*(4), e1001206. https://doi.org/10.1371/journal.pmed.1001206

Chen, N., Wang, H., Huang, Q., Jiang, L., Yan, J., He, D., Fan, C., & Song, H. (2014). Long-term effects of nanoparticles on nutrition and metabolism. *Small, 10*(18), 3603–3611. https://doi.org/10.1002/smll.201303635

Chen, J., Guo, Y., Zhang, X., Liu, J., Gong, P., Su, Z., Fan, L., & Li, G. (2023). Emerging nanoparticles in food: Sources, application, and safety. *Journal of Agricultural and Food Chemistry, 71*(8), 3564–3582. https://doi.org/10.1021/acs.jafc.2c06740

Cota-Ruiz, K., Delgado-Rios, M., Martínez-Martínez, A., Núñez-Gastelum, J. A., Peralta-Videa, J. R., & Gardea-Torresdey, J. L. (2018). Current findings on terrestrial plants – Engineered nanomaterial interactions: Are plants capable of phytoremediating nanomaterials from soil? *Current Opinion in Environmental Science & Health, 6*, 9–15. https://doi.org/10.1016/j.coesh.2018.06.005

Daigle, C. C., Chalupa, D., Gibb, F. R., Morrow, P. E., Oberdörster, G., Utell, M. J., & Frampton, M. W. (2003). Ultrafine particle deposition in humans during rest and exercise. *Inhalation Toxicology, 15*(6), 539–552. https://doi.org/10.1080/08958370304468

De Matteis, V. (2017). Exposure to inorganic nanoparticles: Routes of entry, immune response, biodistribution and in vitro/in vivo toxicity evaluation. *Toxics, 5*, 29. https://doi.org/10.3390/toxics5040029

De Matteis, V., & Rinaldi, R. (2018). Toxicity assessment in the nanoparticle era. In *Advances in experimental medicine and biology* (pp. 1–19). Springer. https://doi.org/10.1007/978-3-319-72041-8_1

Ding, G., Zhang, N., Wang, C., Li, X., Zhang, J., Li, W., Li, R., et al. (2018). Effect of the size on the aggregation and sedimentation of graphene oxide in seawaters with different salinities. *Journal of Nanoparticle Research, 20*. https://doi.org/10.1007/s11051-018-4421-1

Donaldson, K., Cassee, F. R., Newby, D. E., Duffin, R., & Mills, N. L. (2017). Inhaled nanoparticles accumulate at sites of vascular disease. *ACS Nano, 11*(5), 4542–4552. https://doi.org/10.1021/acsnano.6b08551

Ema, M., Kobayashi, N., Naya, M., Hanai, S., & Nakanishi, J. (2010). Reproductive and developmental toxicity studies of manufactured nanomaterials. *Reproductive Toxicology, 30*(3), 343–352. https://doi.org/10.1016/j.reprotox.2010.06.002

Ema, M., Hougaard, K. S., Kishimoto, A., & Honda, K. (2015). Reproductive and developmental toxicity of carbon-based nanomaterials: A literature review. *Nanotoxicology, 10*(4), 391–412. https://doi.org/10.3109/17435390.2015.1073811

Ermolin, M. S., & Fedotov, P. S. (2016). Separation and characterization of environmental nanoandsubmicron particles. *Reviews in Analytical Chemistry, 35*, 185. https://doi.org/10.1515/revac-2016-0006

Ferdous, Z., & Nemmar, A. (2020). Health impact of silver nanoparticles: A review of the biodistribution and toxicity following various routes of exposure. *International Journal of Molecular Sciences, 21*(7), 2375. https://doi.org/10.3390/ijms21072375

Fraser, T. W. K., Reinardy, H. C., Shaw, B. J., Henry, T. B., & Handy, R. D. (2011). Dietary toxicity of single-walled carbon nanotubes and fullerenes (C_{60}) in rainbow trout (*Oncorhynchus mykiss*). *Nanotoxicology, 5*(1), 98–108. https://doi.org/10.3109/17435390.2010.502978

Frohlich, K. L., & Potvin, L. (2008). Transcending the known in public health practice: The inequality paradox: The population approach and vulnerable populations. *American Journal of Public Health, 98*(2), 216–221. https://doi.org/10.2105/AJPH.2007.114777

Fu, P. P., Xia, Q., Hwang, H., Ray, P. C., & Yu, H. (2014). Mechanisms of nanotoxicity: Generation of reactive oxygen species. *Journal of Food and Drug Analysis, 22*(1), 64–75. https://doi.org/10.1016/j.jfda.2014.01.005

Gangwar, R. S., Bevan, G. H., Palanivel, R., Das, L., & Rajagopalan, S. (2020). Oxidative stress pathways of air pollution mediated toxicity: Recent insights. *Redox Biology, 34*, 101545. https://doi.org/10.1016/j.redox.2020.101545

Ganguly, P., Breen, A., & Pillai, S. C. (2018). Toxicity of nanomaterials: Exposure, pathways, assessment, and recent advances. *ACS Biomaterials Science & Engineering, 4*, 2237–2275. https://doi.org/10.1021/acsbiomaterials.8b00068

García-Mouton, C., Hidalgo, A., Cruz, A., & Pérez-Gil, J. (2019). The Lord of the lungs: The essential role of pulmonary surfactant upon inhalation of nanoparticles. *European Journal of Pharmaceutics and Biopharmaceutics, 144*, 230–243. https://doi.org/10.1016/j.ejpb.2019.09.020

Gee, G. C., & Payne-Sturges, D. C. (2004). Environmental health disparities: A framework integrating psychosocial and environmental concepts. *Environmental Health Perspectives, 112*(17), 1645–1653. https://doi.org/10.1289/ehp.7074

Geiser, M., Casaulta, M., Kupferschmid, B., Schulz, H., Semmler-Behnke, M., & Kreyling, W. G. (2008). The role of macrophages in the clearance of inhaled ultrafine titanium dioxide particles. *American Journal of Respiratory Cell and Molecular Biology, 38*(3), 371–376. https://doi.org/10.1165/rcmb.2007-0138oc

Ghoshdastidar, A. J., & Ariya, P. A. (2019). The existence of airborne mercury nanoparticles. *Scientific Reports, 9*, 10733. https://doi.org/10.1038/s41598-019-47086-8

Giraldo, J. P., Wu, H., Newkirk, G. M., & Kruss, S. (2019). Nanobiotechnology approaches for engineering smart plant sensors. *Nature Nanotechnology, 14*(6), 541–553. https://doi.org/10.1038/s41565-019-0470-6

Goswami, L., Kim, K. H., Deep, A., Das, P., Bhattacharya, S. S., Kumar, S., & Adelodun, A. A. (2017). Engineered nano particles: Nature, behavior, and effect on the environment. *Journal of Environmental Management, 196*, 297–315. https://doi.org/10.1016/j.jenvman.2017.01.011

Griffin, S., Masood, M. I., Nasim, M. J., Sarfraz, M., Ebokaiwe, A. P., Schäfer, K. H., Keck, C. M., & Jacob, C. (2018). Natural nanoparticles: A particular matter inspired by nature. *Antioxidants*. https://doi.org/10.3390/antiox7010003

Gupta, P. (2023). *Nanotoxicology in nanobiomedicine*. Springer eBooks. https://doi.org/10.1007/978-3-031-24287-8

Gupta, R., & Xie, H. (2018). Nanoparticles in daily life: Applications, toxicity and regulations. *Journal of Environmental Pathology, Toxicology and Oncology, 37*(3), 209–230. https://doi.org/10.1615/JEnvironPatholToxicolOncol.2018026009

Hagens, W. I., Oomen, A. G., De Jong, W. H., Cassee, F. R., Sips, A. J., & a. M. (2007). What do we (need to) know about the kinetic properties of nanoparticles in the body? *Regulatory Toxicology and Pharmacology, 49*(3), 217–229. https://doi.org/10.1016/j.yrtph.2007.07.006

Hartland, A., Lead, J. R., Slaveykova, V. I., O'Carroll, D., & Valsami-Jones, E. (2013). The environmental significance of natural nanoparticles. *Nature Education Knowledge, 4*(8), 7.

Hashem, N. M., & Sallam, S. M. (2020). Reproductive performance of goats treated with free gonadorelin or nanoconjugated gonadorelin at estrus. *Domestic Animal Endocrinology, 71*, 106390. https://doi.org/10.1016/j.domaniend.2019.106390

Herzog, E., Casey, A., Lyng, F., Chambers, G., Byrne, H., & Davoren, M. (2007). A new approach to the toxicity testing of carbon-based nanomaterials—The clonogenic assay. *Toxicology Letters, 174*(1–3), 49–60. https://doi.org/10.1016/j.toxlet.2007.08.009

Hochella, M. F., Mogk, D. W., Ranville, J., Allen, I. C., Luther, G. W., Marr, L. C., McGrail, B. P., Murayama, M., Qafoku, N. P., Rosso, K. M., Sahai, N., Schroeder, P. A., Vikesland, P., Westerhoff, P., et al. (2019). Natural, incidental, and engineered nanomaterials and their impacts on the Earth system. *Science, 363*, 6434. https://doi.org/10.1126/science.aau8299

Hoet, P., Brüske-Hohlfeld, I., & Salata, O. V. (2004). Nanoparticles—Known and unknown health risks. *Journal of Nanobiotechnology, 2*(1), 12. https://doi.org/10.1186/1477-3155-2-12

Holmannová, D., Borský, P., Švadláková, T., Borská, L., & Fiala, Z. (2022). Reproductive and developmental nanotoxicity of carbon nanoparticles. *Nanomaterials, 12*(10), 1716. https://doi.org/10.3390/nano12101716

Huang, Y. W., Cambre, M., & Lee, H. J. (2017). The toxicity of nanoparticles depends on multiple molecular and physicochemical mechanisms. *International Journal of Molecular Sciences, 18*, 2702. https://doi.org/10.3390/ijms18122702

Hutchison, G. R., & Malone, E. M. (2011). *Nanotoxicity* (pp. 419–434). Springer eBooks. https://doi.org/10.1007/978-1-4419-6956-9_18

Iavicoli, I., Leso, V., Beezhold, D. H., & Shvedova, A. A. (2017). Nanotechnology in agriculture: Opportunities, toxicological implications, and occupational risks. *Toxicology and Applied Pharmacology, 329*, 96–111. https://doi.org/10.1016/j.taap.2017.05.025

Jain, I., & Patel, P. (2021). Effects and mechanism of nanotoxicity: An overview. *Journal of Science and Technology, 6*(4), 2456–5660. https://doi.org/10.46243/jst.2021.v6.i04.pp58-68

Kah, M. (2015). Nanopesticides and nanofertilizers: Emerging contaminants or opportunities for risk mitigation? *Frontiers in Chemistry, 3*. https://doi.org/10.3389/fchem.2015.00064

Kan, H., Pan, D., & Castranova, V. (2018). Engineered nanoparticle exposure and cardiovascular effects: The role of a neuronal-regulated pathway. *Inhalation Toxicology, 30*(9–10), 335–342. https://doi.org/10.1080/08958378.2018.1535634

Keller, A. A., Adeleye, A. S., Conway, J. R., Garner, K. L., Zhao, L., Cherr, G. N., et al. (2017). Comparative environmental fate and toxicity of copper nanomaterials. *NanoImpact, 7*, 28–40. https://doi.org/10.1016/j.impact.2017.05.003

Kendall, M., & Holgate, S. T. (2012). Health impact and toxicological effects of nanomaterials in the lung. *Respirology, 17*(5), 743–758. https://doi.org/10.1111/j.1440-1843.2012.02171.x

Khan, S., Rehman, S., Zeb Khan, A., Amjad Khan, M., & Tahir Shah, M. (2010). Soil and vegetables enrichment with heavy metals from geological sources in Gilgit, northern Pakistan. *Ecotoxicology and Environmental Safety, 73*(7), 1820–1827. https://doi.org/10.1016/j.ecoenv.2010.08.016

Khan, S. H., Fulekar, M. H., & Pathak, B. (2015). Nanotoxicology—Health and environmental impacts: A review. *Journal of Environmental Nanotechnology, 4*(3), 55–73. https://doi.org/10.13074/jent.2015.09.153165

Kim, S., Jeon, S., Lee, D., Lee, S., Jeong, J., Kim, J. S., & Cho, W. (2019). The early onset and persistent worsening pulmonary alveolar proteinosis in rats by indium oxide nanoparticles. *Nanotoxicology, 14*(4), 468–478. https://doi.org/10.1080/17435390.2019.1694184

Kirchner, C., Liedl, T., Kudera, S., Pellegrino, T., Muñoz Javier, A., Gaub, H. E., et al. (2005). Cytotoxicity of colloidal CdSe and CdSe/ZnS nanoparticles. *Nano Letters, 5*(2), 331–338. https://doi.org/10.1021/nl047996m

Korin, N., Kanapathipillai, M., & Ingber, D. E. (2013). Shear-responsive platelet mimetics for 1064 targeted drug delivery. *Israel Journal of Chemistry, 53*, 610–615. https://doi.org/10.1002/ijch.201300052

Kumah, E. A., Fopa, R. D., Harati, S., Boadu, P., Zohoori, F. V., & Pak, T. (2023). Human and environmental impacts of nanoparticles: A scoping review of the current literature. *BMC Public Health, 23*(1), 1059. https://doi.org/10.1186/s12889-023-15958-4

Larese Filon, F., Mauro, M., Adami, G., Bovenzi, M., & Crosera, M. (2015). Nanoparticles skin absorption: New aspects for a safety profile evaluation. *Regulatory Toxicology and Pharmacology, 72*, 310–322. https://doi.org/10.1016/j.yrtph.2015.05.005

Liao, H., Chung, Y., Lai, C., Wang, S., Chiang, H., Li, L., Tsou, T., Li, W., Lee, H., Wu, W., Lin, M., Hsu, J. R. C., Ho, J. A., Chen, C., Shih, T., Lin, C., & Liou, S. (2013). Six-month follow-up study of health markers of nanomaterials among workers handling engineered nanomaterials. *Nanotoxicology, 8*(supl), 100–110. https://doi.org/10.3109/17435390.2013.858793

Liou, S., Tsai, C. S., Pelclová, D., Schubauer-Berigan, M. K., & Schulte, P. A. (2015). Assessing the first wave of epidemiological studies of nanomaterial workers. *Journal of Nanoparticle Research, 17*(10), 1–19. https://doi.org/10.1007/s11051-015-3219-7

Liu, Q., Guan, J., Song, R., Zhang, X., & Mao, S. (2022). Physicochemical properties of nanoparticles affecting their fate and the physiological function of pulmonary surfactants. *Acta Biomaterialia, 140*, 76–87. https://doi.org/10.1016/j.actbio.2021.11.034

Lungu, M., Neculae, A., Bunoiu, M., & Biris, C. (2015). *Nanoparticles' promises and risks, nanoparticles' promises and risks: Characterization, manipulation, and potential hazards to humanity and the environment*. Springer International Publishing. https://doi.org/10.1007/978-3-319-11728-7

Ma, Z., Yin, X., Ji, X., Yue, J. Q., Zhang, L., Qin, J. J., Valiyaveettil, S., et al. (2016). Evaluation and removal of emerging nanoparticle contaminants in water treatment: A review. *Desalination and Water Treatment, 57*, 11221–11232. https://doi.org/10.1080/19443994.2015.1038734

Malakar, A., & Snow, D. D. (2020). Nanoparticles as inorganic pollutant in water. In P. Devi, P. Singh, & S. K. Kansal (Eds.), *Inorganic pollutants in water* (p. 440). Elsevier.

Malakar, A., Snow, D. D., & Ray, C. (2019). Irrigation water quality-A contemporary perspective. *Water, 11*, 1482. https://doi.org/10.3390/w11071482

Malakar, A., Kaiser, M., Snow, D. D., Walia, H., Panda, B., & Ray, C. (2020). Ferrihydrite reduction increases arsenic and uranium bioavailability in unsaturated soil. *Environmental Science & Technology*. https://doi.org/10.1021/acs.est.0c02670

Magdolénová, Z., Collins, A., Kumar, A., Dhawan, A., Stone, V., & Dušinská, M. (2013). Mechanisms of genotoxicity. A review of*in vitro*and*in vivo*studies with engineered nanoparticles. *Nanotoxicology, 8*(3), 233–278. https://doi.org/10.3109/17435390.2013.773464

Mantecca, P., Kasemets, K., Deokar, A., Perelshtein, I., Gedanken, A., Bahk, Y. K., Kianfar, B., et al. (2017). Airborne nanoparticle release and toxicological risk from metal-oxide-coated textiles: Toward a multiscale safe-by-design approach. *Environmental Science & Technology, 51*, 9305–9317. https://doi.org/10.1021/acs.est.7b02390

McClements, D. J., & Xiao, H. (2017). Is nano safe in foods? Establishing the factors impacting the gastrointestinal fate and toxicity of organic and inorganic food-grade nanoparticles. *npj Science of Food, 1*, 6. https://doi.org/10.1038/s41538-017-0005-1

Miller, M. R., Raftis, J. B., Langrish, J. P., McLean, S. G., Samutrtai, P., Connell, S. P., Wilson, S., Vesey, A. T., Fokkens, P. H. B., Boere, A. J. F., Krystek, P., Campbell, C. J., Hadoke, P. W. F., Donaldson, K., Cassee, F. R., Newby, D. E., Duffin, R., et al. (2017). Inhaled anoparticles accumulate at sites of vascular disease. *ACS Nano, 11*, 4542–4552. https://doi.org/10.1021/acsnano.6b08551

Nangia, S., & Sureshkumar, R. (2012). Effects of nanoparticle charge and shape anisotropy on translocation through cell membranes. *Langmuir, 28*, 17666–17671. https://doi.org/10.1021/la303449d

Novotná, B., Pelclová, D., Rössnerová, A., ŽdíMal, V., Ondráček, J., Lischková, L., Vlčková, Š., Fenclová, Z., Klusáčková, P., Závodná, T., Topinka, J., Komarc, M., Dvořáčková, Š., & Rössner, P. (2020). The genotoxic effects in the leukocytes of workers handling nanocomposite materials. *Mutagenesis, 35*(4), 331–340. https://doi.org/10.1093/mutage/geaa016

Oberdörster, G., & Utell, M. J. (2002). Ultrafine particles in the urban air: to the respiratory tract—and beyond? *Environmental Health Perspectives, 110*(8), A440–A441. https://doi.org/10.1289/ehp.110-1240959

Oberdörster, G., Oberdörster, E., & Oberdörster, J. (2005). Nanotoxicology: An emerging discipline evolving from studies of ultrafine particles. *Environmental Health Perspectives, 113*(7), 823–839. https://doi.org/10.1289/ehp.7339

Parisi, C., Vigani, M., & Rodríguez-Cerezo, E. (2015). Agricultural nanotechnologies: What are the current possibilities? *Nano Today, 10*(2), 124–127. https://doi.org/10.1016/j.nantod.2014.09.009

Paul, E. N., Franco-Montoya, M., Paineau, E., Angeletti, B., Vibhushan, S., Ridoux, A., Tiendrébéogo, A. J. F., Salomé, M., Hesse, B., Vantelon, D., Rose, J., Canouï-Poitrine, F., Boczkowski, J., Lanone, S., Delacourt, C., & Pairon, J. (2017). Pulmonary exposure to metallic nanomaterials during pregnancy irreversibly impairs lung development of the offspring. *Nanotoxicology, 11*(4), 484–495. https://doi.org/10.1080/17435390.2017.1311381

Perry, M. J., Arrington, S., Freisthler, M. S., Ibe, I. N., McCray, N. L., Neumann, L. M., Tajanlangit, P., & Trejo Rosas, B. M. (2021). Pervasive structural racism in environmental epidemiology. *Environmental Health, 20*, 119. https://doi.org/10.1186/s12940-021-00801-3

Phillips, D. H., & Arlt, V. M. (2009). Genotoxicity: Damage to DNA and its consequences. *EXS, 99*, 87–110. https://doi.org/10.1007/978-3-7643-8336-7_4

Priyam, A., Singh, P. P., & Gehlout, S. (2018). Role of endocrine-disrupting engineered nanomaterials in the pathogenesis of type 2 diabetes mellitus. *Frontiers in Endocrinology, 9*, 704. https://doi.org/10.3389/fendo.2018.00704

Rai, P. K., Kumar, V., Lee, S., Raza, N., Kim, K.-H., Ok, Y. S., & Tsang, D. C. W. (2018). Nanoparticle-plant interaction: Implications in energy, environment, and agriculture. *Environment International, 119*, 1–19. https://doi.org/10.1016/j.envint.2018.06.012

Rajput, V., Minkina, T., Mazarji, M., Shende, S., Sushkova, S., Mandzhieva, S., et al. (2020a). Accumulation of nanoparticles in the soil-plant systems and their effects on human health. *Annals of Agricultural Science, 65*(2), 137–143. https://doi.org/10.1016/j.aoas.2020.08.001

Rajput, V., Minkina, T., Sushkova, S., Behal, A., Maksimov, A., Blicharska, E., et al. (2020b). ZnO and CuO nanoparticles: A threat to soil organisms, plants, and human health. *Environmental Geochemistry and Health, 42*(1), 147–158. https://doi.org/10.1007/s10653-019-00317-3

Ray, P. C., Yu, H., & Fu, P. P. (2009). Toxicity and environmental risks of nanomaterials: Challenges and future needs. *Journal of Environmental Science and Health. Part C, Environmental Carcinogenesis & Ecotoxicology Reviews, 27*(1), 1–35. https://doi.org/10.1080/10590500802708267

Ray, P., Haideri, N., Haque, I., Mohammed, O., Chakraborty, S., Banerjee, S., Quadir, M., Brinker, A., & Banerjee, S. (2021). The impact of nanoparticles on the immune system: A gray zone of Nanomedicine. *Journal of Immunological Sciences, 5*(1), 19–33. https://doi.org/10.29245/2578-3009/2021/1.1206

Sahu, S. C., Zheng, J., Graham, L., Chen, L., Ihrie, J., Yourick, J. J., & Sprando, R. L. (2014). Comparative cytotoxicity of nanosilver in human liver HepG2 and colon Caco2 cells in culture. *Journal of Applied Toxicology, 34*, 1155–1166. https://doi.org/10.1002/jat.2994

Sahu, S. C., Njoroge, J., Bryce, S. M., Zheng, J., & Ihrie, J. (2016a). Flow cytometric evaluation of the contribution of ionic silver to genotoxic potential of nanosilver in human liver HepG2 and colon Caco2 cells. *Journal of Applied Toxicology, 36*, 521–531. https://doi.org/10.1002/jat.3276

Sahu, S. C., Roy, S., Zheng, J., & Ihrie, J. (2016b). Contribution of ionic silver to genotoxic potential of nanosilver in human liver HepG2 and colon Caco2 cells evaluated by the cytokinesis-block micronucleus assay. *Journal of Applied Toxicology, 36*, 532–542. https://doi.org/10.1002/jat.3279

Sajid, M., Ilyas, M., Basheer, C., Tariq, M., Daud, M., Baig, N., & Shehzad, F. (2015). Impact of nanoparticles on human and environment: Review of toxicity factors, exposures, control strategies, and future prospects. *Environmental Science and Pollution Research, 22*, 4122–4143. https://doi.org/10.1007/s11356-014-3994-1

Schirmer, K. (2014). Mechanisms of Nanotoxicity. *Frontiers of Nanoscience, 7*, 195–221. https://doi.org/10.1016/b978-0-08-099408-6.00006-2

Senut, M. C., Zhang, Y., Liu, F., Sen, A., Ruden, D. M., & Mao, G. (2016). Size-dependent toxicity of gold nanoparticles on human embryonic stem cells and their neural derivatives. *Small, 12*, 631–646. https://doi.org/10.1002/smll.201502346

Shaw, B. J., & Handy, R. D. (2011). Physiological effects of nanoparticles on fish: A comparison of nanometals versus metal ions. *Environment International, 37*(6), 1083–1097. https://doi.org/10.1016/j.envint.2011.03.009

Simeonidis, K., Mourdikoudis, S., Kaprara, E., Mitrakas, M., & Polavarapu, L. (2016). Inorganic engineered nanoparticles in drinking water treatment: a critical review. *Environmental Science, 2*(1), 43–70. https://doi.org/10.1039/c5ew00152h

Simeonidis, K., Martínez-Boubeta, C., Zamora-Pérez, P., Rivera-Gil, P., Kaprara, E., Kokkinos, E., & Mitrakas, M. (2018). Implementing nanoparticles for competitive drinking water purification. *Environmental Chemistry Letters, 17*(2), 705–719. https://doi.org/10.1007/s10311-018-00821-5

Singh, S., Kumar, V., Datta, S., Singh, S., Dhanjal, D. S., Garg, R., Kaur, P., & Sharma, K. (2020). *Challenges and future perspectives of nanotoxicology* (pp. 451–466). Springer eBooks. https://doi.org/10.1007/978-981-15-1702-0_22

Stern, S. T., & McNeil, S. E. (2007). Nanotechnology safety concerns revisited. *Toxicological Sciences, 101*(1), 4–21. https://doi.org/10.1093/toxsci/kfm169

Strong, D. R., & Frank, K. T. (2010). Human involvement in food webs. *Annual Review of Environment and Resources, 35*, 1–23. https://doi.org/10.1146/annurev-environ-031809-133103

Strojny, B., Kurantowicz, N., Sawosz, E., Grodzik, M., Jaworski, S., Kutwin, M., Wierzbicki, M., Hotowy, A., Lipińska, L., & Chwalibog, A. (2015). Long term influence of carbon nanoparticles on health and liver status in rats. *PLoS One, 10*(12), e0144821. https://doi.org/10.1371/journal.pone.0144821

Sukhanova, A., Bozrova, S., Sokolov, P., Berestovoy, M., Karaulov, A., & Nabiev, I. (2018). Dependence of nanoparticle toxicity on their physical and chemical properties. *Nanoscale Research Letters, 13*, 44. https://doi.org/10.1186/s11671-018-2457-x

Sufian, M. M., Khattak, J. Z. K., Yousaf, S., & Rana, M. S. (2017). Safety issues associated with the use of nanoparticles in human body. *Photodiagnosis and Photodynamic Therapy, 19*, 67–72. https://doi.org/10.1016/j.pdpdt.2017.05.012

Sykes, E. A., Chen, J., Zheng, G., & Chan, W. C. W. (2014). Investigating the impact of nanoparticle size on active and passive tumor targeting efficiency. *ACS Nano, 8*, 5696–5706. https://doi.org/10.1021/nn500299p

Teleanu, D. M., Chircov, C., Grumezescu, A. M., Volceanov, A., & Teleanu, R. I. (2018). Impact of nanoparticles on brain health: An up to date overview. *Journal of Clinical Medicine, 7*(12), 490. https://doi.org/10.3390/jcm7120490

Thomas, S. P., Al-Mutairi, E. M., & De, S. K. (2013). Impact of nanomaterials on health and environment. *Arabian Journal for Science and Engineering, 38*(3), 457–477. https://doi.org/10.1007/s13369-012-0324-0

Thu, H. E., Haider, M., Khan, S., Sohail, M., & Hussain, Z. (2023). Nanotoxicity induced by nanomaterials: A review of factors affecting nanotoxicity and possible adaptations. *OpenNano, 14*, 100190. https://doi.org/10.1016/j.onano.2023.100190

Tulinská, J., Mašánová, V., Líšková, A., Mikušová, M. L., Rollerová, E., Krivošíková, Z., Štefíková, K., Uhnáková, I., UrsíNyová, M., Bábíčková, J., Bábelová, A., Bušová, M., Tóthová, Ľ., Wsólová, L., Dušinská, M., Sojka, M., Horváthová, M., Aláčová, R., Večeřa, Z., & Dočekal, B. (2020). Six-week inhalation of CdO nanoparticles in mice: The effects on immune response, oxidative stress, antioxidative defense, fibrotic response, and bones. *Food and Chemical Toxicology, 136*, 110954. https://doi.org/10.1016/j.fct.2019.110954

Ursini, C. L., Fresegna, A. M., Ciervo, A., Maiello, R., Del Frate, V., Folesani, G., Galetti, M., Poli, D., Buresti, G., Di Cristo, L., Sabella, S., Iavicoli, S., & Cavallo, D. (2020). Occupational exposure to graphene and silica nanoparticles. Part II: Pilot study to identify a panel of sensitive biomarkers of genotoxic, oxidative and inflammatory effects on suitable biological matrices. *Nanotoxicology, 15*(2), 223–237. https://doi.org/10.1080/17435390.2020.1850903

van der Meer, L., Barsties, L. S., Daalderop, L. A., Waelput, A. J. M., Steegers, E. A. P., & Bertens, L. C. M. (2022). Social determinants of vulnerability in the population of reproductive age: A systematic review. *BMC Public Health, 22*(1), 1252. https://doi.org/10.1186/s12889-022-13651-6

Viswanath, B., & Kim, S. (2016). Influence of nanotoxicity on human health and environment: The alternative strategies. In *Reviews of environmental contamination and toxicology* (pp. 61–104). Springer. https://doi.org/10.1007/398_2016_12

Wang, B., Zhou, X., Chai, Z., Zhao, Y., Feng, W., Yin, J.-J., & Kurash, I. (2013). Physicochemical origin for free radical generation of iron oxide nanoparticles in biomicroenvironment: Catalytic activities mediated by surface chemical states. *Journal of Physical Chemistry C, 117*, 383–392. https://doi.org/10.1021/jp3101392

Wang, L., Yan, L., Liu, J., Chen, C., & Zhao, Y. (2018). Quantification of nanomaterial/nanomedicine trafficking in vivo. *Analytical Chemistry, 90*, 589–614. https://doi.org/10.1021/acs.analchem.7b04765

Westmeier, D., Solouk-Saran, D., Vallet, C., Siemer, S., Docter, D., Götz, H., Männ, L., Hasenberg, A., Hahlbrock, A., Erler, K., Reinhardt, C., Schilling, O., Becker, S., Gunzer, M., Hasenberg, M., Knauer, S. K., et al. (2018). Nanoparticle decoration impacts airborne fungal pathobiology. *Proceedings of the National Academy of Sciences, 115*, 7087–7092. https://doi.org/10.1073/pnas.1804542115

Xiong, T., Dumat, C., Dappe, V., Vezin, H., Schreck, E., Shahid, M., et al. (2017). Copper oxide nanoparticle foliar uptake, phytotoxicity, and consequences for sustainable urban agriculture. *Environmental Science & Technology, 51*(9), 5242–5251. https://doi.org/10.1021/acs.est.6b05546

Yah, C., Iyuke, S., & Simate, G. (2012). A review of nanoparticles toxicity and their routes of exposures. *Iranian Journal of Pharmaceutical Sciences, 8*(1), 299–314.

Yousef, M. I., Roychoudhury, S., Jafaar, K. S., Slama, P., Kesari, K. K., & Kamel, M. A. E. (2022). Aluminum oxide and zinc oxide induced nanotoxicity in rat brain, heart, and lung. *Physiological Research, 71*, 677–694. https://doi.org/10.33549/physiolres.934831

Zergui, A. (2023). Metallic nanoparticles in food: An overview on consumers' health effects. *Food and Humanity, 1*, 51–56. https://doi.org/10.1016/j.foohum.2023.04.004

Chapter 13
Risk Assessment and Management in Nanotoxicology

Patrick Omoregie Isibor, Ifeoluwa Ihotu Kayode-Edwards, Oluwafemi Adebayo Oyewole, David Osagie Agbontaen, Ogochukwu Oritseweyinmi Onwaeze, Esther Paul, Samuel Adeniyi Oyegbade, Ummulkhair Salamah Ilyasu, and Konjerimam Ishaku Chimbekujwo

13.1 Introduction

Risk assessment and management in nanotoxicology are vital components of ensuring the safe and responsible use of nanoparticles, which are increasingly prevalent in various industries and consumer products. Nanoparticles, due to their unique properties, raise concerns regarding their potential impacts on human health and the environment. As these particles can easily enter biological systems and have distinct chemical and physical characteristics, understanding their risks and implementing effective management strategies are essential. In the realm of nanotoxicology, risk assessment encompasses a comprehensive evaluation of the potential adverse effects of nanoparticles, considering factors such as size, shape, surface properties, and chemical composition. This assessment involves the characterization of nanoparticles, understanding their exposure pathways, evaluating their toxicity through both in vitro and in vivo studies, and establishing dose-response relationships. Identifying uncertainties and variabilities within the data is crucial to develop accurate risk profiles (Tsuji et al., 2006).

P. O. Isibor (✉) · I. I. Kayode-Edwards · O. O. Onwaeze · S. A. Oyegbade
Department of Biological Sciences, College of Science and Technology, Covenant University, Ota, Nigeria
e-mail: patrick.isibor@covenantuniversity.edu.ng

O. A. Oyewole · E. Paul · U. S. Ilyasu
Department of Microbiology, Federal University of Technology Minna, Minna, Nigeria

D. O. Agbontaen
Department of Public Health, University of South Wales, Pontypridd, UK

K. I. Chimbekujwo
Department of Microbiology, Modibbo Adama University, Yola, Nigeria

P. O. Isibor et al. (eds.), *Environmental Nanotoxicology*,
https://doi.org/10.1007/978-3-031-54154-4_13

Risk management strategies, on the other hand, focus on mitigating the identified risks. This includes the implementation of stringent regulations and guidelines specific to nanomaterials, the adoption of engineering controls and personal protective equipment in workplaces, and the development of safer nanomaterials through continuous research and innovation. Monitoring and surveillance mechanisms help in tracking exposure levels and assessing the effectiveness of implemented safety measures. Public awareness campaigns and international collaboration further enhance our collective ability to manage the risks associated with nanoparticles (Warheit & Sayes, 2015).

In the face of rapid technological advancements and increasing applications of nanotechnology, a proactive and adaptive approach to risk assessment and management in nanotoxicology is essential. By integrating scientific research, regulatory frameworks, and public awareness initiatives, navigating the complexities of nanotoxicology and ensuring the benefits of nanotechnology while minimizing potential harms to human health and the environment are ensured. Nemmar et al. (2013) report that increases in mortality and visitations to emergency rooms are linked with exposure to particulate matter (PM). When inhaled, these particles penetrate into the respiratory tract from which they target various anatomical sites based on their aerodynamic size and other properties.

The characterization of nanoparticles is crucial in understanding their properties, behavior, and potential applications. Nanoparticles (NPs) are materials with dimensions on the nanometer scale, typically ranging from 1 to 100 nanometers. Characterization methods provide valuable information about their size, shape, surface area, surface charge, composition, and other physicochemical attributes (Warheit et al., 2008). These properties can significantly influence their toxicity. Some techniques used for the characterization of nanoparticles include scanning electron microscopy (SEM) for high-resolution images to visualize the surface morphology and size distribution of nanoparticles (Akhtar et al., 2018); transmission electron microscopy (TEM) for detailed imaging at the atomic scale of the particle size, shape, and crystallinity of nanoparticles (Dachraoui et al., 2021); atomic force microscopy (AFM) for mapping the surface topography, particularly for biological nanoparticles to provide data on their size, shape, and surface roughness (Marrese et al., 2017; Sbarigia et al., 2022); dynamic light scattering (DLS) for measuring the hydrodynamic size of nanoparticles in a liquid medium to provide information on particle size distribution and polydispersity (Langevin et al., 2018; Karmakar, 2019); X-ray diffraction (XRD) for determining the crystalline structure of nanoparticles to characterize the composition and identify the crystal phases present in nanoparticles (Lu et al., 2015); X-ray photoelectron spectroscopy (XPS) for analyzing the elemental composition and oxidation states of nanoparticles by providing information on the surface composition and chemical state of nanoparticles (Sarma et al., 2013); Fourier transform infrared spectroscopy (FTIR) for studying the functional groups of nanoparticles to provide information on the chemical bonds, surface modifications, and presence of specific functional groups in nanoparticles (Jang & Lim, 2010; Behboudi et al., 2023); nuclear magnetic resonance (NMR) spectroscopy for studying organic nanoparticles and their interactions with other molecules (Salassa & Bürgi, 2018); surface area and porosity analysis (using Brunauer-Emmett-Teller (BET) analysis, Barrett-Joyner-Halenda (BJH) method, and the modern density functional theory (DFT)) for studying the surface area, pore size distribution, and surface energy of nanoparticles (Wani et al., 2011; Singh & Dutta, 2018; Bardestani et al., 2019); and zeta potential measurement to determine the surface charge of nanoparticles in a liquid medium (Sikora et al., 2015; Serrano-Lotina et al., 2023).

The toxicity profiles of nanoparticles are influenced by a variety of factors, making the study of nanotoxicology complex and multifaceted. These factors as shown in Fig. 13.2 contribute to the diverse and sometimes unpredictable effects of nanoparticles on living organisms and the environment. Some key factors that influence the toxicity of nanoparticles include particle size, surface chemistry, shape and aspect ratio, composition, agglomeration and aggregation, solubility and dissolution rate, biological interactions, route of exposure, and environmental factors.

Nanoparticles exhibit unique size-dependent properties, affecting their interaction with biological systems. Smaller nanoparticles can penetrate cellular structures more easily, potentially causing greater harm (Zhao et al., 2011). Additionally, coatings and functionalization can alter the surface chemistry of nanoparticles, influencing their reactivity and interactions with biological molecules (Albanese et al., 2012). The surface charge of nanoparticles (zeta potential) also affects their stability and cellular uptake as positively or negatively charged nanoparticles can interact differently with cell membranes and proteins (Fröhlich, 2012). Nanoparticles come in various shapes as shown in Fig. 13.1, such as spherical, rod-shaped, bar-shaped (nanobar), star-shaped, cubical (nanocube), polyhedral, dendritic, wire-like (nanowires), needle-shaped (nanoneedles), pyramid-shaped, prism-shaped (nanoprism), flower-shaped, tripodal, and tetrapodal, impacting their biological interactions

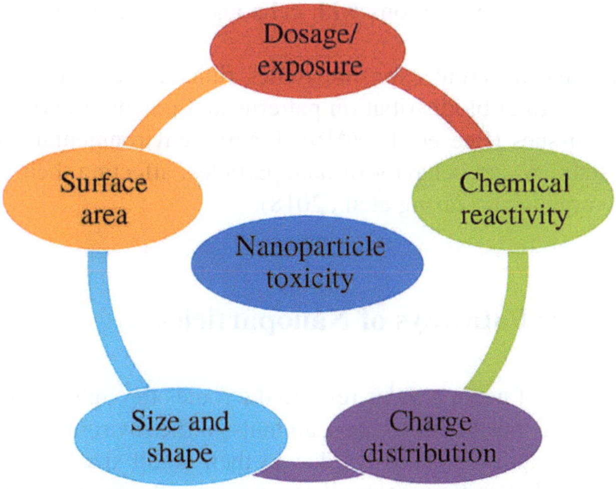

Fig. 13.1 Physicochemical factors affecting the toxicity of nanoparticles (Naseer et al., 2018)

(Tzitzios et al., 2008; Zhang et al., 2008; Al-Sherbini, 2010; Paul et al., 2010; Sun et al., 2012; Woo et al., 2012; Li et al., 2014; Melnikau et al., 2014; Sun et al., 2014; Gayathri et al., 2015; Sayed & Polshettiwar, 2015; Demille et al., 2018; Khodashenas & Ghorbani, 2019; Castillo-Henríquez et al., 2020; Liu et al., 2020a, b). Needle-like or high-aspect-ratio nanoparticles may cause more damage due to their sharp edges (Lin et al., 2018).

Different materials (metals, oxides, polymers, etc.) exhibit varying toxicities. For instance, certain metal nanoparticles like silver and titanium dioxide can have distinct biological effects including neurotoxicity, immunotoxicity, genotoxicity, cytotoxicity, and inflammation causing damage to the skin, lungs, kidneys, and liver. Additionally, the addition of different elements to nanoparticles (doping) or the creation of alloys can modify their physical and chemical properties and toxicity (Mahmoud et al., 2019).

Aggregation and agglomeration, which involve the formation of nanoparticle clusters, can impact their sedimentation, mobility, and bioavailability in aquatic environments (Cross et al., 2015; Xu, 2018; Abbas et al., 2020). The solubility and dissolution rate of nanoparticles can also affect their toxicity. Soluble nanoparticles can release ions in biological fluids, influencing their toxicity (Zhong et al., 2017). On the other hand, insoluble nanoparticles might dissolve slowly, leading to long-term exposure to toxic ions, and this may cause physical damage in tissues (Misra et al., 2012). Figure 13.1 summarizes the various physicochemical factors that affect the dissolution of nanoparticles, some of which could also independently affect the toxicity of these particles. The biological interactions of nanoparticles, such as cellular uptake and protein corona formation, also affect their toxicity (Liu et al., 2020a, b). Cellular uptake refers to the ability of nanoparticles to enter cells and accumulate in organelles which can lead to various cellular responses, impacting toxicity (Adjei et al., 2014). Protein corona refers to the layer of proteins that adsorb onto nanoparticles' surfaces upon exposure to biological fluids, and the formation of these can alter their interactions with cells and potentially their toxicity (Obst et al., 2017).

Furthermore, the different exposure routes (inhalation, ingestion, and dermal contact) lead to distinct biodistribution patterns and potential toxic effects in specific organs or tissues (Lee et al., 2016). Finally, environmental conditions also influence the stability and behavior of nanoparticles, affecting their bioavailability and toxicity to organisms (Zhang et al., 2018).

13.2 Exposure Pathways of Nanoparticles

Exposure pathways of nanoparticles refer to the routes through which these minute particles come into contact with humans, animals, or the environment. Nanoparticles can enter various exposure pathways due to their small size and unique properties (Teleanu et al., 2018). Some common exposure pathways include:

(i) Inhalation: Nanoparticles can be inhaled, particularly those airborne particles generated during manufacturing processes or product use. Once inhaled, nanoparticles can potentially reach the respiratory system and even penetrate deep into the lungs, posing health risks.
(ii) Ingestion: Nanoparticles might enter the body through ingestion, either unintentionally through contaminated food, beverages, or particles released during dental procedures or intentionally through the use of nanoparticle-based supplements or medications.
(iii) Dermal contact: Skin contact with products containing nanoparticles can lead to dermal exposure. Nanoparticles in personal care products, cosmetics, or textiles can potentially penetrate the skin and enter the bloodstream, although the extent of this penetration varies based on particle characteristics and skin conditions.
(iv) Medical applications: Nanoparticles used in medical treatments, such as drug delivery systems or imaging agents, can be deliberately introduced into the body through injections, implants, or medical procedures.
(v) Environmental exposure: Nanoparticles released into the environment through industrial activities, waste disposal, or product usage can enter soil, water bodies, or air. These particles may subsequently impact ecosystems and organisms, affecting environmental health.

Understanding these exposure pathways is crucial in assessing the potential risks associated with nanoparticle exposure and in developing safety measures to minimize any adverse effects on human health and the environment.

13.2.1 Environment Exposure

Nanoparticles can enter the environment through various pathways, leading to potential ecological impacts. Understanding these exposure pathways is essential for assessing and managing the environmental risks associated with nanomaterials. The primary exposure pathways of nanoparticles to the environment include natural processes, airborne emissions, water contamination, soil pollution, direct discharges, and atmospheric deposition.

Nanoparticles, such as inorganic oxides or fullerenes, can be introduced into the environment through various natural processes from sources like volcanic eruptions, forest fires, and weathering (Nowack & Bucheli, 2007). Volcanic eruptions are responsible for releasing vast quantities of nanoparticles into the atmosphere (Sonwani et al., 2021). These volcanic nanoparticles are typically in the form of volcanic ash, which contains a mixture of minerals and nanoscale particles (Fedotov et al., 2023). Forest fires, both natural and human-induced, can also produce nanoparticles (Costa et al., 2018). The intense heat generated during a wildfire can lead to the formation of nanoscale particles from the combustion of organic material (Rose & Ruppel, 2015). Weathering processes can break down larger minerals and rocks into nanoparticles over geological timescales (Hochella et al., 2012). These naturally occurring nanoparticles can enter the environment, particularly aquatic systems, through erosion and transportation by water. While natural processes contribute to nanoparticle exposure in the environment, the scale of human-induced nanoparticle production and release is more significant (Kumah et al., 2023).

Airborne emissions of nanoparticles primarily occur through industrial activities, combustion processes, and the use of nanomaterial-containing products (Thanachoksawang et al., 2022). For instance, manufacturing facilities release nanoparticles into the atmosphere during production, leading to their dispersion in the air, and combustion sources, such as vehicle exhaust and power plants, generate nanoparticles as byproducts. These airborne nanoparticles can be carried by wind and dispersed over wide areas. Nanoparticles can contribute to air pollution and have detrimental effects on air quality and respiratory health in humans (Kumar, 2013; Sonwani et al., 2021).

The inadvertent exposure of nanoparticles to the environment through water contamination has also become a significant concern. This contamination primarily arises from industrial discharges, runoff from land, improper disposal of nanomaterial-containing products, and the transformation of larger materials into nanoparticles in aquatic systems (Brar et al., 2010; Bundschuh et al., 2018; Bradford et al., 2022). Industrial processes often release nanoparticles into water bodies, such as rivers and oceans, where they can accumulate and disperse downstream. Inadequate management and disposal practices for products containing nanoparticles, like electronics or textiles, can also result in these particles ending up in wastewater treatment facilities and eventually entering natural water systems. Furthermore, some nanoparticles can be formed naturally in water through processes like weathering and corrosion, posing additional challenges. The environmental consequences of nanoparticle water contamination are multifaceted. Nanoparticles can disrupt aquatic ecosystems by altering the chemistry of water and sediments, affecting the health of aquatic organisms (Turan et al., 2019). They may also accumulate in the food chain, posing potential risks to human health (Kahlon et al., 2018).

Nanoparticles enter the soil environment primarily through industrial processes, through the use of nanomaterial-containing products, and through atmospheric deposition (Ahmed et al., 2021; Bose, 2022). Industrial activities can release nanoparticles directly into the soil during production and manufacturing processes. Improper waste disposal and leaching from landfills can also introduce these particles into the soil environment. Additionally, nanomaterial-containing products like pesticides and fertilizers, when applied to crops, can lead to nanoparticle accumulation in the soil. The presence of nanoparticles in soil can have far-reaching ecological implications. It can alter soil chemistry, affect nutrient availability, and potentially harm soil-dwelling organisms (Ahmed et al., 2021). Another concern is regarding the potential uptake of nanoparticles by plants, which could result in their entry into the food chain and subsequent human exposure. However, it is important to note that the current scientific consensus is that there is no evidence to suggest that NP uptake by plants poses a significant risk to human health (Azim et al., 2023).

Understanding and mitigating these exposure pathways are essential to prevent or minimize the environmental impact of nanoparticles. Regulations, proper waste management practices, responsible manufacturing practices, responsible consumer behavior, and ongoing research are crucial to ensuring the accountable use of nanomaterials and develop effective strategies for minimizing environmental contamination.

13.2.2 Human Exposure

Nanoparticles can enter the human body through various exposure pathways as presented in Fig. 13.2. Understanding these pathways is crucial for assessing potential health risks associated with nanomaterials. The primary exposure pathways of nanoparticles to humans include inhalation, ingestion, and dermal contact (Khan & Shanker, 2015; Sajid et al., 2015).

Inhalation is a significant exposure pathway for nanoparticles, particularly in occupational settings and urban environments. Nanoparticles, due to their small size, can remain suspended in the air for extended periods, making them easily inhalable (Guo et al., 2021). Nho (2020) and Fröhlich and Salar-Behzadi (2014) report that, when inhaled, NPs can penetrate deep into the respiratory system, reaching the lungs' alveoli. From the alveoli, they may enter the bloodstream, posing potential health risks. Occupational exposure occurs in industries involving nanomaterial production, handling, or processing (Ding et al., 2017), while environmental exposure can result from vehicle emissions, industrial processes, or consumer products (Coelho et al., 2012; Kumar et al., 2012). Proper ventilation, personal protective equipment, and regulatory measures are crucial to minimizing the inhalation risks associated with nanoparticles and safeguarding public health.

Ingestion is another exposure pathway for nanoparticles, particularly through contaminated food, water, or consumer products (Bergin & Witzmann, 2013). Nanoparticles can unintentionally enter the human body when ingested orally. In the food industry, nanoparticles are utilized for various purposes, such as in the production of food packaging, potentially leading to their presence in edible items (Polat et al., 2019; Chaudhary et al., 2020). Additionally, nanoparticles might contaminate water sources due to industrial discharges or environmental pollution. Once ingested, nanoparticles can interact with the digestive system, potentially entering the bloodstream and spreading throughout the body (Yang et al., 2022). Ensuring the safety of food products, regulating nanoparticles in consumables, and monitoring water sources are crucial measures to mitigate the risks associated with nanoparticle ingestion and safeguard public health.

Dermal contact is also an exposure pathway for nanoparticles, occurring when nanomaterials come into direct contact with the skin (Khater et al., 2021). Nanoparticles are present in various consumer products like cosmetics, textiles, and sunscreen, making skin exposure common (Puolamaa, 2006; NIST, 2018). Due to their small size, nanoparticles can penetrate the skin, potentially reaching deeper

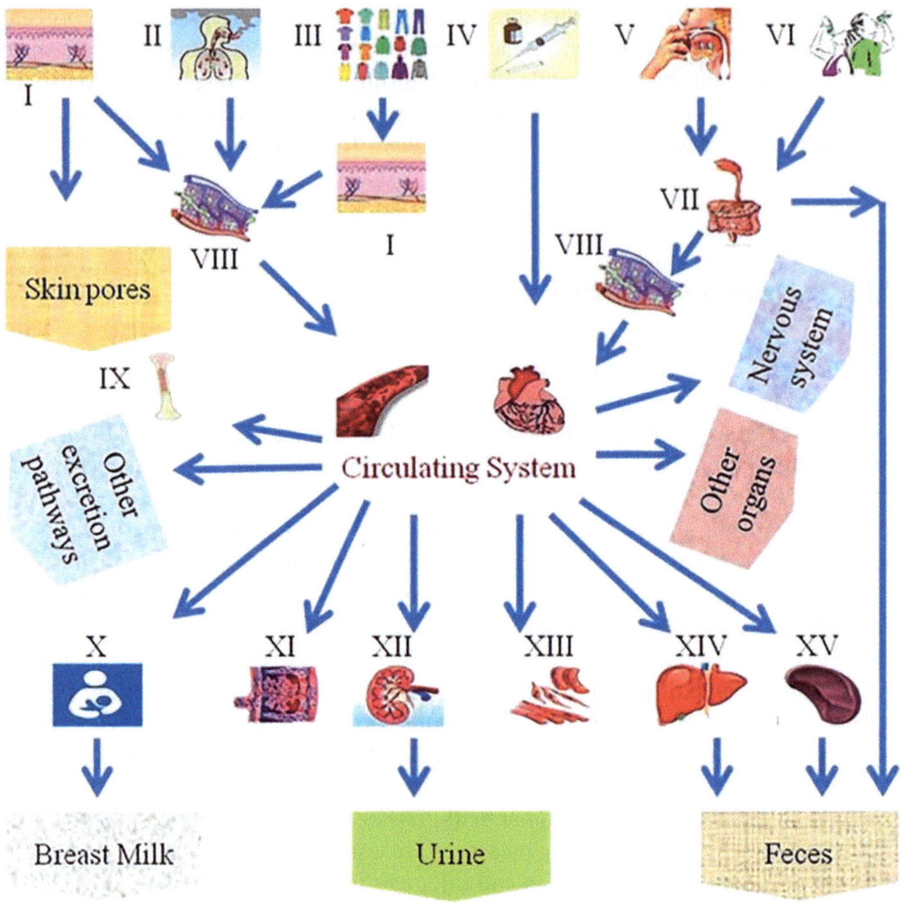

Fig. 13.2 Exposure of nanoparticles through different routes in the human body and distribution and excretion of the nanoparticles from the body. (I) Skin, (II) inhalation, (III) fabric, (VI) intravenous injection, (V) food intake, (VI) water intake, (VII) gastrointestinal tract, (VIII) lymph, (IX) bone marrow, (X) breast milk, (XI) placenta, (XII) kidney, (XIII) muscles, (XIV) liver, and (XV) spleen (Brohi et al., 2017)

tissues (Liang et al., 2013). Occupational exposure also happens in industries where workers handle nanomaterials directly (Bergamaschi, 2009). Dermal contact raises concern because nanoparticles might induce skin irritation, allergies, or systemic effects, especially if they carry toxic substances (Filon et al., 2016; Zaiter et al., 2022).

Additionally, nanoparticles are used in medical imaging, drug delivery systems, and diagnostic tools. Patients might be exposed during medical procedures involving these materials. However, the exposure level is generally low and is not considered harmful (Patra et al., 2018). Some pharmaceutical formulations also contain nanoparticles, and patients can be exposed through ingestion or injection (Moghimi, 2018).

13.3 Toxicity Assessment

The National Institute for Occupational Safety and Health (NIOSH) has proposed a recommended exposure limit (REL) of 1 μg/m^3 elemental carbon as a respirable mass 8 h time-weighted average (TWA) concentration for carbon nanotubes and nanofibers. NIOSH has also derived a recommended exposure limit (REL) for silver nanomaterials (< 100 nm primary particle size) of 0.9 micrograms per cubic meter (μg/m^3) as an airborne respirable 8 h time-weighted average (TWA) concentration (NIOSH, 2021).

Toxicity assessment of nanoparticles is the systematic evaluation of the potential adverse effects these particles may have on living organisms and the environment. It involves in vitro and in vivo studies to determine the extent of harm caused by nanoparticles. In vitro experiments, using cell cultures, help understand cellular responses to nanoparticles, while in vivo studies, utilizing animal models, provide insights into organ toxicity, immune responses, and other physiological effects (Zhang et al., 2011). Common endpoints include assessing cell viability, genotoxicity, inflammation, and the impact on organs or tissues. Such assessments aim to identify the dose-response relationships, thresholds for toxicity, and underlying mechanisms. Toxicity assessments play a vital role in ensuring the safe development and application of nanomaterials, informing regulatory decisions, and protecting both human health and the environment from the potential hazards associated with nanoparticles (Sufian et al., 2017).

Cell culture experiments to understand the cellular response to nanomaterials have been investigated to assess endpoints such as cell viability, oxidative stress, inflammation, and genotoxicity. Various cell lines and primary cultures are employed to model specific cell types or tissues. For example, researchers often use human cell lines to simulate human responses to nanomaterials. There are numerous in vitro experiments to assess the endpoints of nanotoxicity. An MTT or MTS assay is commonly used to measure cell metabolic activity, providing insights into the cytotoxic effects of nanoparticles. Comet assay or micronucleus assay is used to detect genetic damage and chromosomal aberrations. Tumor necrosis factor (TNF)-alpha or interleukin (IL)-6 is used to measure the release of pro-inflammatory cytokines from immune cells or lung epithelial cells exposed to nanoparticles. Fluorescent probes and biochemical assays can be used for oxidative stress assays to evaluate reactive oxygen species (ROS) production. Fluorescence and electron microscopy can be used to visualize the uptake, intracellular localization, and distribution of nanoparticles within cells. Other studies involved in in vitro experiments are mechanistic studies, nanomaterial-drug interaction studies, biodistribution and pharmacokinetics, organ-on-a-chip models, and nanoparticle-protein interaction studies (Zoeller & Vandenberg, 2015).

Sani et al. (2021), reported the toxicology of gold nanoparticles (AuNPs), with emphasis on oxidative damage to in vitro cell lines which aided observations of toxicity to the kidney, spleen, and liver. However, in vitro studies also have limitations as they do not capture the complexity of the whole organism and the dynamic interactions that occur in a living system. Therefore, in vivo studies, epidemiological data, and environmental assessments are also necessary to provide a comprehensive understanding of the potential risks and benefits associated with nanomaterials.

In scientific investigations, animal models have been effectively employed to examine the consequences of exposure to nanomaterials, enabling the evaluation of organ toxicity, immune responses, inflammation, oxidative stress, genotoxicity, and other pertinent endpoints. The utilization of in vivo studies offers a closer emulation of the natural routes of exposure, such as inhalation, ingestion, or injection, thereby facilitating a comprehensive comprehension of the impacts of nanomaterials on the human body. Notably, although inhaled particles, including nanoparticles, primarily affect the respiratory system, systemic effects, chronic, acute, or subacute in nature, can also be impacted (Nemmar et al., 2013). Furthermore, a multitude of in vivo experiments have been conducted to assess a wide range of nanotoxicity endpoints. Radiolabeling and fluorescent markers are implemented in pharmacokinetics and biodistribution investigations. Studies focusing on immunotoxicity involve the examination of immunosuppression, allergic reactions, and variations in immune cell populations. In vivo research spans various domains, encompassing toxicity evaluations, studies specific to particular organs, research related to cancer, investigations into reproductive and developmental aspects, assessments of chronic and long-term effects, studies of neurotoxicity, examinations of metabolic and nutritional impacts, and investigations into environmental ramifications (Sufian et al., 2017).

Investigations conducted in vivo provide essential revelations regarding the internal distribution of nanoparticles, their propensity to accumulate within distinct organs, and their capacity to elicit systemic reactions. Nevertheless, in vivo research is not without constraints. Ethical apprehensions concerning animal well-being often arise, and outcomes derived from animal-based studies may not invariably extrapolate directly to human responses. Consequently, it is imperative to amalgamate in vivo findings with in vitro analyses, epidemiological data, and environmental appraisals to achieve a comprehensive assessment of the safety profile of nanomaterials.

13.4 Potential Hazards of Nanoparticles

Nanoparticles, characterized by their exceptional attributes and diverse utilization, have introduced noteworthy health, environmental, economic, and social risks that warrant thorough evaluation. These minute particles have permeated numerous sectors, encompassing medicine, electronics, and consumer goods. However, it is imperative not to disregard their potential ramifications.

13.4.1 Human Health Hazards

Nanoparticles can travel through the bloodstream to the organs after they have entered the body. Moreover, they have the ability to pass through the blood-brain barrier or travel by axonal transport via the olfactory nerve, contingent on factors like size, shape, and chemical reactivity (Simko & Mattsson, 2014). The small size and distinctive characteristics of nanoparticles can give rise to possible threats to human health and can lead to or contribute to the onset of a range of illnesses, as outlined in Fig. 13.3. The precise perils are contingent upon an array of factors, including the nanoparticle type, levels of exposure, exposure duration, and individual susceptibility. A few conceivable risks associated with nanoparticles encompass respiratory complications, cardiovascular repercussions, dermal issues, neurotoxicity, immunotoxicity, genotoxicity, challenges linked to reproduction and development, and unanticipated risks (Sharma et al., 2020).

Respiratory complications arising from nanoparticles are attributable to their capacity to penetrate deeply into the respiratory system, where they can initiate or exacerbate a spectrum of health concerns. Inhalation of nanoparticles can result in

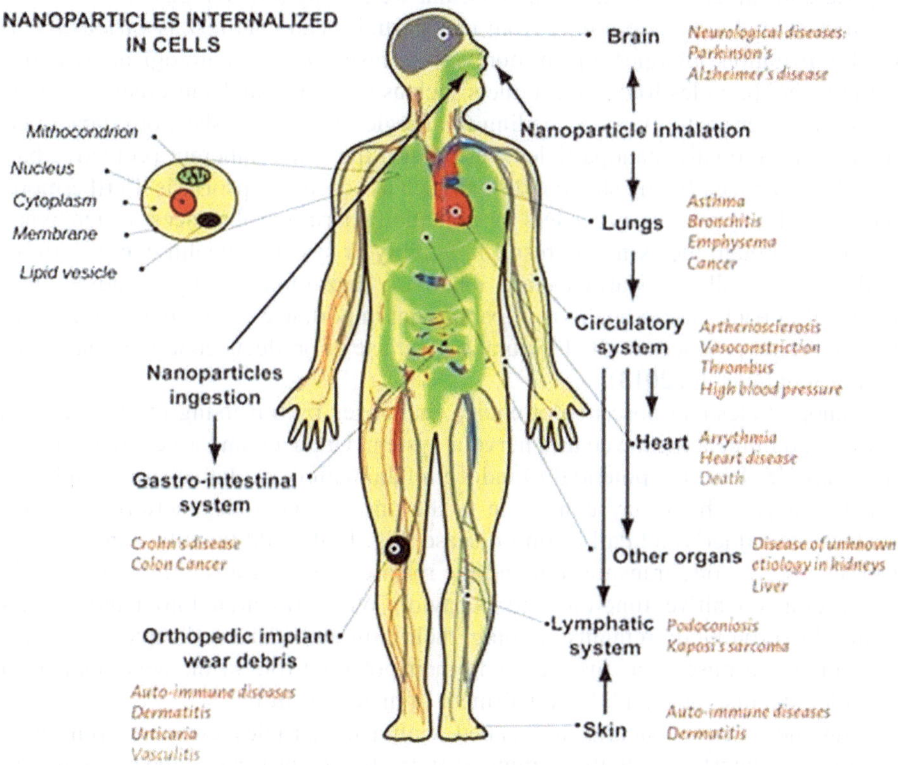

Fig. 13.3 Schematics of the exposure pathways and major diseases associated with nanoparticle exposure from epidemiological and clinical studies (Buzea et al., 2007)

lung inflammation, with potential implications for conditions like asthma, bronchitis, and chronic obstructive pulmonary disease (COPD). It can further manifest as fibrosis and compromised pulmonary functions. The diminutive size and physicochemical attributes of nanoparticles afford them the ability to circumvent the body's innate defenses, affording access to vulnerable lung tissues and the potential for tissue damage. Furthermore, nanoparticles can carry adsorbed toxic substances, exacerbating respiratory issues (Schulte et al., 2014).

Nanoparticles can migrate from the lungs to other organs, including the bloodstream, possibly contributing to cardiovascular disturbances. These particles may exert substantial cardiovascular repercussions by infiltrating the bloodstream, potentially implicating heart and circulatory system maladies. Inhalation or absorption of nanoparticles may incite systemic inflammation and oxidative stress, culminating in vascular damage and an augmented risk of atherosclerosis and heart disease. Certain nanoparticles may also interfere with blood coagulation, heightening the potential for thrombosis or stroke. Additionally, these minuscule particles can impair the endothelium's function, the inner lining of blood vessels, further imperiling cardiovascular health. Although ongoing research strives to elucidate the full extent of these impacts, it underscores the significance of reducing nanoparticle exposure to safeguard overall cardiovascular well-being (OSHA, 2013).

Skin-related issues stemming from nanoparticles can manifest as various forms of skin irritation, allergies, or, in more severe instances, dermatological concerns. Certain nanoparticles found in products such as cosmetics and sunscreens have the capacity to permeate the skin, leading to allergic reactions, rashes, or contact dermatitis. Additionally, nanoparticles can perturb the skin's inherent protective barrier, thereby exacerbating skin conditions such as eczema or psoriasis. Furthermore, nanoparticles may serve as carriers for potentially harmful substances, which, when absorbed through the skin, can result in toxicity or systemic health complications. Although not all nanoparticles are inherently hazardous, comprehending their potential impact on skin health is of paramount importance, necessitating the adoption of secure practices in the formulation and use of products containing nanoparticles (Schulte et al., 2013).

Nanoparticles can pose neurotoxicity challenges by traversing the blood-brain barrier and influencing the central nervous system. Upon gaining access to the brain, nanoparticles have the potential to induce inflammation, oxidative stress, and neuronal damage, which may contribute to the onset of neurodegenerative ailments such as Alzheimer's and Parkinson's disease (Fig. 13.4). Additionally, nanoparticles can disrupt synaptic transmission and the release of neurotransmitters, potentially influencing cognitive function and behavior. While research into nanoparticle-induced neurotoxicity remains ongoing, their diminutive size and capacity to infiltrate the brain raise concerns regarding their potential role in the development of neurological disorders and the impairment of mental health.

Immunotoxicity concerns associated with nanoparticles emanate from their capability to interface with the immune system, thereby instigating the possibility of adverse consequences. Nanoparticles have the potential to incite inflammation, suppress immune responses, or incite allergic reactions. These immunotoxic effects

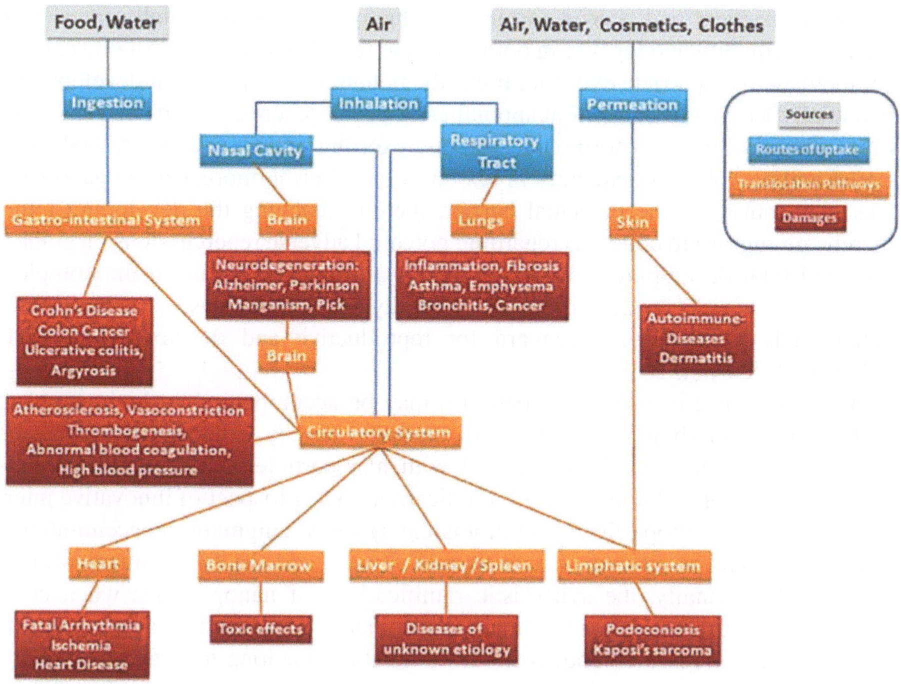

Fig. 13.4 The predominant routes of NP exposure, uptake, translocation, and potential risks (Zoroddu et al., 2014)

may compromise the body's resilience against infections and augment the susceptibility to autoimmune disorders. Given the diminutive size and unique attributes of nanoparticles, they may pose a greater challenge for the immune system to manage, potentially eliciting atypical responses. A comprehensive comprehension of nanoparticle-induced immunotoxicity holds paramount significance, especially in the context of assessing their safety across diverse applications, spanning from medical treatments to consumer products. Moreover, it is instrumental in the formulation of strategies aimed at mitigating potential risks to human health and the integrity of the immune system (Nasrullah et al., 2023).

Genotoxicity issues precipitated by nanoparticles result from their capacity to inflict damage upon an organism's DNA, thereby heightening the prospect of genetic mutations and an escalated susceptibility to cancer. Upon entering cells, nanoparticles have the potential to generate reactive oxygen species and disrupt cellular machinery, thereby causing DNA strand breaks, mutations, and chromosomal aberrations. Such genetic harm can enhance the likelihood of cancer onset and other genetic anomalies. It is imperative to acknowledge that the genotoxic consequences of nanoparticles hold a particular concern, as they may be inheritable and have implications for future generations (Murdock et al., 2008).

Reproductive and developmental concerns arising from nanoparticles pose a substantive area of apprehension. Nanoparticles exhibit the capacity to infiltrate biological barriers and accrue within reproductive organs, potentially interfering with sperm quality and fertility in both male and female individuals. Among expectant mothers, nanoparticle exposure holds the potential to impact fetal development, potentially leading to congenital anomalies or developmental disorders. These particles can also perturb hormonal systems, exerting influence over reproductive cycles and provoking imbalances in sex hormones. Furthermore, certain nanoparticles may transcend the placental barrier, thereby reaching the developing fetus, thereby giving rise to concerns regarding potential adverse repercussions on embryonic and fetal development. Research efforts are ongoing to unveil the complete scope of these risks, underscoring the necessity for comprehensive evaluations of nanoparticle safety as a safeguard for reproductive and developmental well-being (Park & Choi, 2010).

The emergence of novel nanoparticles may be accompanied by an incomplete understanding of their potential health hazards, thereby ushering in unforeseen risks. These unknown risks associated with nanoparticles encompass their protracted ecological and human health implications. The prospect of innovative interactions between nanoparticles and biological systems, enigmatic bioaccumulation trends, and their persistency within tissues, organs, and the environment warrant concern. Additionally, the synergistic ramifications of nanoparticles when combined with other contaminants remain inadequately apprehended, amplifying the call for sustained research endeavors. In summation, the long-term and cumulative health consequences of nanoparticle exposure necessitate further research to ensure the preservation of public safety (Schluesener & Schluesener, 2013).

13.4.2 Environmental Health Hazards

Nanoparticles, given their diminutive size and distinctive characteristics, introduce the potential for environmental hazards. Significant concerns pertaining to the environmental footprint of nanoparticles encompass issues such as ecotoxicity, bioaccumulation, biomagnification, soil contamination, air pollution, alterations in ecosystem dynamics, the proliferation of detrimental species, and the uncertainties surrounding their environmental fates.

Ecotoxicity induced by nanoparticles denotes the deleterious impacts of nanoscale particles on ecosystems and the organisms inhabiting them. Due to their minute size and distinctive attributes, nanoparticles possess the capacity to infiltrate cell membranes, disrupt biological processes, and accrue within living organisms (Ochoa, 2018). This capacity can culminate in detrimental ecological consequences, notably including disruptions in food chains and alterations in microbial communities within soil and water ecosystems. In aquatic environments, nanoparticles have the potential to inflict harm upon aquatic life, affecting both primary producers, such as phytoplankton and algae, and organisms occupying higher trophic levels, including fish and invertebrates, via mechanisms of bioaccumulation and biomagnification (Zhao et al., 2023). In soil ecosystems, nanoparticles can bring about changes in microbial communities, potentially impacting nutrient cycling and thereby influencing plant health. Moreover, these particles may gain entry into plants through their roots, subsequently raising the possibility of human exposure upon the consumption of contaminated crops. Consequently, it is imperative to grasp and mitigate the ecotoxic repercussions of nanoparticles to ensure the preservation of environmental well-being and the sustainability of ecosystems amidst mounting nanoparticle exposure (Murdock, 2017).

Soil contamination arising from nanoparticles is the consequence of the introduction of these minuscule particles into the soil environment, typically occurring through industrial processes, improper disposal practices, or the utilization of nanomaterial-containing products. Upon entering the soil matrix, nanoparticles have the capacity to modify its physical and chemical characteristics, thereby influencing nutrient availability and soil structure. These transformations can have detrimental effects on soil microorganisms, plants, and soil-dwelling organisms, thereby disturbing the intricate equilibrium of terrestrial ecosystems. Additionally, there is a mounting apprehension concerning the prospect of nanoparticle uptake by plants, which could facilitate their entry into the food chain, thereby impacting not only ecosystems but also human health via the consumption of contaminated crops. Soil contamination by nanoparticles poses a substantial threat to ecosystems and agricultural productivity and may have far-reaching consequences for environmental health and human well-being (Nafisi & Maibach, 2018).

Air pollution induced by nanoparticles arises from the dispersion of these minute particles into the atmosphere. The diminutive size and unique attributes of nanoparticles contribute to their adverse effects. These particles have the propensity to remain suspended in the air for protracted durations and possess the ability to penetrate deeply into the respiratory system upon inhalation. This capability to infiltrate lung tissues and potentially access the bloodstream raises concerns regarding their health implications. Exposure to airborne nanoparticles has been correlated with respiratory complications, cardiovascular challenges, and systemic inflammation. Furthermore, these minuscule particles can carry an array of toxic substances, augmenting the complexity of their potential health hazards (Kuempel et al., 2012). Vulnerable demographic segments, including children, the elderly, and individuals with preexisting health conditions, are particularly susceptible. To mitigate air pollution stemming from nanoparticles, regulatory interventions, cleaner industrial methodologies, and advancements in emission control technologies hold pivotal significance (Macko et al., 2022).

Alterations in ecosystem dynamics precipitated by nanoparticles ensue from the introduction of these ultrasmall particles into natural environments. The perturbation of aquatic and terrestrial ecosystems due to nanoparticle contamination can result in a diminution of biodiversity, impacting the stability and functioning of these ecosystems. Moreover, nanoparticles may impede nutrient cycles, leading to imbalances and influencing the growth of plants and other organisms. The potential for nanoparticles to interfere with nutrient cycles and create imbalances is particularly noteworthy. In addition, the bioaccumulation of nanoparticles, especially in apex predators, can perturb predator-prey relationships. Both this bioaccumulation and the conceivable genotoxic impacts of nanoparticles have the potential to induce shifts in population dynamics over time (Lundquist & Artursson, 2016).

Nanoparticles have the potential to exacerbate the proliferation of detrimental species, including specific pathogens and invasive organisms, through various mechanisms. Initially, their minuscule size and extensive surface area render nanoparticles alluring vehicles for an assortment of substances, encompassing nutrients, pollutants, and toxins. This quality can amplify the availability and conveyance of these noxious substances in the environment, creating a more hospitable milieu for their proliferation. Additionally, nanoparticles have the capacity to disrupt the natural equilibrium of ecosystems, with potential implications for beneficial species and the facilitation of harmful species. Finally, nanoparticles may directly influence the physiology and biology of certain organisms, potentially conferring advantages upon detrimental species. For instance, specific nanoparticles can stimulate the proliferation of harmful algae, thereby engendering algal blooms that can deplete oxygen levels in aquatic systems, consequently compromising aquatic life (Ezealisiji et al., 2019).

The fate of nanoparticles in the environment is uncertain owing to their peculiar attributes. They can exhibit high mobility and may not conform to the behaviors of larger particles, influencing their distribution and accumulation within ecosystems. Furthermore, nanoparticles can engage in interactions with other environmental contaminants, potentially yielding uncharted synergistic effects and the creation of novel compounds with distinct properties and toxicity. The long-term persistence, transformations, and the plausible accumulation of nanoparticles in varied environmental compartments, spanning air, water, soil, and even biological tissues, remain areas of incomplete comprehension. Moreover, there are concerns regarding their potential to bioaccumulate within food chains, impacting both wildlife and human health. Consequently, the intricate and sometimes unpredictable environmental behaviors of nanoparticles impart critical questions regarding their enduring repercussions on ecosystems, underscoring the necessity for further research and regulatory considerations (Feitshans, 2017).

From an economic vantage point, nanoparticles can introduce potential perils. The fervor to embrace nanotechnology without a thorough comprehension and the enforcement of regulations can give rise to substantial financial risks. The manufacturing and market introduction of nanomaterials, if not vigilantly managed, can lead to costly product recalls, legal disputes, and reputational damage to companies. Enterprises must allocate resources toward research, safety protocols, and adherence to evolving regulatory frameworks, potentially escalating operational expenses. Additionally, the increased health issues stemming from nanoparticle exposure may contribute to augmented healthcare costs and place an economic burden upon societies (Winkler et al., 2013; Nikam et al., 2021).

One of the most disconcerting social challenges associated with nanoparticles centers on their potential health implications. Occupational exposure to nanoparticles, prevalent in industries like manufacturing, construction, and healthcare, may expose workers to risks. Such occupational exposure can potentially translate to reduced productivity and heightened absenteeism among affected labor forces. The social ramifications of nanoparticles also extend to ethical considerations. The integration of nanoparticles into consumer products, including cosmetics and sunscreens, has given rise to apprehensions about their safety, particularly when these products are disseminated widely devoid of adequate regulation and labeling. Ethical dilemmas emerge pertaining to issues of transparency and informed consumer choices. Consumer anxieties regarding the safety of nanoparticles can exert an impact on consumer confidence in an array of products and industries associated with nanoparticle use (Saifi et al., 2018).

13.4.3 Risk Management Strategies

In the field of nanotoxicology, risk assessment and risk management are indispensable processes with the primary objective of comprehending and alleviating potential hazards linked to nanoparticles. Nanotoxicology encompasses the systematic investigation of the deleterious impacts of nanoparticles on both biological organisms and the surrounding ecosystem. In light of the swift progression of nanotechnology and the escalating utilization of nanoparticles in diverse applications, it becomes imperative to evaluate the perils they engender and institute efficacious measures for safeguarding human well-being and the environment. Proficiently handling the risks associated with nanoparticles necessitates a combination of regulatory protocols, research endeavors, and best practices across various industrial sectors. Noteworthy strategies for the effective management of nanoparticle-related risks encompass regulatory supervision, research and development initiatives, occupational and consumer safety measures, environmental preservation efforts, collaborative research ventures and information exchange, and the application of life cycle assessment (LCA) methodologies (Warheit, 2018).

The development of regulations and their subsequent adherence constitute a central and indispensable facet of the effective management of nanoparticle-related risks. Governments and regulatory bodies must undertake the formulation of comprehensive guidelines to govern the safe utilization and disposal of nanomaterials. These regulations establish benchmarks for research practices, manufacturing processes, and product labeling, thereby mitigating potential hazards to human health and the environment (Lakshmi et al., 2022). Companies must conform to these regulatory standards not only to maintain public confidence but also to avert legal liabilities. Compliance necessitates the rigorous implementation of testing protocols and safety measures across the entire life cycle of nanoparticles, spanning from their production to ultimate disposal. In this manner, regulation development and compliance function as pivotal risk management strategies, ensuring that the advantages of nanotechnology are realized while concurrently mitigating potential detrimental impacts on society and the natural world.

Furthermore, the strategies of labeling and reporting also hold significance as fundamental risk management approaches in the context of nanoparticles. Properly labeling products containing nanoparticles empowers consumers and workers to make informed choices, thereby minimizing their exposure risks (Geiser et al., 2017). Meanwhile, reporting requirements for manufacturers and researchers serve to enhance transparency and facilitate the collection of data concerning nanoparticle safety. The timely reporting of adverse incidents or accidents involving nanoparticles enables regulatory agencies to respond expeditiously. Collectively, these strategies heighten safety and facilitate more robust risk assessment procedures (Feitshans, 2017).

An additional crucial risk management strategy in the context of nanoparticles pertains to research and development efforts, encompassing the exploration of toxicity studies and the generation of safer nanoparticle alternatives. Toxicity studies serve as a means to assess the potential health and environmental risks linked to nanoparticle exposure. They yield indispensable data concerning the safety and repercussions of nanoparticles, thereby contributing to the establishment of regulatory frameworks and safety guidelines (Karunarathne et al., 2022). By identifying potential hazards and defining safe exposure limits, these studies play a pivotal role in safeguarding the well-being of workers, consumers, and ecosystems. A comprehensive exploration of nanoparticle toxicity fosters responsible and well-informed decision-making within both industries and governmental bodies, enabling the realization of the advantages of nanotechnology while concurrently minimizing risks to human health and the environment. It represents an indispensable element in guaranteeing the secure utilization of nanoparticles across a spectrum of applications (Graham et al., 2017).

Furthermore, the adoption of life cycle assessment (LCA) stands as an effective risk management strategy for nanoparticles. LCA entails a holistic examination of the entire life cycle of products containing nanoparticles, ranging from the extraction of raw materials to their eventual disposal. The objective is to pinpoint potential environmental and health risks inherent at each stage of the life cycle. This approach aids in evaluating the ecological footprint, energy consumption, and waste generation associated with nanoparticles, thereby facilitating the development of environmentally friendly and safer alternatives. LCA champions sustainability by guiding industries toward responsible manufacturing practices and reducing adverse impacts. By scrutinizing the complete life cycle of nanoparticle-based products, LCA ensures that decision-making is informed, potential hazards are minimized, and the integration of nanoparticles into products and processes transpires in a manner that is environmentally and socially responsible (Date et al., 2016).

Additionally, the identification and promotion of safer alternatives stand as pivotal risk management strategies for nanoparticles. This approach entails the research and development of nanomaterials characterized by reduced toxicity and heightened safety profiles. By substituting hazardous nanoparticles with more benign options, industries can effectively mitigate potential health and environmental hazards. Safer alternatives not only serve to protect the well-being of workers and consumers but also address ethical considerations. Furthermore, they frequently engender innovative solutions that either maintain or enhance product performance while concurrently reducing the likelihood of harm. This proactive strategy aligns technological progress with responsible practices, thereby ensuring the sustainable and secure utilization of nanotechnology across various sectors, encompassing but not limited to healthcare and manufacturing (Liu et al., 2014; Egbuna et al., 2021).

The implementation of workplace safety protocols represents another potent risk management strategy for nanoparticles. These protocols establish guidelines for the proper handling, storage, and disposal of nanomaterials, which include provisions for engineering controls, personal protective equipment, and worker training. These measures effectively reduce the risks associated with nanoparticle exposure in diverse industrial settings. Adequate training and adherence to safety protocols are instrumental in safeguarding workers against potential health hazards, such as respiratory issues or skin contact reactions. Additionally, these safety protocols guarantee the secure disposal of nanoparticles, thus preventing environmental contamination. Strict compliance with these measures is imperative to protect employees and forestall workplace accidents. By fostering a culture of safety, workplace safety protocols adeptly manage the risks linked to nanoparticles, thus enabling their responsible and secure use in research, manufacturing, and other sectors (De Matteis, 2017).

Furthermore, workplace monitoring and surveillance constitute additional risk management strategies for nanoparticles. Regular monitoring serves the purpose of tracking airborne nanoparticle levels to ensure that exposure remains within safe thresholds (WHO, 2021). Surveillance encompasses health assessments and the early detection of potential adverse effects in workers who are exposed to nanoparticles (Hassan & Hasary, 2022). These measures are vital for identifying and mitigating risks, safeguarding the health of employees, and facilitating prompt intervention when necessary. Through proactive monitoring and surveillance in workplaces, it is possible to effectively manage and minimize the risks associated with nanoparticles, thereby ensuring a safer and healthier work environment for employees in industries where nanomaterials are utilized (Palmer & DeLouise, 2016).

13.5 Conclusion

Consumer safety emerges as a paramount risk management strategy in the context of nanoparticles embedded in various products. The proper labeling and disclosure of nanoparticle utilization enable consumers to make informed choices, thereby minimizing potential health hazards. Stringent regulatory measures and adherence to compliance requirements ensure that products undergo thorough testing and meet established safety standards. By placing consumer safety at the forefront, individuals can be shielded from potential adverse effects, such as skin irritation or exposure to toxic substances. This approach fosters trust and confidence in products containing nanoparticles.

Concurrently, environmental protection constitutes a pivotal risk management strategy for nanoparticles. Prudent disposal of nanomaterials is imperative to avert environmental contamination and safeguard public health. Specialized waste handling protocols are instrumental in preventing the infiltration of nanoparticles into ecosystems, thus mitigating the potential disruption of natural equilibrium. This strategy serves to minimize the enduring perils associated with nanoparticle waste and acts as a barrier against their unintended release into the environment. Additionally, environmental monitoring, which constitutes another risk management strategy for nanoparticles, entails the regular assessment of nanoparticle presence and impact in ecosystems, air, water, and soil. This surveillance is instrumental in the early identification of potential environmental contamination and ecological disturbances stemming from nanoparticle discharge. By monitoring these effects, timely corrective measures can be implemented, and comprehensive regulations to restrict nanoparticle exposure can be devised. This approach plays a pivotal role in preserving the environment, protecting biodiversity, and ensuring the sustainable utilization of nanomaterials.

Collaborative research and information sharing stand as yet another vital risk management strategy for nanoparticles. In an interconnected global landscape, the sharing of data and findings concerning the safety and potential hazards of nanoparticles assumes critical importance. By pooling resources and collective knowledge, scientists, regulatory authorities, and industrial sectors can collaboratively assess risks, discern best practices, and formulate comprehensive safety guidelines. Collaboration fosters transparency and harmonizes safety standards, thereby promoting responsible utilization of nanoparticles. Moreover, the dissemination of shared information aids in the early identification of hazards, risk mitigation, and well-informed decision-making, all of which are fundamental for protecting public health and the environment. This cooperative approach ensures a safer and more sustainable integration of nanotechnology across various sectors. By amalgamating these strategies, regulatory bodies, industries, and researchers can effectively cooperate in the management of risks associated with nanoparticles, thereby ensuring their secure utilization and minimizing potential harm to human health and the environment.

References

Abbas, Q., Yousaf, B., Ali, M. U., Munir, M. A. M., El-Naggar, A., Rinklebe, J., & Naushad, M. (2020). Transformation pathways and fate of engineered nanoparticles (ENPs) in distinct interactive environmental compartments: A review. *Environment International, 138*, 105646.

Adjei, I. M., Sharma, B., & Labhasetwar, V. (2014). Nanoparticles: Cellular uptake and cytotoxicity. In *Nanomaterial: Impacts on cell biology and medicine* (pp. 73–91). Springer.

Ahmed, B., Rizvi, A., Ali, K., Lee, J., Zaidi, A., Khan, M. S., & Musarrat, J. (2021). Nanoparticles in the soil–plant system: A review. *Environmental Chemistry Letters, 19*, 1545–1609.

Akhtar, K., Khan, S. A., Khan, S. B., & Asiri, A. M. (2018). Scanning electron microscopy: Principle and applications in nanomaterials characterization. In *Handbook of materials characterization* (pp. 113–145). Springer.

Albanese, A., Tang, P. S., & Chan, W. C. (2012). The effect of nanoparticle size, shape, and surface chemistry on biological systems. *Annual Review of Biomedical Engineering, 14*, 1–16.

Al-Sherbini, A. S. A. (2010). Thermal reshaping of gold nanorods in micellar solution of water/glycerol mixtures. *Journal of Nanomaterials, 2010*, 1–6.

Azim, Z., Singh, N. B., Singh, A., Amist, N., Niharika, K. S., Yadav, R. K., Bano, C., & Yadav, V. (2023). A review summarizing uptake, translocation and accumulation of nanoparticles within the plants: Current status and future prospectus. *Journal of Plant Biochemistry and Biotechnology, 32*(2), 211–224.

Bardestani, R., Patience, G. S., & Kaliaguine, S. (2019). Experimental methods in chemical engineering: Specific surface area and pore size distribution measurements—BET, BJH, and DFT. *The Canadian Journal of Chemical Engineering, 97*(11), 2781–2791.

Behboudi, A., Mohammadi, T., & Ulbricht, M. (2023). High performance antibiofouling hollow fiber polyethersulfone nanocomposite membranes incorporated with novel surface-modified silver nanoparticles suitable for membrane bioreactor application. *Journal of Industrial and Engineering Chemistry, 119*, 298–314.

Bergamaschi, E. (2009). Occupational exposure to nanomaterials: Present knowledge and future development. *Nanotoxicology, 3*(3), 194–201.

Bergin, I. L., & Witzmann, F. A. (2013). Nanoparticle toxicity by the gastrointestinal route: Evidence and knowledge gaps. *International Journal of Biomedical Nanoscience and Nanotechnology, 3*(1–2), 163–210.

Bose, P. (2022). *Nanoparticles in soil: What is the risk?* AZoNano.com. Available at: https://www.azonano.com/article.aspx?ArticleID=6046. Accessed on 12 Oct 2023.

Bradford, S. A., Shen, C., Kim, H., Letcher, R. J., Rinklebe, J., Ok, Y. S., & Ma, L. (2022). Environmental applications and risks of nanomaterials: An introduction to CREST publications during 2018–2021. *Critical Reviews in Environmental Science and Technology, 52*(21), 3753–3762.

Brar, S. K., Verma, M., Tyagi, R. D., & Surampalli, R. Y. (2010). Engineered nanoparticles in wastewater and wastewater sludge–evidence and impacts. *Waste Management, 30*(3), 504–520.

Brohi, R. D., Wang, L., Talpur, H. S., Wu, D., Khan, F. A., Bhattarai, D., Rehman, Z. U., Farmanullah, F., & Huo, L. J. (2017). Toxicity of nanoparticles on the reproductive system in animal models: A review. *Frontiers in Pharmacology, 148*, 606.

Bundschuh, M., Filser, J., Lüderwald, S., McKee, M. S., Metreveli, G., Schaumann, G. E., Schulz, R., & Wagner, S. (2018). Nanoparticles in the environment: Where do we come from, where do we go to? *Environmental Sciences Europe, 30*(1), 1–17.

Buzea, C., Pacheco, I. I., & Robbie, K. (2007). Nanomaterials and nanoparticles: Sources and toxicity. *Biointerphases, 2*(4), MR17–MR71.

Castillo-Henríquez, L., Alfaro-Aguilar, K., Ugalde-Álvarez, J., Vega-Fernández, L., Montes de Oca-Vásquez, G., & Vega-Baudrit, J. R. (2020). Green synthesis of gold and silver nanoparticles from plant extracts and their possible applications as antimicrobial agents in the agricultural area. *Nanomaterials, 10*(9), 1763.

Chaudhary, P., Fatima, F., & Kumar, A. (2020). Relevance of nanomaterials in food packaging and its advanced future prospects. *Journal of Inorganic and Organometallic Polymers and Materials, 30*, 5180–5192.

Coelho, M. C., Torrao, G., & Emami, N. (2012). Nanotechnology in automotive industry: Research strategy and trends for the future—Small objects, big impacts. *Journal of Nanoscience and Nanotechnology, 12*(8), 6621–6630.

Costa, M. A. M., Fogarin, H. M., Costa, A. F., Pires, L. O., Silva, D. D., Lima-Souza, M., & Dussán, K. J. (2018). Nanoparticles emitted by biomass burning: Characterization and monitoring of risks. In *Nanomaterials: Ecotoxicity, safety, and public perception* (pp. 253–279). Springer.

Cross, R. K., Tyler, C., & Galloway, T. S. (2015). Transformations that affect fate, form and bioavailability of inorganic nanoparticles in aquatic sediments. *Environmental Chemistry, 12*(6), 627–642.

Dachraoui, W., Keller, D., Henninen, T. R., Ashton, O. J., & Erni, R. (2021). Atomic mechanisms of nanocrystallization via cluster-clouds in solution studied by liquid-phase scanning transmission electron microscopy. *Nano Letters, 21*(7), 2861–2869.

Date, A. A., Hanes, J., & Ensign, L. M. (2016). Nanoparticles for oral delivery: Design, evaluation and state-of-the-art. *Journal of Controlled Release: Official Journal of the Controlled Release Society, 240*, 504–526. https://doi.org/10.1016/j.jconrel.2016.06.016

De Matteis, V. (2017). Exposure to inorganic nanoparticles: Routes of entry, immune response, biodistribution and in vitro/in vivo toxicity evaluation. *Toxics, 5*(4), 29. https://doi.org/10.3390/toxics5040029

Demille, T. B., Hughes, R. A., Preston, A. S., Adelung, R., Mishra, Y. K., & Neretina, S. (2018). Light-mediated growth of noble metal nanostructures (Au, Ag, Cu, Pt, Pd, Ru, Ir, Rh) from micro-and nanoscale ZnO tetrapodal backbones. *Frontiers in Chemistry, 6*, 411.

Ding, Y., Kuhlbusch, T. A., Van Tongeren, M., Jiménez, A. S., Tuinman, I., Chen, R., Alvarez, I. L., Mikolajczyk, U., Nickel, C., Meyer, J., & Kaminski, H. (2017). Airborne engineered nanomaterials in the workplace—A review of release and worker exposure during nanomaterial production and handling processes. *Journal of Hazardous Materials, 322*, 17–28.

Egbuna, C., Parmar, V. K., Jeevanandam, J., Ezzat, S. M., Patrick-Iwuanyanwu, K. C., Adetunji, C. O., Khan, J., Onyeike, E. N., Uche, C. Z., Akram, M., & Ibrahim, M. S. (2021). Toxicity of nanoparticles in biomedical application: Nanotoxicology. *Journal of Toxicology, 2021*, 1–21.

Ezealisiji, K. M., Siwe-Noundou, X., Maduelosi, B., Nwachukwu, N., & Krause, R. W. M. (2019). Green synthesis of zinc oxide nanoparticles using Solanum torvum (L) leaf extract and evaluation of the toxicological profile of the ZnO nanoparticles–hydrogel composite in Wistar albino rats. *International Nano Letters, 9*(2), 99–107. https://doi.org/10.1007/s40089-018-0263-1

Fedotov, P. S., Ermolin, M. S., & Ivaneev, A. I. (2023). Study of elemental composition and properties of volcanic ash and urban dust nanoparticles. In *Advances in geochemistry, analytical chemistry, and planetary sciences: 75th Anniversary of the Vernadsky institute of the Russian academy of Sciences* (pp. 133–143). Springer International Publishing.

Feitshans, I. L. (2017). Remember the ladies: Gender equity for corporate compliance programs using nanotechnology. *International Journal of Women's Health Care, 15*, 2573–9506.

Filon, F. L., Bello, D., Cherrie, J. W., Sleeuwenhoek, A., Spaan, S., & Brouwer, D. H. (2016). Occupational dermal exposure to nanoparticles and nano-enabled products: Part I—Factors affecting skin absorption. *International Journal of Hygiene and Environmental Health, 219*(6), 536–544.

Fröhlich, E. (2012). The role of surface charge in cellular uptake and cytotoxicity of medical nanoparticles. *International Journal of Nanomedicine, 7*, 5577–5591.

Fröhlich, E., & Salar-Behzadi, S. (2014). Toxicological assessment of inhaled nanoparticles: Role of in vivo, ex vivo, in vitro, and in silico studies. *International Journal of Molecular Sciences, 15*(3), 4795–4822.

Gayathri, T., Sundaram, N. M., & Kumar, R. A. (2015). Gadolinium oxide nanoparticles for magnetic resonance imaging and cancer theranostics. *Journal of Bionanoscience, 9*(6), 409–423.

Geiser, M., Jeannet, N., Fierz, M., & Burtscher, H. (2017). Evaluating adverse effects of inhaled nanoparticles by realistic in vitro technology. *Nanomaterials (Basel, Switzerland), 7*(2), 49. https://doi.org/10.3390/nano7020049

Graham, U. M., Jacobs, G., Yokel, R. A., Davis, B. H., Dozier, A. K., Birch, M. E., Tseng, M. T., Oberdörster, G., Elder, A., & DeLouise, L. (2017). From dose to response: In vivo nanoparticle processing and potential toxicity. In *Modelling the toxicity of nanoparticles* (pp. 71–100). Springer.

Guo, Y., Bera, H., Shi, C., Zhang, L., Cun, D., & Yang, M. (2021). Pharmaceutical strategies to extend pulmonary exposure of inhaled medicines. *Acta Pharmaceutica Sinica B, 11*(8), 2565–2584.

Hassan, D. S., & Hasary, H. J. (2022). Nanotoxicology: An integrative environmental challenge. *Al-Rafidain Journal of Medical Sciences, 3*, 41–47. https://doi.org/10.54133/ajms.v3i.80

Hochella, M. F., Aruguete, D., Kim, B., & Madden, A. S. (2012). Naturally occurring inorganic nanoparticles: General assessment and a global budget for one of earth's last unexplored major geochemical components. In *Nature's nanostructures* (pp. 1–31). Pan Stanford.

Jang, J. H., & Lim, H. B. (2010). Characterization and analytical application of surface modified magnetic nanoparticles. *Microchemical Journal, 94*(2), 148–158.

Kahlon, S. K., Sharma, G., Julka, J. M., Kumar, A., Sharma, S., & Stadler, F. J. (2018). Impact of heavy metals and nanoparticles on aquatic biota. *Environmental Chemistry Letters, 16*, 919–946.

Karmakar, S. A. N. A. T. (2019). Particle size distribution and zeta potential based on dynamic light scattering: Techniques to characterize stability and surface charge distribution of charged colloids. In *Recent trends in materials physics and chemistry* (pp. 117–159). Studium Press.

Karunarathne, P., Pocquet, N., Labbé, P., & Milesi, P. (2022). BioRssay: An R package for analyses of bioassays and probit graphs. *Parasites & Vectors, 15*(1), 35.

Khan, H. A., & Shanker, R. (2015). Toxicity of nanomaterials. In *BioMed research international*. Hindawi.

Khater, D., Nsairat, H., Odeh, F., Saleh, M., Jaber, A., Alshaer, W., Al Bawab, A., & Mubarak, M. S. (2021). Design, preparation, and characterization of effective dermal and transdermal lipid nanoparticles: A review. *Cosmetics, 8*(2), 39.

Khodashenas, B., & Ghorbani, H. R. (2019). Synthesis of silver nanoparticles with different shapes. *Arabian Journal of Chemistry, 12*(8), 1823–1838.

Kuempel, E. D., Geraci, C. L., & Schulte, P. A. (2012). Risk assessment and risk management of nanomaterials in the workplace: Translating research to practice. *The Annals of Occupational Hygiene, 56*(5), 491–505. https://doi.org/10.1093/annhyg/mes040

Kumah, E. A., Fopa, R. D., Harati, S., Boadu, P., Zohoori, F. V., & Pak, T. (2023). Human and environmental impacts of nanoparticles: A scoping review of the current literature. *BMC Public Health, 23*(1), 1–28.

Kumar, P. (2013). Explainer: Nanoparticles in air pollution. *The Conversation*. Available at: https://theconversation.com/explainer-nanoparticles-in-air-pollution-16013. Accessed on 12 Oct 2023.

Kumar, P., Kumar, A., & Lead, J. R. (2012). Nanoparticles in the Indian environment: Known, unknowns and awareness. *Environmental Science & Technology, 46*, 7071.

Lakshmi, T., Geetha, R. V., Sreekanth, K. M., Ezhilarasan, D., Royapuram, P. P., Nazmul, H. S., Kamal, D., Dinesh, K. C., Sri, R. B., & Ujjal, K. B. (2022). Role of nanoparticles in environmental remediation: An insight into heavy metal pollution from dentistry. In *Bioinorganic chemistry and applications* (pp. 1–13). Hindawi. https://doi.org/10.1155/2022/1946724

Langevin, D., Raspaud, E., Mariot, S., Knyazev, A., Stocco, A., Salonen, A., Luch, A., Haase, A., Trouiller, B., Relier, C., & Lozano, O. (2018). Towards reproducible measurement of nanoparticle size using dynamic light scattering: Important controls and considerations. *NanoImpact, 10*, 161–167.

Lee, I. C., Ko, J. W., Park, S. H., Shin, N. R., Shin, I. S., Moon, C., Kim, J. H., Kim, H. C., & Kim, J. C. (2016). Comparative toxicity and biodistribution assessments in rats following subchronic oral exposure to copper nanoparticles and microparticles. *Particle and Fibre Toxicology, 13*(1), 1–16.

Li, Y., Bian, T., Du, J., Xiong, Y., Zhan, F., Zhang, H., & Yang, D. (2014). Facile synthesis of high-quality Pt nanostructures with a controlled aspect ratio for methanol electro-oxidation. *CrystEngComm, 16*(36), 8340–8343.

Liang, X. W., Xu, Z. P., Grice, J., Zvyagin, A. V., Roberts, M. S., & Liu, X. (2013). Penetration of nanoparticles into human skin. *Current Pharmaceutical Design, 19*(35), 6353–6366.

Lin, N., Berton, P., Moraes, C., Rogers, R. D., & Tufenkji, N. (2018). Nanodarts, nanoblades, and nanospikes: Mechano-bactericidal nanostructures and where to find them. *Advances in Colloid and Interface Science, 252*, 55–68.

Liu, Y., Xia, Q., Liu, Y., Zhang, S., Cheng, F., Zhong, Z., & Xiao, K. (2014). Genotoxicity assessment of magnetic iron oxide nanoparticles with different particle sizes and surface coatings. *Nanotechnology, 25*(42), 425101.

Liu, N., Tang, M., & Ding, J. (2020a). The interaction between nanoparticles-protein corona complex and cells and its toxic effect on cells. *Chemosphere, 245*, 125624.

Liu, Z., Fan, T., Zhang, Y., Ren, X., Wang, Y., Ma, H., & Wei, Q. (2020b). Electrochemical assay of ampicillin using Fe 3 N-Co 2 N nanoarray coated with molecularly imprinted polymer. *Microchimica Acta, 187*, 1–9.

Lu, P. J., Huang, S. C., Chen, Y. P., Chiueh, L. C., & Shih, D. Y. C. (2015). Analysis of titanium dioxide and zinc oxide nanoparticles in cosmetics. *Journal of Food and Drug Analysis, 23*(3), 587–594.

Lundquist, P., & Artursson, P. (2016). Oral absorption of peptides and nanoparticles across the human intestine: Opportunities, limitations and studies in human tissues. *Advanced Drug Delivery Reviews, 106*, 256–276. https://doi.org/10.1016/j.addr.2016.07.007

Macko, M., Antoš, J., Božek, F., Konečný, J., Huzlík, J., Hegrová, J., & Kuřitka, I. (2022). Development of new health risk assessment of nanoparticles: EPA health risk assessment revised. *Nanomaterials, 13*(1), 20.

Mahmoud, A., Echabaane, M., Omri, K., El Mir, L., & Chaabane, R. B. (2019). Development of an impedimetric non enzymatic sensor based on ZnO and Cu doped ZnO nanoparticles for the detection of glucose. *Journal of Alloys and Compounds, 786*, 960–968.

Marrese, M., Guarino, V., & Ambrosio, L. (2017). Atomic force microscopy: A powerful tool to address scaffold design in tissue engineering. *Journal of functional biomaterials, 8*(1), 7.

Melnikau, D., Savateeva, D., & Rakovich, Y. P. (2014). Coupling effect in hybrid system of plasmonic nanoparticles and J-aggregates yields double Rabi splitting. In *2014 16th International Conference on Transparent Optical Networks (ICTON)* (pp. 1–4). IEEE.

Misra, S. K., Dybowska, A., Berhanu, D., Luoma, S. N., & Valsami-Jones, E. (2012). The complexity of nanoparticle dissolution and its importance in nanotoxicological studies. *Science of the Total Environment, 438*, 225–232.

Moghimi, S. M. (2018). Nanomedicine safety in preclinical and clinical development: Focus on idiosyncratic injection/infusion reactions. *Drug Discovery Today, 23*(5), 1034–1042.

Murdock, G. A. (2017). *Cryomilling of high magnesium aluminum alloy with boron carbide reinforcement*. Doctoral dissertation, California State University, Sacramento.

Murdock, R. C., Braydich-Stolle, L., Schrand, A. M., Schlager, J. J., & Hussain, S. M. (2008). Characterization of nanomaterial dispersion in solution prior to in vitro exposure using dynamic light scattering technique. *Toxicological Sciences: An Official Journal of the Society of Toxicology, 101*(2), 239–253. https://doi.org/10.1093/toxsci/kfm240

Nafisi, S., & Maibach, H. I. (2018). Skin penetration of nanoparticles. In R. Shegokar & E. B. Souto (Eds.), *Emerging nanotechnologies in immunology* (pp. 47–88). Elsevier.

Naseer, B., Srivastava, G., Qadri, O. S., Faridi, S. A., Islam, R. U., & Younis, K. (2018). Importance and health hazards of nanoparticles used in the food industry. *Nanotechnology Reviews, 7*(6), 623–641.

Nasrullah, M., Sundaram, D. N. M., Claerhout, J., Ha, K., Demirkaya, E., & Uludag, H. (2023). Nanoparticles and cytokine response. *Frontiers in Bioengineering and Biotechnology, 11*, 1243651.

National Institute of Standards and Technology (NIST). (2018). *Nanomaterials in consumer products*. Available at: https://www.nist.gov/programs-projects/nanomaterials-consumer-products. Accessed on 19 Oct 2023.

Nemmar, A., Holme, J. A., Rosas, I., Schwarze, P. E., & Alfaro-Moreno, E. (2013). Recent advances in particulate matter and nanoparticle toxicology: A review of the in vivo and in vitro studies. In *BioMed research international*. Hindawi.

Nho, R. (2020). Pathological effects of nano-sized particles on the respiratory system. *Nanomedicine: Nanotechnology, Biology and Medicine, 29*, 102242.

NIOSH. (2021). *Current intelligence bulletin 70: health effects of occupational exposure to silver nanomaterials*. By Kuempel E, Roberts JR, Roth G, Dunn KL, Zumwalde R, Drew N, Hubbs A, Trout D, and Holdsworth G. Cincinnati, OH: U.S. Department of Health and Human Services, Centers for Disease Control and Prevention, National Institute for Occupational Safety and Health, DHHS (NIOSH) Publication No. 2021-112. https://doi.org/10.26616/NIOSHPUB2021112.

Nowack, B., & Bucheli, T. D. (2007). Occurrence, behavior and effects of nanoparticles in the environment. *Environmental Pollution, 150*(1), 5–22.

Obst, K., Yealland, G., Balzus, B., Miceli, E., Dimde, M., Weise, C., Eravci, M., Bodmeier, R., Haag, R., Calderón, M., & Charbaji, N. (2017). Protein corona formation on colloidal polymeric nanoparticles and polymeric nanogels: Impact on cellular uptake, toxicity, immunogenicity, and drug release properties. *Biomacromolecules, 18*(6), 1762–1771.

Occupational Safety and Health Administration (OSHA). (2013). *Working safely with nanomaterials*. U.S. Department of Labor. Available at: https://www.osha.gov/sites/default/files/publications/OSHA_FS-3634.pdf. Accessed on 19 Oct 2023.

Ochoa, R. (2018). Design of studies and risk management in toxicologic pathology: Addressing risks in product discovery and development. In *Fundamentals of toxicologic pathology* (pp. 105–122). Academic Press.

Palmer, B., & DeLouise, L. (2016). Nanoparticle-enabled transdermal drug delivery systems for enhanced dose control and tissue targeting. *Molecules (Basel, Switzerland), 21*(12), 1719. https://doi.org/10.3390/molecules21121719

Park, S. Y., & Choi, J. (2010). Geno-and ecotoxicity evaluation of silver nanoparticles in freshwater crustacean Daphnia magna. *Environmental Engineering Research, 15*(1), 23–27.

Patra, J. K., Das, G., Fraceto, L. F., Campos, E. V. R., Rodriguez-Torres, M. D. P., Acosta-Torres, L. S., Diaz-Torres, L. A., Grillo, R., Swamy, M. K., Sharma, S., & Habtemariam, S. (2018). Nano based drug delivery systems: Recent developments and future prospects. *Journal of Nanobiotechnology, 16*(1), 1–33.

Paul, K. B., Athanasopoulos, G. I., Doumanidis, C. C., & Rebholz, C. (2010). Growth and characterization of nanowires. In *Advances in condensed matter physics*. Hindawi.

Polat, S., Ağçam, E., Dündar, B., & Akyildiz, A. (2019). Nanoparticles in food packaging: Opportunities and challenges. In *Health and safety aspects of food processing technologies* (pp. 577–611). Springer.

Nikam, A. P., Anubhav, D., & Ratan, G. (2021). Nanoparticles: An overview. *Drugs and Cell Therapies in Haematology, 10*(1), 2281–4876.

Puolamaa, M. (2006). *The appropriateness of existing methodologies to assess the potential risks associated with engineered and adventitious products of nanotechnologies*. European Asylum Support Office. Malta. Retrieved from https://policycommons.net/artifacts/2053365/the-appropriateness-of-existing-methodologies-to-assess-the-potential-risks-associated-with-engineered-and-adventitious-produc/2806456/ on 19 Oct, 2023.

Rose, N. L., & Ruppel, M. (2015). Environmental archives of contaminant particles. In *Environmental contaminants: Using natural archives to track sources and long-term trends of pollution* (pp. 187–221). Springer.

Saifi, M. A., Khan, W., & Godugu, C. (2018). Cytotoxicity of nanomaterials: Using nanotoxicology to address the safety concerns of nanoparticles. *Pharmaceutical Nanotechnology, 6*(1), 3–16. https://doi.org/10.2174/2211738505666171023152928

Sajid, M., Ilyas, M., Basheer, C., Tariq, M., Daud, M., Baig, N., & Shehzad, F. (2015). Impact of nanoparticles on human and environment: Review of toxicity factors, exposures, control strategies, and future prospects. *Environmental Science and Pollution Research, 22*, 4122–4143.

Salassa, G., & Bürgi, T. (2018). NMR spectroscopy: A potent tool for studying monolayer-protected metal nanoclusters. *Nanoscale Horizons, 3*(5), 457–463.

Sani, A., Cao, C., & Cui, D. (2021). Toxicity of gold nanoparticles (AuNPs): A review. *Biochemistry and Biophysics Reports, 26*, 100991.

Sarma, D. D., Santra, P. K., Mukherjee, S., & Nag, A. (2013). X-ray photoelectron spectroscopy: A unique tool to determine the internal heterostructure of nanoparticles. *Chemistry of Materials, 25*(8), 1222–1232.

Sayed, F. N., & Polshettiwar, V. (2015). Facile and sustainable synthesis of shaped iron oxide nanoparticles: Effect of iron precursor salts on the shapes of iron oxides. *Scientific Reports, 5*(1), 9733.

Sbarigia, C., Tacconi, S., Mura, F., Rossi, M., Dinarelli, S., & Dini, L. (2022). High-resolution atomic force microscopy as a tool for topographical mapping of surface budding. *Frontiers in Cell and Developmental Biology, 10*, 975919.

Schluesener, J. K., & Schluesener, H. J. (2013). Nanosilver: Application and novel aspects of toxicology. *Archives of Toxicology, 87*(4), 569–576. https://doi.org/10.1007/s00204-012-1007-z

Schulte, P. A., Geraci, C. L., Hodson, L. L., Zumwalde, R. D., Kuempel, E. D., Murashov, V., Martinez, K. F., & Heidel, D. S. (2013). Overview of risk management for engineered nanomaterials. *Journal of Physics: Conference Series, 429*, 012062. https://doi.org/10.1088/1742-6596/429/1/012062

Schulte, P. A., Geraci, C. L., Murashov, V., Kuempel, E. D., Zumwalde, R. D., Castranova, V., et al. (2014). Occupational safety and health criteria for responsible development of nanotechnology. *Journal of Nanoparticle Research: An Interdisciplinary Forum for Nanoscale Science and Technology, 16*(1), 1–17. https://doi.org/10.1007/s11051-013-2153-9

Serrano-Lotina, A., Portela, R., Baeza, P., Alcolea-Rodriguez, V., Villarroel, M., & Ávila, P. (2023). Zeta potential as a tool for functional materials development. *Catalysis Today, 423*, 113862.

Sharma, V., Aneja, B., Yata, V. K., Malakar, D., & Mohanty, A. K. (2020). Systemic nanotoxicity and its assessment in animal models. *Nanopharmaceuticals: Principles and Applications, 3*, 201–243.

Sikora, A., Bartczak, D., Geißler, D., Kestens, V., Roebben, G., Ramaye, Y., Varga, Z., Palmai, M., Shard, A. G., Goenaga-Infante, H., & Minelli, C. (2015). A systematic comparison of different techniques to determine the zeta potential of silica nanoparticles in biological medium. *Analytical Methods, 7*(23), 9835–9843.

Simko, M., & Mattsson, M.-O. (2014). Interactions between nanosized materials and the brain. *Current Medicinal Chemistry, 21*(37), 4200–4214. Retrieved from https://www.ingentaconnect.com/content/ben/cmc/2014/00000021/00000037/art00002

Singh, R., & Dutta, S. (2018). Synthesis and characterization of solar photoactive TiO2 nanoparticles with enhanced structural and optical properties. *Advanced Powder Technology, 29*(2), 211–219.

Sonwani, S., Madaan, S., Arora, J., Suryanarayan, S., Rangra, D., Mongia, N., Vats, T., & Saxena, P. (2021). Inhalation exposure to atmospheric nanoparticles and its associated impacts on human health: A review. *Frontiers in Sustainable Cities, 3*, 690444.

Sufian, M. M., Khattak, J. Z. K., Yousaf, S., & Rana, M. S. (2017). Safety issues associated with the use of nanoparticles in human body. *Photodiagnosis and Photodynamic Therapy, 19*, 67–72. https://doi.org/10.1016/j.pdpdt.2017.05.012

Sun, S., Kong, C., You, H., Song, X., Ding, B., & Yang, Z. (2012). Facet-selective growth of Cu–Cu 2 O heterogeneous architectures. *CrystEngComm, 14*(1), 40–43.

Sun, J., Ge, J., Liu, W., Lan, M., Zhang, H., Wang, P., Wang, Y., & Niu, Z. (2014). Multi-enzyme co-embedded organic–inorganic hybrid nanoflowers: Synthesis and application as a colorimetric sensor. *Nanoscale, 6*(1), 255–262.

Teleanu, D. M., Chircov, C., Grumezescu, A. M., Volceanov, A., & Teleanu, R. I. (2018). Impact of nanoparticles on brain health: An up to date overview. *Journal of Clinical Medicine, 7*(12), 490.

Thanachoksawang, C., Navasumrit, P., Hunsonti, P., Chompoobut, C., Chaisatra, K., Autrup, H., & Ruchirawat, M. (2022). Exposure to airborne iron oxide nanoparticles induces oxidative DNA damage and inflammatory responses: A pilot study in welders and in human lung epithelial cell line. *Toxicology and Environmental Health Sciences, 14*(4), 339–349.

Tsuji, J. S., Maynard, A. D., Howard, P. C., James, J. T., Lam, C.-W., Warheit, D. B., & Santamaria, A. B. (2006). Research strategies for safety evaluation of nanomaterials, part IV: Risk assessment of nanoparticles. *Toxicological Sciences: An Official Journal of the Society of Toxicology, 89*(1), 42–50. https://doi.org/10.1093/toxsci/kfi339

Turan, N. B., Erkan, H. S., Engin, G. O., & Bilgili, M. S. (2019). Nanoparticles in the aquatic environment: Usage, properties, transformation and toxicity—A review. *Process Safety and Environmental Protection, 130*, 238–249.

Tzitzios, V., Georgakilas, V., Zafiropoulou, I., Boukos, N., Basina, G., Niarchos, D., & Petridis, D. (2008). A general chemical route for the synthesis of capped nanocrystalline materials. *Journal of Nanoscience and Nanotechnology, 8*(6), 3117–3122.

Wani, I. A., Ganguly, A., Ahmed, J., & Ahmad, T. (2011). Silver nanoparticles: Ultrasonic wave assisted synthesis, optical characterization and surface area studies. *Materials Letters, 65*(3), 520–522.

Warheit, D. B. (2018). Hazard and risk assessment strategies for nanoparticle exposures: How far have we come in the past 10 years? *F1000Research, 7*, 376. https://doi.org/10.12688/f1000research.12691

Warheit, D. B., & Sayes, C. M. (2015). Routes of exposure to nanoparticles. In P. I. Dolez (Ed.), *Nanoengineering* (pp. 41–54). Elsevier.

Warheit, D. B., Sayes, C. M., Reed, K. L., & Swain, K. A. (2008). Health effects related to nanoparticle exposures: Environmental, health and safety considerations for assessing hazards and risks. *Pharmacology & Therapeutics, 120*(1), 35–42. https://doi.org/10.1016/j.pharmthera.2008.07.001

WHO Global Air Quality Guidelines. (2021). *World health organization*. Available at: https://www.who.int/news-room/questions-and-answers/item/who-global-air-quality-guidelines. Accessed on 12 Oct 2023.

Winkler, D. A., Mombelli, E., Pietroiusti, A., Tran, L., Worth, A., Fadeel, B., & McCall, M. J. (2013). Applying quantitative structure–activity relationship approaches to nanotoxicology: Current status and future potential. *Toxicology, 313*(1), 15–23. https://doi.org/10.1016/j.tox.2012.11.005

Woo, H., Kang, H., Kim, A., Jang, S., Park, J. C., Park, S., Kim, B. S., Song, H., & Park, K. H. (2012). Azide-alkyne huisgen [3 + 2] cycloaddition using CuO nanoparticles. *Molecules, 17*(11), 13235–13252.

Xu, F. (2018). Review of analytical studies on TiO2 nanoparticles and particle aggregation, coagulation, flocculation, sedimentation, stabilization. *Chemosphere, 212*, 662–677.

Yang, Z., Deng, W., Zhang, X., An, Y., Liu, Y., Yao, H., & Zhang, Z. (2022). Opportunities and challenges of nanoparticles in digestive tumours as anti-angiogenic therapies. *Frontiers in Oncology, 11*, 789330.

Zaiter, T., Cornu, R., El Basset, W., Martin, H., Diab, M., & Béduneau, A. (2022). Toxicity assessment of nanoparticles in contact with the skin. *Journal of Nanoparticle Research, 24*(7), 149.

Zhang, J., Meng, L., Zhao, D., Fei, Z., Lu, Q., & Dyson, P. J. (2008). Fabrication of dendritic gold nanoparticles by use of an ionic polymer template. *Langmuir, 24*(6), 2699–2704.

Zhang, J., Wu, L., Chan, H.-K., & Watanabe, W. (2011). Formation, characterization, and fate of inhaled drug nanoparticles. *Advanced Drug Delivery Reviews, 63*(6), 441–455. https://doi.org/10.1016/j.addr.2010.11.002

Zhang, J., Guo, W., Li, Q., Wang, Z., & Liu, S. (2018). The effects and the potential mechanism of environmental transformation of metal nanoparticles on their toxicity in organisms. *Environmental Science: Nano, 5*(11), 2482–2499.

Zhao, F., Zhao, Y., Liu, Y., Chang, X., Chen, C., & Zhao, Y. (2011). Cellular uptake, intracellular trafficking, and cytotoxicity of nanomaterials. *Small, 7*(10), 1322–1337.

Zhao, M., Zhou, G., Wang, J., Zhang, Y., Xue, J., Liu, J., et al. (2023). MiR-5622-3p inhibits ZCWPW1 to induce apoptosis in silica-exposed mice and spermatocyte cells. *Nanotoxicology, 17*(4), 372–384. https://doi.org/10.1080/17435390.2023.2223632

Zhong, L., Yu, Y., Lian, H. Z., Hu, X., Fu, H., & Chen, Y. J. (2017). Solubility of nano-sized metal oxides evaluated by using in vitro simulated lung and gastrointestinal fluids: Implication for health risks. *Journal of Nanoparticle Research, 19*, 1–10.

Zoeller, R. T., & Vandenberg, L. N. (2015). Assessing dose–response relationships for endocrine disrupting chemicals (EDCs): A focus on non-monotonicity. *Environmental Health, 14*, 1–5.

Zoroddu, M. A., Medici, S., Ledda, A., Nurchi, V. M., Lachowicz, J. I., & Peana, M. (2014). Toxicity of nanoparticles. *Current Medicinal Chemistry, 21*(33), 3837–3853.

Chapter 14
Regulations and Policy Considerations for Nanoparticle Safety

Patrick Omoregie Isibor

14.1 Introduction

Regulating nanoparticle safety is crucial due to their unique properties and potential risks they pose to human health and the environment. Nanoparticles, due to their small size and high reactivity, can interact with biological systems in ways that larger particles do not. Regulating their safety ensures protection against potential health hazards such as respiratory issues, skin irritation, and systemic toxicity arising from exposure.

Establishing regulations helps in identifying and mitigating potential risks associated with nanoparticle exposure. This includes preventing adverse health effects arising from inhalation, ingestion, or skin contact with nanoparticles. Nanoparticles can impact ecosystems, wildlife, and overall environmental health (Patel et al., 2021). Regulations aid in minimizing nanoparticle release into the environment, preventing contamination of air, water, and soil. Ensuring the safety of products containing nanoparticles, such as cosmetics, food packaging, and electronics, is vital to safeguard consumers from potential health risks associated with their use. Regulations provide standardized guidelines for nanoparticle safety assessment, allowing for consistency in testing methodologies and safety evaluations across industries and jurisdictions. Regulatory frameworks encourage research and development in understanding nanoparticle behavior, toxicity, and safe handling practices. This helps in advancing scientific knowledge and improving safety assessments (Patel et al., 2021).

Stringent regulations enhance public trust by assuring consumers that products containing nanoparticles have undergone rigorous safety assessments, thereby

P. O. Isibor (✉)
Department of Biological Sciences, College of Science and Technology, Covenant University, Ota, Nigeria
e-mail: patrick.isibor@covenantuniversity.edu.ng

P. O. Isibor et al. (eds.), *Environmental Nanotoxicology*,
https://doi.org/10.1007/978-3-031-54154-4_14

increasing confidence in their use. Establishing regulatory standards promotes international cooperation and harmonization, facilitating trade while ensuring consistent safety measures globally. Regulations protect workers from potential occupational hazards associated with nanoparticle handling by setting guidelines for safe working conditions and protective measures. As nanotechnology evolves, regulations need to adapt to new advancements, ensuring that safety standards remain current and effective. Overall, regulating nanoparticle safety is essential for ensuring responsible and sustainable utilization of nanotechnology while minimizing potential risks to human health and the environment and facilitating innovation in a safe and ethical manner (Saif et al., 2018).

The scope of regulatory policies on nanotoxicity encompasses a range of measures aimed at assessing, mitigating, and managing the potential risks associated with nanoparticles. Regulatory policies focus on conducting comprehensive risk assessments to evaluate the potential hazards posed by nanoparticles. This includes identifying exposure routes, assessing toxicity, determining safe exposure levels, and understanding the potential impacts on human health and the environment. Establishing standardized testing methodologies and protocols for assessing nanotoxicity is crucial. Regulatory frameworks outline specific guidelines for toxicity testing, ensuring consistency and comparability of results across studies. Defining criteria for classifying different types of nanomaterials based on their physicochemical properties, such as size, shape, surface area, and composition, helps in categorizing their potential risks and setting appropriate safety standards. Regulatory policies provide guidelines for safe handling, storage, transportation, and disposal of nanoparticles. These guidelines aim to minimize occupational exposure risks and prevent environmental contamination. Mandating clear labeling and disclosure of nanomaterials in consumer products or industrial applications ensures transparency and allows consumers, workers, and regulators to make informed decisions regarding their use.

Establishing occupational safety standards specific to nanoparticle handling environments ensures the protection of workers from potential health hazards associated with exposure. Policies include measures to safeguard the environment by regulating the release of nanoparticles into air, water, and soil. This involves setting limits on nanoparticle discharge and implementing pollution prevention strategies. Implementing monitoring programs to track nanoparticle exposure levels in various environmental compartments aids in early detection of potential contamination and facilitates risk management strategies.

Regulatory policies govern the approval, manufacturing, and marketing of products containing nanoparticles, ensuring compliance with safety standards before their introduction into the market. Efforts to align regulatory policies on nanotoxicity across countries and regions promote international cooperation and harmonization, facilitating consistency in safety evaluations and trade practices.

The scope of regulatory policies on nanotoxicity is continually evolving to keep pace with advancements in nanotechnology, ensuring the safe and responsible development and use of nanomaterials while protecting human health and the environment.

14.2 Regulatory Framework

Various regulatory bodies, such as the United States Environmental Protection Agency (EPA), the Food and Drug Administration (FDA), and the Health and Consumer Protection Directorate of the European Commission, play a pivotal role in guiding the safe handling and disposal practices of potentially hazardous materials across diverse environments. Their primary objective is to curtail human exposure to these substances (Bundschuh et al., 2018). These organizations undertake comprehensive risk assessments to make informed decisions regarding risk management strategies. However, the existing regulatory frameworks fall short in addressing all concerns specifically associated with nanoparticles (NPs). Nanoparticles introduce a distinct realm of potential risks, for which present risk assessment methodologies are not optimally tailored.

Presently, only a limited number of regulatory bodies globally are adopting risk-based methodologies to evaluate and control exposure, particularly concerning potential hazards associated with nanoparticles (NPs) and their corresponding exposure scenarios. For instance, the European Union's Registration, Evaluation, Authorization, and Restriction of Chemicals (REACH) initiative mandates an evaluation of engineered nanomaterials during the registration of the parent substance. Notably, all manufacturers and importers dealing with carbon nanoproducts are obliged to furnish comprehensive health and safety data. Similarly, the US Food and Drug Administration (FDA) convenes periodic forums dedicated to scrutinizing and regulating various substances, necessitating safety assessments for each application involving nanotechnology. Despite the evaluation of new substances on an individual basis within global regulatory frameworks, there is presently no overarching international regulation specifically addressing nanotechnology. Moreover, there exists a notable absence of universally agreed-upon definitions or standardized protocols for toxicity testing of nanoparticles.

In practical terms, risk management aims to institute measures that effectively reduce exposures to nanoparticles (NPs) below hazardous thresholds. Mitigating the risks associated with environmental nanoparticle exposure necessitates the implementation of occupational safety strategies, especially targeting workers facing the highest risks. These measures encompass alterations in manufacturing processes, engineering controls, and the provision of personal protective equipment, all pivotal for ensuring worker safety in industries employing nanomaterials. Furthermore, regulatory frameworks need to strike a balance between the rapid technological advancements in nanoparticles and environmental conservation. Environmental risk assessment of nanomaterials involves a comprehensive investigation into their fate, transport, and potential ecological impacts, which in turn could amplify the risks of human NP exposure. Addressing these risks requires the deployment of new remediation technologies, formulation of pertinent policies, and bolstering public awareness. Employing risk analysis, encompassing thorough risk assessment, management, and effective communication, emerges as the most suitable approach for handling nanomaterials. It is imperative to establish robust risk assessment techniques, regulatory frameworks, and effective risk management strategies to ensure the safe integration of nanotechnology and nanomaterials across diverse industries (Saif et al., 2018).

The European Chemicals Agency (ECHA) is an agency of the European Union (EU) responsible for managing the technical, scientific, and administrative aspects of chemical substances within the EU. Established in 2007, ECHA operates under the framework of the REACH Regulation (Registration, Evaluation, Authorization, and Restriction of Chemicals), which aims to ensure the safe use of chemicals while protecting human health and the environment. ECHA operates databases, including the REACH database, which stores information on registered substances, and the Candidate List of SVHC, listing substances subject to authorization. It also maintains the Classification and Labeling Inventory and the Biocidal Products Regulation (BPR) database. Overall, ECHA plays a pivotal role in managing and regulating chemicals to ensure their safe use, protect human health, and safeguard the environment across the European Union.

The Organization for Economic Co-operation and Development (OECD) is an international organization comprising 38 member countries, dedicated to promoting policies that improve economic and social well-being globally. Established in 1961, the OECD serves as a platform for countries to discuss, develop, and coordinate policies across various domains, including economics, trade, social issues, and environmental concerns. The OECD covers a diverse array of policy areas, including economics, finance, trade, education, health, environment, innovation, governance, and development. Its mission is to provide a platform for countries to work together to develop evidence-based policies that promote sustainable economic growth, social progress, and environmental protection worldwide.

The World Health Organization (WHO) is a specialized agency of the United Nations (UN) responsible for international public health. Established in 1948, the WHO operates as a leading authority on global health issues, setting norms and standards, providing technical support and guidance, and coordinating international efforts to address health challenges worldwide. The WHO operates through its headquarters in Geneva, Switzerland, and a network of regional and country offices worldwide. It works in close collaboration with member states and partners to promote health, prevent diseases, and improve the quality of life for people around the globe, guided by the principle that health is a fundamental human right (Runa et al., 2022). The International Organization for Standardization (ISO) is an independent, nongovernmental international organization that develops and publishes voluntary international standards across various industries and sectors. Established in 1947, ISO facilitates global standardization to enhance product quality, safety, efficiency, and interoperability, promoting international trade and fostering innovation. ISO standards cover a wide range of areas, including quality management, environmental management, information security, food safety, healthcare, energy management, and many more. These standards are developed through a consensus-based process, involving technical committees and experts from member countries who contribute their expertise to draft, review, and finalize standards. ISO standards are not legally binding but serve as guidelines and best practices that organizations and industries can voluntarily adopt to improve their processes, products, and services. Adoption of ISO standards can provide a competitive edge, facilitate market access, and demonstrate compliance with internationally recognized benchmarks for quality and safety (Bakand & Hayes, 2016; Okeke, 2022; Bocca et al., 2023).

14.3 Key Safety Considerations

When assessing nanoparticle properties for safety considerations, several key factors are evaluated to determine their potential risks and impact on human health and the environment. Nanoparticles exhibit unique properties based on their size and shape. Their small size can influence their behavior within biological systems and their ability to penetrate barriers such as cell membranes or the blood-brain barrier (Semenzin et al., 2019).

Nanoparticles often have a high surface area-to-volume ratio, affecting their reactivity and interactions with biological systems. Surface modifications and coatings can alter their behavior, stability, and potential toxicity. The elemental composition and chemical structure of nanoparticles play a significant role in determining their toxicity. Certain metals or compounds used in nanoparticles may pose health risks or induce adverse effects (Armstead & Li, 2016).

The solubility and stability of nanoparticles in various environments, such as biological fluids or water systems, can influence their bioavailability and potential for accumulation in tissues. Assessing the toxicological effects of nanoparticles involves understanding their potential to induce cytotoxicity, inflammation, oxidative stress, genotoxicity, or other adverse biological responses. Biocompatibility studies aim to evaluate the compatibility of nanoparticles with biological systems. Understanding how nanoparticles interact with cells, tissues, organs, and biological processes is crucial. This includes their uptake, distribution, metabolism, and excretion pathways within living organisms (Gao et al., 2015).

Evaluating the behavior of nanoparticles in environmental systems, their persistence, bioaccumulation potential, and effects on ecosystems, aquatic life, and soil health is essential for understanding their environmental impact. Assessing the relationship between nanoparticle exposure levels and biological responses helps determine safe exposure limits and potential hazards associated with different concentrations. Developing standardized testing protocols and methodologies for assessing nanoparticle safety is vital to ensure consistency and comparability of results across studies.

Considering these factors, comprehensive risk assessments and toxicity studies are conducted to determine the safety profiles of nanoparticles. Such assessments guide regulatory decisions, risk management strategies, and the development of safe handling practices for nanoparticles in various industries and applications.

14.3.1 Measurement Techniques and Standards

Measurement techniques and standards for nanoparticles play a critical role in accurately characterizing their properties and ensuring consistency in assessment methodologies. Several techniques are commonly used to measure and analyze nanoparticles, and various standards have been established to guide these measurements. Dynamic light scattering (DLS) measures particle size distribution based on the fluctuations in light scattering caused by Brownian motion of particles in suspension. Transmission electron microscopy (TEM) provides high-resolution imaging to visualize individual nanoparticles and determine their size, shape, and morphology. Scanning electron microscopy (SEM) offers detailed surface imaging and characterization of nanoparticles. Atomic force microscopy (AFM) maps surface topography and measures particle dimensions with atomic-scale resolution (Armstead & Li, 2016).

BET (Brunauer-Emmett-Teller) method measures specific surface area and porosity of nanoparticles by nitrogen gas adsorption. X-ray diffraction (XRD) identifies crystalline structure and phase composition of nanoparticles. X-ray photoelectron spectroscopy (XPS) analyzes surface chemistry and elemental composition of nanoparticles. Energy-dispersive X-ray spectroscopy (EDS/EDX) provides elemental analysis of nanoparticles. Electrophoretic light scattering (ELS) determines the zeta potential, measuring the surface charge and stability of nanoparticles in suspension. Cytotoxicity, genotoxicity, oxidative stress, and other cellular responses to nanoparticles are evaluated using various cell-based assays. In vivo studies, animal models are used to assess nanoparticle toxicity, biodistribution, and effects on organs and tissues (Mohanta & Ahmaruzzaman, 2020).

Standards organizations such as the International Organization for Standardization (ISO) and American Society for Testing and Materials (ASTM) have developed guidelines and standards for nanoparticle characterization and testing methods. ISO 19430:2016 is employed in determination of particle size distribution by electrical sensing zone method (Coulter principle). ISO/TR 13014:2012 provides guidance on physicochemical characterization of engineered nanoscale materials for toxicologic assessment. By standard practice, ASTM E2490–09(2018) is a reference guide for measurement of particle size distribution of nanomaterials in suspension by photon correlation spectroscopy (PCS), while ASTM E2456–06(2018) is standard terminology relating to nanotechnology. These standards provide guidance on terminology, testing procedures, quality control, and reporting protocols, ensuring consistency and reliability in the characterization and assessment of nanoparticles across different laboratories and industries. Compliance with these standards helps in fostering comparability and accuracy in nanoparticle measurements, supporting safe and responsible use of nanomaterials (Siegrist et al., 2019).

14.3.2 Risk Evaluation

Dose-response relationships in nanoparticle toxicology refer to the correlation between the dose or concentration of nanoparticles and the biological response or toxic effect elicited in living organisms. Understanding these relationships is crucial for assessing the potential risks and hazards associated with nanoparticle exposure (Stern & McNeil, 2007; Asare et al., 2012; Tsang et al., 2017).

Generally, as the concentration or dose of nanoparticles increases, the biological response tends to intensify. This relationship helps determine the threshold at which adverse effects become evident. Some nanoparticles may exhibit threshold effects, where no observable adverse effects occur below a certain concentration. Beyond this threshold, the biological response begins to manifest. Dose-response relationships for nanoparticles may not always follow a linear pattern. At higher doses, biological responses might escalate disproportionately or plateau, indicating complex interactions and saturation effects (Thu et al., 2023).

Nanoparticles can induce biological responses even at low doses that might not produce apparent effects immediately but could have cumulative or long-term consequences. Variations in individual susceptibility, differences in nanoparticle characteristics (size, shape, surface properties), and exposure routes can lead to varied responses among different organisms or cell types (Thannickal & Fanburg, 2000). Dose-response relationships might change over time, particularly for chronic exposures, where prolonged or repeated exposure to low doses may accumulate and lead to different outcomes compared to acute exposures. In some instances, the response to nanoparticles might display a biphasic or U-shaped curve, where both low and high doses can elicit adverse effects, while intermediate doses may have minimal impact (Bakand & Hayes, 2016).

Quantifying dose-response relationships involves conducting systematic studies where organisms or cells are exposed to varying concentrations of nanoparticles, observing and analyzing the resulting biological responses. These studies help establish dose-response curves, enabling the determination of safe exposure limits, identification of threshold levels, and characterization of potential hazards associated with nanoparticle exposure. Applying knowledge of dose-response relationships is fundamental in setting regulatory guidelines, establishing safety thresholds, and informing risk assessments to ensure safe handling and use of nanoparticles in diverse applications while mitigating potential health and environmental risks (Mohanta & Ahmaruzzaman, 2020).

The developing immune systems in children and infants may make them more vulnerable to the effects of nanoparticles. Their higher metabolic rates and greater intake of air, food, and water per unit of body weight can lead to increased exposure levels. Aging populations might have compromised immune function or underlying health conditions, making them more susceptible to the adverse effects of nanoparticles (Bocca et al., 2023). Nanoparticle exposure during pregnancy can impact fetal development. Nanoparticles might cross the placental barrier, potentially affecting the developing fetus and increasing the risk of adverse birth outcomes. People with preexisting respiratory conditions (e.g., asthma, chronic obstructive pulmonary disease) or cardiovascular diseases might experience exacerbated symptoms due to nanoparticle exposure (Singh et al., 2020). Workers in industries involving the production, handling, or disposal of nanoparticles are at higher risk due to potential direct and prolonged exposure in occupational settings. Certain communities might face higher exposure to nanoparticles due to living in areas with increased industrial activities or pollution, contributing to greater health risks. Genetic variations among individuals might lead to differential responses to nanoparticle exposure, making some more prone to adverse effects (Singh et al., 2020).

Identifying vulnerable populations is essential for targeted risk assessments, regulatory considerations, and implementing protective measures to minimize exposure. Protecting these groups often involves comprehensive risk assessments, strict adherence to safety protocols in occupational settings, and public health interventions aimed at reducing overall nanoparticle exposure. Additionally, specific guidelines may be developed to protect vulnerable populations, ensuring their safety in diverse environments where nanoparticles are present.

Ecotoxicity studies of nanoparticles aim to understand their potential effects on ecosystems and organisms in the environment by assessing the impact of nanoparticles at various levels, from individual organisms to entire ecosystems. It also aims at assessing the impact of nanoparticles on individual organisms such as aquatic species (fish, algae, zooplankton), soil organisms (earthworms, microbes), and terrestrial organisms (plants, insects). The changes in growth, reproduction, behavior, and mortality rates due to nanoparticle exposure are also of keen interest. Furthermore, understanding the uptake, accumulation, and transformation of nanoparticles within organisms and along the food chain is crucial for a complete knowledge. Bioaccumulation determine the extent to which nanoparticles are absorbed and stored in tissues, potentially leading to biomagnification in higher trophic levels, investigating how nanoparticles affect entire ecosystems by studying interactions among different species, changes in biodiversity, and ecosystem functioning. Alterations in community structures and ecosystem services due to nanoparticle exposure, elucidating the mechanisms through which nanoparticles exert toxicity in organisms, such as oxidative stress, inflammation, genotoxicity, and disruption of cellular functions are central to nanotoxocology. A full understanding of these mechanisms helps predict and assess potential adverse effects (Rodríguez-Ibarra et al., 2017), evaluating the behavior of nanoparticles in environmental compartments like water, soil, and air. Knowledge on the persistence, mobility, transformation, and potential of nanomaterials to leach into different environmental matrices is crucial to developing standardized methods for ecotoxicity testing of nanoparticles to ensure consistency and comparability of results across studies and laboratories. These protocols aid in regulatory assessments and risk management strategies.

Ecotoxicity studies involving nanoparticles face challenges due to their unique properties, including small size, high surface area, and reactivity. Researchers aim to bridge knowledge gaps and develop methodologies to accurately assess the environmental impact of nanoparticles. Additionally, considering the diverse range of nanoparticles and their applications, conducting comprehensive and standardized ecotoxicity assessments becomes crucial for regulatory decision-making and ensuring the safe use of nanomaterials while minimizing environmental risks.

14.4 Challenges and Emerging Issues

In regulatory policies pertaining to nanoparticle safety, several data gaps and uncertainties persist, necessitating further exploration and refinement to ensure comprehensive risk assessment and effective governance. These gaps encompass diverse facets, ranging from understanding nanoparticle-specific characteristics to evaluating their potential effects on human health and the environment (Crist et al., 2021).

One prominent data gap pertains to the limited understanding of the physicochemical properties and behaviors of diverse nanoparticles. Variability in nanoparticle characteristics, such as size, shape, surface modifications, and coating materials, poses challenges in standardizing characterization protocols and comprehensively assessing their safety profiles. Addressing these gaps requires robust characterization methods that encompass the diverse range of nanoparticles and their dynamic interactions within biological and environmental matrices. Furthermore, uncertainties exist regarding the toxicological effects and potential hazards posed by different types of nanoparticles. Variations in toxicity profiles, modes of action, and mechanisms of cellular responses across various nanoparticle types underscore the need for comprehensive toxicity assessments and elucidation of structure-toxicity relationships. Establishing standardized testing methodologies and frameworks for evaluating nanoparticle toxicity and elucidating their potential adverse effects remains an imperative (Singh et al., 2020).

Data gaps also encompass limited knowledge concerning nanoparticle exposure routes, dose-response relationships, and their fate within biological systems. Understanding the routes of exposure, kinetics of absorption, distribution, metabolism, and excretion of nanoparticles aids in delineating their bioavailability and potential for accumulation, crucial for assessing their safety thresholds and establishing exposure limits. Moreover, there are uncertainties in assessing the potential environmental impacts of nanoparticles, including their fate in ecosystems, interactions with biota, and long-term ecological effects. Evaluating nanoparticle behavior in environmental matrices and elucidating their potential effects on ecosystems requires comprehensive studies that encompass diverse environmental compartments and ecological interactions. Resolving these data gaps and uncertainties in regulatory policies necessitates concerted efforts in research, standardization of methodologies, interdisciplinary collaborations, and the development of robust risk assessment frameworks. Bridging these knowledge gaps is crucial for ensuring the safe development, utilization, and governance of nanomaterials across various industries and applications while safeguarding human health and the environment.

14.4.1 Advancements in Nanotechnology

Rapid advancements in nanotechnology have propelled transformative innovations across diverse fields, owing to the unprecedented control and manipulation of matter at the nanoscale. These advancements signify a paradigm shift in scientific and technological capabilities, enabling novel applications with significant societal, economic, and scientific implications. In materials science, nanotechnology has ushered in a new era by offering engineered materials with tailored properties and functionalities. Nanomaterials, characterized by their unique size-dependent properties, find applications in diverse sectors, from electronics and energy storage to medicine and environmental remediation (De Jong, 2019). Their remarkable mechanical, optical, electrical, and catalytic properties have revolutionized material design and fabrication processes. In the realm of medicine and healthcare, nanotechnology has introduced groundbreaking approaches for drug delivery, diagnostics, and therapeutics. Nanoscale drug delivery systems offer precise targeting, controlled release, and enhanced bioavailability, optimizing treatment efficacy while minimizing side effects. Nanoparticle-based diagnostic tools enable early disease detection with heightened sensitivity and specificity, transforming clinical practices (Xuan et al., 2023).

Moreover, nanotechnology has revolutionized electronics and computing, enabling the development of smaller, faster, and more energy-efficient devices. Nanoelectronics leverage the unique properties of nanomaterials to design high-performance electronic components, paving the way for advancements in information technology and communication systems. In the energy sector, nanotechnology offers innovative solutions for energy generation, storage, and conservation. Nanomaterial-based photovoltaics, energy storage devices, and catalytic systems exhibit enhanced efficiency and sustainability, contributing to the development of cleaner and more efficient energy technologies. Environmental applications of nanotechnology encompass pollution remediation, water purification, and sustainable materials. Nanomaterial-based solutions facilitate efficient removal of contaminants, water purification, and the development of eco-friendly materials, addressing critical environmental challenges.

The rapid progress in nanotechnology is propelled by interdisciplinary collaborations, enabling the convergence of expertise from diverse fields such as physics, chemistry, biology, engineering, and materials science. The ability to engineer and manipulate materials at the nanoscale continues to inspire innovative solutions, driving the evolution of technology and shaping our future across multiple domains.

14.4.2 International Harmonization of Regulatory Policies

The international harmonization of regulatory policies for nanoparticle safety represents a pivotal endeavor aimed at establishing cohesive frameworks and standards to govern the safe development, utilization, and oversight of nanomaterials on a global scale. This concerted effort addresses the complexities and challenges surrounding nanotechnology and its potential implications for human health, the environment, and various industries (Halamoda-Kenzaoui et al., 2019). Harmonizing regulatory policies involves aligning standards, guidelines, and methodologies across different countries or regions to ensure consistency in the assessment, management, and regulation of nanomaterials (Devasahayam, 2019). This collaborative approach seeks to promote mutual recognition of safety assessments, facilitate trade, foster innovation, and uphold safety standards while avoiding unnecessary barriers to market entry. International harmonization efforts aim to establish common definitions, terminology, and classification systems for nanomaterials, enhancing clarity and consistency in regulatory frameworks. Standardized characterization protocols and testing methodologies for evaluating the safety profiles of nanomaterials enable comparability of data and assessments across jurisdictions (Allan et al., 2021). Moreover, harmonization initiatives strive to develop risk assessment frameworks and exposure guidelines that account for the unique properties and behaviors of nanomaterials. These frameworks consider diverse aspects, including toxicity assessments, dose-response relationships, exposure routes, fate in biological and environmental systems, and potential hazards, ensuring comprehensive evaluation and regulation of nanomaterials. Collaborative platforms and international forums facilitate knowledge sharing, information exchange, and the establishment of best practices among regulatory authorities, scientific communities, industry stakeholders, and policymakers. These platforms foster dialogue, promote transparency, and enable consensus-building on regulatory approaches, enhancing the effectiveness and efficiency of regulatory oversight (Dobrovolskaia & McNeil, 2013).

International harmonization of regulatory policies for nanoparticle safety also addresses challenges associated with emerging technologies and ensures regulatory agility to keep pace with advancements. It encourages adaptive regulatory strategies that can accommodate evolving scientific understanding and technological innovations while maintaining stringent safety standards. Ultimately, international harmonization fosters a cohesive global regulatory landscape that promotes innovation, ensures consumer and environmental safety, facilitates trade, and instills public confidence in the safe and responsible development and utilization of nanomaterials across diverse applications and industries.

14.4.3 Balancing Innovation with Safety

Balancing innovation with the safety of nanoparticle applications represents a critical challenge in harnessing the immense potential of nanotechnology while ensuring the protection of human health, the environment, and societal well-being. This delicate equilibrium necessitates a comprehensive approach that integrates innovation-driven progress with robust safety considerations, fostering responsible development and utilization of nanomaterials (Agrahari & Hiremath, 2017; Đorđević et al., 2021).

Innovation in nanotechnology drives transformative advancements across various sectors, offering unprecedented opportunities for improved products, processes, and technologies. The unique properties of nanomaterials enable novel applications in medicine, electronics, energy, and materials science, propelling advancements in efficiency, performance, and sustainability. However, this innovation needs to be coupled with rigorous safety assessment and regulatory scrutiny to mitigate potential risks associated with nanomaterials. Ensuring the safety of nanoparticle applications requires a proactive and multidisciplinary approach that incorporates comprehensive risk assessment, toxicity evaluation, and life cycle analysis. Rigorous characterization of nanomaterials, including their physicochemical properties, potential hazards, and interactions within biological and environmental systems, is crucial to understanding their safety profiles. Integration of safety considerations at the early stages of nanomaterial design and development facilitates the creation of safer nanoproducts and processes. This involves employing safer-by-design principles, utilizing less hazardous materials, and engineering nanomaterials with reduced toxicity without compromising their functionalities or performance. Transparent communication, information sharing, and collaboration among stakeholders, including researchers, industries, regulatory bodies, and the public, are fundamental in addressing safety concerns while fostering innovation. Stakeholder engagement promotes responsible innovation, facilitates risk-aware decision-making, and enhances public trust in the safety of nanotechnology applications (Allan et al., 2021).

Regulatory frameworks need to evolve in tandem with technological advancements, ensuring adaptive and robust oversight while enabling the pace of innovation. Harmonization of regulatory policies, international collaboration, and the establishment of standardized safety assessment protocols aid in ensuring consistency, comparability, and global acceptance of safety evaluations for nanomaterials. The pursuit of innovation in nanotechnology must be underpinned by ethical considerations, societal values, and a commitment to responsible research and development. Balancing innovation with safety necessitates a proactive, precautionary, and science-based approach, striving to harness the transformative potential of nanotechnology while safeguarding human health and the environment (Forest et al., 2022).

The EU has implemented a regulatory framework specifically addressing nanomaterials' safety and risk assessment. The Registration, Evaluation, Authorization, and Restriction of Chemicals (REACH) regulation has provisions requiring registration and assessment of nanomaterials, treating them as distinct substances due to their unique properties. The EU Observatory for Nanomaterials provides information and guidance to support risk assessment and regulation of nanomaterials. In the USA, the Food and Drug Administration (FDA) provides guidance and regulations for nanotechnology-based products, particularly in the areas of pharmaceuticals, cosmetics, and food. The FDA conducts risk assessments and evaluates the safety of nanomaterials used in consumer products. Health Canada has established guidelines for nanomaterials used in various products, including drugs and cosmetics. The government agency conducts risk assessments and monitors the safety of nanomaterials through the New Substances Notification Regulations (NSNR). The Australian government, through the National Industrial Chemicals Notification and Assessment Scheme (NICNAS), evaluates and regulates industrial chemicals, including nanomaterials. NICNAS requires notification and assessment of new and existing nanomaterials to ensure their safe use.

Several countries in the Asia-Pacific region, such as Japan and South Korea, have regulatory frameworks for nanotechnology. Japan has implemented guidelines and safety assessments for nanomaterials used in various industries, including electronics and healthcare. South Korea's Ministry of Food and Drug Safety (MFDS) regulates nanotechnology-based products, emphasizing safety evaluations and risk (Gaur et al., 2020).

Concerns arose regarding the potential toxicity of titanium dioxide nanoparticles (TiO_2 NPs), commonly used in sunscreens, cosmetics, and food products, due to their ability to generate reactive oxygen species under certain conditions. European Chemicals Agency (ECHA) carried out a regulatory response to the incidence by reviewing the safety of TiO_2 NPs and recommended its classification as a suspected carcinogen when inhaled. Consequently, France banned the use of TiO_2 NPs in food products, primarily food coloring additives (E171) which took effect from January 2020.

Concerns have also been raised about the potential health risks associated with carbon nanotubes (CNTs), particularly their resemblance to asbestos fibers and potential to cause respiratory issues. Consequently, the US National Institute for Occupational Safety and Health (NIOSH) issued recommendations for the safe handling of CNTs in workplaces. In Japan, the Ministry of Health, Labour and Welfare released guidelines recommending precautions when handling CNTs to minimize exposure and potential health risks. These incidents prompted regulatory bodies worldwide to reassess the safety profiles of certain nanoparticles, leading to revised guidelines, recommendations, or restrictions on their use to mitigate potential health and environmental risks. These responses highlight the importance of ongoing assessment and regulation of nanomaterials to ensure their safe utilization across various industries.

14.5 Future Directions

Addressing the following future directions will be instrumental in advancing the safe and responsible development of nanotechnology, harnessing its potential while mitigating potential risks to human health, the environment, and society.

14.5.1 Safer Nanomaterial Design

Efforts in advancing safer-by-design principles within nanomaterial engineering aim to integrate methodologies that mitigate toxicity risks while preserving intended functionalities. This strategic approach emphasizes the proactive design and synthesis of nanomaterials with reduced inherent toxicity or adverse effects on biological systems (He et al., 2019). Central to this initiative is the development of predictive models and computational tools that allow for a comprehensive assessment of nanomaterial safety during the design phase. These sophisticated models incorporate intricate physicochemical characteristics of nanomaterials, such as size, shape, surface properties, and composition, to forecast their interactions with biological entities (Galey et al., 2020). By integrating predictive modeling and computational simulations, researchers and engineers can anticipate the behavior and potential toxicity of nanomaterials prior to their synthesis or application. This preemptive evaluation empowers the identification of potential hazards or adverse effects, enabling informed modifications to the material's design to enhance safety. The objective is to establish a systematic framework that guides the design and development of nanomaterials, prioritizing safety considerations without compromising their intended functionalities. These efforts not only aim to minimize potential risks associated with nanomaterial exposure but also foster innovation by aligning safety assessments with the early stages of material design and synthesis (Hedberg et al., 2021).

14.5.2 Advanced Characterization Techniques

Recent strides in nanomaterial characterization methodologies have revolutionized our ability to scrutinize and comprehend the intricate physicochemical traits and behaviors of a diverse array of nanomaterials (Gao et al., 2015). These advancements have culminated in the development and implementation of cutting-edge techniques that afford a comprehensive understanding of nanomaterial dynamics. High-resolution imaging techniques, coupled with state-of-the-art spectroscopic methods, stand at the forefront of these advancements, enabling real-time monitoring of nanomaterial interactions within complex biological and environmental systems. These sophisticated imaging modalities, such as transmission electron microscopy (TEM), scanning electron microscopy (SEM), and atomic force microscopy (AFM), offer unprecedented spatial resolution, allowing visualization of nanomaterial morphology, size, and distribution with exceptional detail (Janoušek et al., 2023).

Complementing imaging techniques, advanced spectroscopic methodologies, including surface-enhanced Raman spectroscopy (SERS), X-ray photoelectron spectroscopy (XPS), and nuclear magnetic resonance (NMR), provide valuable insights into nanomaterial properties and interactions at molecular and atomic levels. These spectroscopic tools facilitate the elucidation of chemical composition, surface functionalization, and structural modifications, offering a nuanced understanding of nanomaterial behavior in diverse environments (Giannakou et al., 2022).

The integration of these cutting-edge characterization techniques not only enables a comprehensive assessment of nanomaterial physicochemical properties but also affords real-time monitoring of their interactions with biological entities and environmental matrices. This capability is pivotal in unraveling the complex interplay between nanomaterials and biological systems, thereby informing safer design strategies and enhancing our ability to predict and manage potential risks associated with their applications (Gaur et al., 2020).

14.5.3 Risk Assessment and Toxicity Studies

Conducting long-term toxicity studies represents a crucial facet in elucidating the chronic implications and potential accumulation patterns of nanoparticles within living organisms. These investigations delve into the extended effects of nanoparticle exposure over prolonged periods, shedding light on their persistent impact on biological systems (Leong et al., 2019).

Moreover, exploring the consequences of nanoparticle combinations and intricate mixtures serves to capture the complexities inherent in real-world exposure scenarios. In contrast to singular nanoparticle assessments, these studies consider the amalgamation of various nanoparticles or their interactions within complex mixtures, mimicking the diverse exposures encountered in practical environments. Through comprehensive long-term investigations, researchers scrutinize the cumulative effects and potential bioaccumulation of nanoparticles within organisms. These studies offer insights into how extended exposure durations may influence biological systems, potentially unraveling chronic implications, such as adverse health outcomes or altered physiological responses (Abudayyak et al., 2016).

Furthermore, delving into the effects of nanoparticle combinations and complex mixtures encompasses a more holistic approach to assess the intricate interplay among diverse nanoparticles. This approach enables a more accurate representation of the multifaceted exposure scenarios prevalent in natural settings, providing a deeper understanding of their collective impact on biological entities and ecosystems (Subramanian et al., 2016).

14.5.4 Transparency and Stakeholder Engagement

Promoting transparent communication and fostering collaborative networks among researchers, industrial stakeholders, regulatory entities, and the public stands as a cornerstone in addressing safety apprehensions and nurturing responsible innovation within the realm of nanotechnology. This approach prioritizes open dialogue, ensuring that findings, advancements, and safety considerations regarding nanomaterials are widely shared and comprehensively discussed among diverse stakeholders. Engaging in transparent communication cultivates a shared understanding of the benefits, risks, and ethical implications associated with nanotechnological developments (Limaye et al., 2018).

Moreover, involving stakeholders in decision-making processes facilitates the integration of societal values and ethical considerations into the formulation and implementation of regulatory policies. By incorporating diverse perspectives, these policies can be crafted to align with societal needs, values, and ethical frameworks, ensuring that safety and ethical dimensions remain integral components of the regulatory landscape governing nanotechnology. This collaborative approach serves to harmonize viewpoints, bridge knowledge gaps, and establish consensus on safety guidelines and ethical standards. In doing so, it reinforces responsible innovation in nanotechnology by fostering an environment where safety concerns are addressed proactively, societal values are upheld, and ethical principles are integrated into regulatory frameworks (Mohanta & Ahmaruzzaman, 2020).

14.5.5 Ethical, Legal, and Social Implications

Navigating the ethical, legal, and social implications inherent in nanotechnology necessitates a multifaceted approach that integrates ethical considerations and societal acceptance into governmental frameworks.

Addressing the challenges associated with the ethical, legal, and social implications involves a conscientious examination of the ethical dimensions associated with nanotechnological advancements. This includes ethical assessments of the potential risks and benefits, the equitable distribution of benefits across society, and the implications of these innovations on individual rights, privacy, and autonomy. Furthermore, societal acceptance plays a pivotal role in shaping the governance of nanotechnology, emphasizing the importance of understanding public perceptions, concerns, and values. Incorporating ethical and societal perspectives into regulatory decision-making processes is pivotal. It involves devising mechanisms that not only consider scientific evidence and technical aspects but also integrate ethical analysis and societal values. This integration ensures that regulatory frameworks are comprehensive, addressing the broader implications of nanotechnological developments on society, while upholding ethical principles and aligning with societal norms (Mühlebach, 2018; Paradise, 2019).

This approach strives to establish responsible governance by creating frameworks that are not solely based on scientific merit but are also responsive to ethical considerations and societal expectations. By integrating ethical and societal perspectives into regulatory decision-making, governance structures can better accommodate societal values, ethical norms, and public sentiments, fostering greater acceptance and trust in the governance of nanotechnology.

14.6 Further Thoughts and Remarks

The evolving regulatory landscape surrounding nanotechnology is characterized by ongoing advancements, challenges, and the continual refinement of frameworks to ensure the safe development and utilization of nanomaterials. Integration of nanotechnology-specific regulations into existing frameworks (e.g., REACH in the EU, FDA guidelines in the USA) to address nanomaterials' unique properties. Emphasis on risk-based approaches, considering exposure potential and hazard identification for different nanomaterials. Efforts to harmonize regulatory approaches globally for consistent safety assessment and risk management of nanomaterials. Collaboration among regulatory agencies, such as the OECD's Working Party on Manufactured Nanomaterials (WPMN), fostering information exchange and guideline development. Development of standardized testing protocols and guidelines for assessing the safety and environmental impacts of nanomaterials (Rodríguez-Ibarra et a., 2017).

Emphasis on comprehensive risk assessment, including toxicity, exposure scenarios, and life cycle analysis, to inform regulatory decisions. Promotion of safer-by-design strategies, encouraging the development of nanomaterials with reduced toxicity from the design phase. Integration of predictive models and computational tools to assess nanomaterial safety early in the development process. Engagement of stakeholders, including researchers, industries, regulatory bodies, and the public, in decision-making processes to foster transparency and consensus-building. Enhanced communication of regulatory decisions and risk assessment outcomes to ensure public awareness and understanding. Integration of ethical, legal, and societal implications into regulatory frameworks to address concerns related to societal acceptance, ethics, and responsible governance of nanotechnology (Thu et al., 2023).

Flexibility in regulatory frameworks to adapt to rapidly evolving nanotechnologies and innovative applications. Continuous monitoring of advancements to update regulations and guidelines accordingly may boost attention to the environmental impact, fate, and ecological implications of nanomaterials, and drive efforts toward sustainable nanotechnology. The evolving regulatory landscape reflects a proactive approach to ensure the safe and responsible development of nanomaterials. These approaches have the possibilities to address toxicity challenges through collaborative efforts, innovation, coupled with commitment to scientific rigor and ethical considerations.Research and development in nanosafety focus on comprehensively understanding the potential hazards and risks associated with nanomaterials, ensuring their safe and responsible application across various industries. Studying the toxicity profiles of different nanomaterials across varying sizes, shapes, surface modifications, and composition. Investigating the mechanisms of nanotoxicity, including cellular interactions, oxidative stress, inflammation, and genotoxicity (Abudayyak et al., 2016). Evaluating routes of exposure and determining the levels of human and environmental exposure to nanomaterials across different industries and applications. Studying the fate of nanomaterials in biological systems, including absorption, distribution, metabolism, and excretion. Furthermore, conducting risk assessments to determine potential risks to human health, occupational safety, and the environment posed by different nanomaterial. Developing predictive models and frameworks for quantitative risk assessment and management (Xuan et al., 2023).

Establishing standardized protocols and methodologies for nanomaterial characterization, toxicity testing, and risk assessment to ensure consistency and comparability of results. Advancing high-throughput screening methods to streamline toxicity testing and risk assessment processes. Incorporating safer-by-design principles into nanomaterial development, aiming to engineer nanomaterials with reduced toxicity while maintaining functionality (Eom & Choi, 2010). Designing nanomaterials with specific surface coatings or modifications to enhance biocompatibility and reduce adverse effects. Conducting comprehensive life cycle assessments to evaluate the environmental impacts of nanomaterials throughout their life cycle, from production to disposal. Investigating the ecological implications, persistence, and potential bioaccumulation of nanomaterials in different environmental compartments (Okeke, 2022).

Contributing to the development of regulatory frameworks and policies by providing scientific evidence and data on nanomaterial safety. Collaborating with regulatory agencies to establish guidelines, standards, and safety thresholds for the safe use of nanomaterials. Translating research findings into practical applications, ensuring that safety considerations are integrated into the development and commercialization of nanoproducts. Facilitating knowledge transfer among stakeholders, including academia, industry, policymakers, and the public has the potential to promote informed decision-making and responsible nanotechnology development. Furthermore, research and development efforts in nanosafety ascertains the sustainable and safe advancement of nanotechnology, and balances innovation with risk management to maximize its potential benefits while minimizing potential risks to human health and the environment.

Public awareness and stakeholder engagement also play crucial roles in promoting informed decision-making, fostering transparency, and ensuring responsible development and use of nanotechnology. Conducting educational campaigns, workshops, and public forums to disseminate information about nanotechnology, its applications, benefits, and potential risks. Providing accessible and understandable information tailored for different audiences, including the general public, policymakers, industry stakeholders, and healthcare professionals. Encouraging transparent communication between researchers, regulators, industry, and the public regarding the potential benefits, uncertainties, and safety aspects of nanomaterials.

Establishing platforms for open dialogue and information sharing to address concerns and foster trust among stakeholders is imperative to ascertain sustainable monitoring regimes. There is need to assess risk perceptions and misconceptions surrounding nanotechnology by providing accurate and evidence-based information on the safety and regulatory oversight of nanomaterials. Furthermore, empowering the public to make informed choices through clear communication about the potential benefits and risks associated with nanotechnology applications is imperative. There is also need to engage diverse stakeholders, including community groups, NGOs, academia, industry, regulatory agencies, and policymakers, in decision-making processes related to nanotechnology development, regulation, and governance, while incorporating public input and concerns into policy development and regulatory decision-making to ensure inclusivity and consideration of societal values (Armstead & Li, 2016).

For optimum advancement, ethical considerations and societal values must be integrated into discussions surrounding the development, deployment, and regulation of nanotechnology to address broader societal concerns while considering potential ethical implications, equity, and social justice aspects related to the adoption of nanomaterials. Participatory approaches, such as citizen science initiatives or public consultations, to involve the public in research, risk assessment, and decision-making processes related to nanotechnology need to be adopted while empowering communities to engage actively in discussions about the implications and applications of nanotechnology in their respective contexts. These strategies can be achieved by providing training and capacity-building programs to empower stakeholders, including policymakers, educators, and healthcare professionals, with the knowledge and skills to address nanotechnology-related issues. Public awareness and stakeholder engagement efforts serve as essential components in ensuring the responsible and inclusive development of nanotechnology, fostering trust, and facilitating a balanced dialogue among all involved parties.

14.7 Conclusion

In conclusion, the burgeoning field of nanotechnology presents immense potential for transformative innovations across diverse sectors, offering novel solutions to complex challenges. The unique properties of nanomaterials pave the way for groundbreaking advancements in medicine, electronics, materials science, and beyond. However, this technological frontier also necessitates careful consideration of potential risks to human health, the environment, and societal well-being.

Efforts in nanosafety research, regulatory frameworks, and stakeholder engagement are pivotal in ensuring the safe and responsible development of nanotechnology. Research endeavors focus on understanding the toxicity, exposure pathways, and risk profiles of nanomaterials, guiding the design of safer nanoproducts and informing regulatory decisions. Regulatory agencies globally are evolving their frameworks to adapt to nanotechnology's advancements, harmonizing standards, and emphasizing risk-based assessments. Public awareness and stakeholder engagement initiatives play a crucial role in fostering transparency, informed decision-making, and societal acceptance of nanotechnology. Educating the public, engaging diverse stakeholders, and promoting open communication contribute to addressing concerns, enhancing trust, and integrating ethical considerations into the development and deployment of nanomaterials.

As nanotechnology continues to evolve, it is imperative to strike a balance between innovation and safety, ensuring that the potential benefits are maximized, while potential risks are minimized. Collaboration among researchers, industries, regulatory bodies, policymakers, and the public is key to navigating the opportunities and challenges presented by nanotechnology, paving the way for a sustainable, inclusive, and responsible integration of this groundbreaking technology into society.

References

Abudayyak, M., Guzel, E. E., & Özhan, G. (2016). Copper (II) oxide nanoparticles induced nephrotoxicity in vitro conditions. *Applied In Vitro Toxicology, 2*(3), 157–164.

Agrahari, V., & Hiremath, P. (2017). Challenges associated and approaches for successful translation of nanomedicines into commercial products. *Nanomedicine, 12*, 819–823.

Allan, J., Belz, S., Hoeveler, A., Hugas, M., Okuda, H., Patri, A., Rauscher, H., Silva, P., Slikker, W., Sokull-Kluettgen, B., Tong, W., & Anklam, E. (2021). Regulatory landscape of nanotechnology and nanoplastics from a global perspective. *Regulatory Toxicology and Pharmacology, 122*, 104885.

Armstead, A., & Li, B. (2016). Nanotoxicity: Emerging concerns regarding nanomaterial safety and occupational hard metal (WC-Co) nanoparticle exposure. *International Journal of Nanomedicine, 11*, 6421–6433. https://doi.org/10.2147/ijn.s121238

Asare, N., Instanes, C., & Sandberg, W. J. (2012). Cytotoxic and genotoxic effects of silver nanoparticles in testicular cells. *Toxicology, 291*(1–3), 65–72.

Bakand, S., & Hayes, A. (2016). Toxicological considerations, toxicity assessment, and risk management of inhaled nanoparticles. *International Journal of Molecular Sciences, 17*(6), 929. https://doi.org/10.3390/ijms17060929. PMID: 27314324; PMCID: PMC4926462.

Bocca, B., Leso, V., Battistini, B., Caimi, S., Senofonte, M., Fedele, M., Cavallo, D. M., Cattaneo, A., Lovreglio, P., & Iavicoli, I. (2023). Human biomonitoring and personal air monitoring. An integrated approach to assess exposure of stainless-steel welders to metal-oxide nanoparticles. *Environmental Research, 216*, 114736.

Bundschuh, M., Filser, J., Lüderwald, S., McKee, M. S., Metreveli, G., Schaumann, G. E., Schulz, R., & Wagner, S. (2018). Nanoparticles in the environment: Where do we come from, where do we go to? *Environmental Sciences Europe, 30*(1), 1–17.

Crist, R. M., Dasa, S. S. K., Liu, C. H., Clogston, J. D., Dobrovolskaia, M. A., Stern, A. N. D., & S.T. (2021). Challenges in the development of nanoparticle-based imaging agents: Characterization and biology. *Wiley Interdisciplinary Reviews: Nanomedicine and Nanobiotechnology, 13*, e1665.

De Jong, W. (2019). Safety assessment of nanomaterials as ingredients for cosmetic products. In *Global Summit on Regulatory Science*. Pan Stanford.

Devasahayam, S. (2019). Nanotechnology and nanomedicine in market: A global perspective on regulatory issues. In *Characterization and biology of nanomaterials for drug delivery* (pp. 477–522). Elsevier.

Dobrovolskaia, M. A., & McNeil, S. E. (2013). Understanding the correlation between in vitro and in vivo immunotoxicity tests for nanomedicines. *Journal of Controlled Release, 172*, 456–466.

Đorđević, S., Gonzalez, M. M., Conejos-Sánchez, I., Carreira, B., Pozzi, S., Acúrcio, R. C., Satchi-Fainaro, R., Florindo, H. F., & Vicent, M. J. (2021). Current hurdles to the translation of nanomedicines from bench to the clinic. *Drug Delivery and Translational Research, 12*, 500–525.

Eom, H.-J., & Choi, J. (2010). p38 MAPK activation, DNA damage, cell cycle arrest and apoptosis as mechanisms of toxicity of silver nanoparticles in Jurkat T cells. *Environmental Science & Technology, 44*(21), 8337–8342.

Forest, V., Hochepied, J.-F., & Pourchez, J. (2022). Importance of choosing relevant biological end points to predict nanoparticle toxicity with computational approaches for human health risk assessment. *Chemical Research in Toxicology, 32*, 1320–1326.

Galey, L., Audignon, S., Witschger, O., Bau, S., Judon, N., Lacourt, A., & Garrigou, A. (2020). What does ergonomics have to do with nanotechnologies? A case study. *Applied Ergonomics, 87*, 103116.

Gao, H., Korn, J. M., Ferretti, S., Monahan, J. E., Wang, Y., Singh, M., Zhang, C., Schnell, C., Yang, G., Zhang, Y., Balbin, O. A., Barbe, S., Cai, H., Casey, F., Chatterjee, S., Chiang, D. Y., Chuai, S., Cogan, S. M., Collins, S. D., Dammassa, E., Ebel, N., Embry, M., Green, J., Kauffmann, A., Kowal, C., Leary, R. J., Lehar, J., Liang, Y., Loo, A., Lorenzana, E., Robert McDonald, E., ME, M. L., Merkin, J., Meyer, R., Naylor, T. L., Patawaran, M., Reddy, A., Röelli, C., Ruddy, D. A., Salangsang, F., Santacroce, F., Singh, A. P., Tang, Y., Tinetto, W., Tobler, S., Velazquez, R., Venkatesan, K., Von Arx, F., Wang, H. Q., Wang, Z., Wiesmann, M., Wyss, D., Xu, F., Bitter, H., Atadja, P., Lees, E., Hofmann, F., Li, E., Keen, N., Cozens, R., Jensen, M. R., Pryer, N. K., Williams, J. A., & Sellers, W. R. (2015). High-throughput screening using patient-derived tumor xenografts to predict clinical trial drug response. *Nature Medicine, 21*(11), 1318–1325. https://doi.org/10.1038/nm.3954

Gaur, N., Sharma, N., Dahiya, A., Yadav, P., Ojha, H., & Goyal, R. K. (2020). Toxicity and regulatory concerns for nanoformulations in medicine. In *The ELSI handbook of nanotechnology* (pp. 333–357). Wiley.

Giannakou, C., Geertsma, E. R., de Jong, H. W., van Loveren, H. J., Vandebriel, R., & Park, V. D. Z. M. (2022). Immunotoxicity testing of nanomedicinal products: Possible pitfalls in endotoxin determination. *Current Bionanotechnology, 2*, 95–102.

Halamoda-Kenzaoui, B., Holzwarth, U., Roebben, G., Bogni, A., & Bremer-Hofmann, S. (2019). Mapping of the available standards against the regulatory needs for nanomedicines. *Wiley Interdisciplinary Reviews: Nanomedicine and Nanobiotechnology, 11*, 1–17.

He, H., Liu, L., Morin, E. E., Liu, M., & Schwendeman, A. (2019). Survey of clinical translation of cancer nanomedicines—Lessons learned from successes and failures. *Accounts of Chemical Research, 52*, 2445–2461.

Hedberg, Y. S., Wei, Z., McCarrick, S., Romanovski, V., Theodore, J., Westin, E. M., Wagner, R., Persson, K. A., Karlsson, H. L., & Wallinder, I. O. (2021). Welding fume nanoparticles from solid and flux-cored wires: Solubility, toxicity, and role of fluorides. *Journal of Hazardous Materials, 413*, 0304–3894.

Janoušek, V., Florisbal, L. M., Konopásek, J., Jeřábek, P., Bitencourt, M. F., Gadas, P., Erban, V., & Kopačková-Strnadová, V. (2023). Arc-like magmatism in syn- to post-collisional setting: The Ediacaran Angra Fria Magmatic Complex (NW Namibia) and its cross-Atlantic correlatives in the south Brazilian Florianópolis Batholith. *Journal of Geodynamics, 155*, 0264–3707.

Leong, H. S., Butler, K. S., Brinker, C. J., Azzawi, M., Conlan, S., & Dufès, C. (2019). On the issue of transparency and reproducibility in nanomedicine. *Nature Nanotechnology, 14*, 811.

Limaye, V., Fortwengel, G., & Limaye, D. (2018). Regulatory roadmap for nanotechnology based medicines. *International Journal of Drug Regulatory Affairs, 2*, 33–41.

Mohanta, D., & Ahmaruzzaman, M. (2020). Biogenic synthesis of SnO2 quantum dots encapsulated carbon nanoflakes: An efficient integrated photocatalytic adsorbent for the removal of bisphenol A from aqueous solution. *Journal of Alloys and Compounds, 828*, 0925–8388.

Mühlebach, S. (2018). Regulatory challenges of nanomedicines and their follow-on versions: A generic or similar approach? *Advanced Drug Delivery Reviews, 131*, 122–131.

Okeke, E. N. (2022). Playing defense? Health care in the era of Covid. *Journal of Health Economics, 85*, 0167–6296.

Paradise, J. (2019). Regulating nanomedicine at the food and drug administration. *AMA Journal of Ethics, 21*, 347–355.

Patel, D., Sandefur, J., & Subramanian, A. (2021). The new era of unconditional convergence. *Journal of Development Economics, 152*, 0304–3878.

Rodríguez-Ibarra, C., Déciga-Alcaraz, A., Ispanixtlahuatl-Meráz, O., Medina-Reyes, E. I., Delgado-Buenrostro, N. L., & Chirino, Y. I. (2017). International landscape of limits and recommendations for occupational exposure to engineered nanomaterials. *Toxicology Letters, 322*, 111–119. https://doi.org/10.1016/j.toxlet.2020.01.016. Epub 2020 Jan 22.

Runa, R. A., Maksud, S., Rahman, M. S., Hasan, M., & Alam, M. R. (2022). Impact of drinking of saline water on hemato-biochemical parameters of Black Bengal goats in the selected areas of Bangladesh. *Saudi Journal of Biological Sciences, 29*, 103397.

Saif, M. A., Khan, W., & Godugu, C. (2018). Cytotoxicity of nanomaterials: Using nanotoxicology to address the safety concerns of nanoparticles. *Pharmaceutical Nanotechnology, 6*, 3–16.

Semenzin, E., Subramanian, V., Pizzol, L., Zabeo, A., Fransman, W., Oksel, C., Hristozov, D., & Marcomini, A. (2019). Controlling the risks of nano-enabled products through the life cycle: The case of nano copper oxide paint for wood protection and nano-pigments used in the automotive industry. *Environment International, 131*, 0160–4120.

Siegrist, S., Cörek, E., Detampel, P., Sandström, J., Wick, P., & Huwyler, J. (2019). Preclinical hazard evaluation strategy for nanomedicines. *Nanotoxicology, 13*, 73–99.

Singh, S., Kumar, V., Datta, S., Singh, S., Dhanjal, D. S., Garg, R., Kaur, P., & Sharma, K. (2020). Challenges and future perspectives of nanotoxicology. In *Springer eBooks* (pp. 451–466). https://doi.org/10.1007/978-981-15-1702-0_22

Stern, S. T., & McNeil, S. E. (2007). Nanotechnology safety concerns revisited. *Toxicological Sciences, 101*(1), 4–21. https://doi.org/10.1093/toxsci/kfm169

Subramanian, V., Semenzin, E., Hristozov, D., Zabeo, A., Malsch, I., McAlea, E., & Linkov, I. (2016). Sustainable nanotechnology decision support system: Bridging risk management, sustainable innovation and risk governance. *Journal of Nanoparticle Research, 18*, 1–13.

Thannickal, V. J., & Fanburg, B. L. (2000). Reactive oxygen species in cell signaling. *American Journal of Physiology. Lung Cellular and Molecular Physiology, 279*(6), L1005–L1028.

Thu, N. T., Patra, S., Pranudta, A., Nguyen, T. T., El-Moselhy, M. M., & Padungthon, S. (2023). Desalination of brackish groundwater using self-regeneration hybrid ion exchange and reverse osmosis system (HSIX-RO). *Desalination, 550*, 116378.

Tsang, M. P., Hristozov, D., Zabeo, A., Koivisto, A. J., Jensen, A. C. Ø., Jensen, K. A., Pang, C., Marcomini, A., & Sonnemann, G. (2017). Probabilistic risk assessment of emerging materials: Case study of titanium dioxide nanoparticles. *Nanotoxicology, 11*(4), 558–568.

Xuan, L., Wang, Y., Yang, K., Shao, R., Huang, F., Fan, Z., Chi, P., Xu, Y., Xu, N., Deng, L., Li, X., Liang, X., Luo, X., Shi, P., Liu, H., Wang, Z., Jiang, L., Lin, R., Chen, Y., Tu, S., Zhang, Y., Sun, J., Huang, X., & Liu, Q. (2023). Sorafenib maintenance after allogeneic haemopoietic stem-cell transplantation in patients with FLT3-ITD acute myeloid leukaemia: Long-term follow-up of an open-label, multicentre, randomised, phase 3 trial. *The Lancet Haematology, 10*(8), e600–e611. ISSN 2352-3026.

Chapter 15
Life Cycle Assessment of Nanoparticles

Margaret Ikhiwili Oniha and Stephen Oluwanifise Oyejide

15.1 Introduction

Researchers have publicized the significance of nanotechnology and the role of nanomaterials, particularly the copious applications of nanochitosan in biomedical, health, textile environment, agriculture, electronics, energy, construction, and marine building sectors, among others. However, such processes require feedback systems to circumvent adverse intrusions in the channels to salvage the economies (personal view). Additionally, it is important to comprehensively situate the possible human health and environmental challenges associated with the use of nanomaterials through the early evaluation of the life cycle of novel product formulations. Only a little focus has been paid to the detailed assessment of the environmental conduct and primary aspects of NMs with regard to LCA methodology (Nizam et al., 2021). Nizam et al. (2021) reported the tendency of false information from reviews due to the absence of comprehensive LCA studies on the environmental impacts of nanomaterials until the termination of life stages. The applications of nanomaterials (NMs) are multifaceted, grounded in high-performance materials, significant commercial impacts, energy storage capabilities, transformative potential, cost and energy efficiencies, and diminished environmental footprints.

These utilizations comprise of single, fused, aggregated, tubular, and unidimensional shapes, while the diversified types encase nanotubes, quantum dots, films, plates, and fullerenes (Guerra et al., 2018). Other features include their unique physical, chemical, mechanical, and potent characteristics, as well as the physical behavior that changed from classical physics to quantum physics with decreasing particle

M. I. Oniha (✉)
Department of Biological Sciences, Covenant University, Ota, Ogun State, Nigeria
e-mail: margaret.oniha@covenantuniversity.edu.ng

S. O. Oyejide
Department of Cell Biology and Genetics, University of Lagos, Akoka, Lagos State, Nigeria

P. O. Isibor et al. (eds.), *Environmental Nanotoxicology*,
https://doi.org/10.1007/978-3-031-54154-4_15

size. The size effect of NMs renders high facet energy, an expansive fraction of surface atoms, and spatial confinements (Cucurachi & Rocha, 2019; Kim et al., 2016), all of which have continually improved their use in different areas. The utilization of nanotechnologies in endeavors related to energy efficiency, pollution control and reduction, as well as environmental monitoring and remediation, has recently emerged as a central focus within the scientific community. For instance, reports have indicated their applications in industries such as healthcare, textile, agriculture, and energy. Several applications of nanoparticles are probably due to their varied forms: single, aggregated, tubular, irregular, etc. Another reason could be due to their small size which is related to their high surface energy and quantum effects which affect their optical, magnetic, and electrical behavior (Bratovcic, 2019). While this is applaudable, there is the other side of ensuring the effects of these nanomaterials are not detrimental to the environment and public health. The synthesis of nanomaterials is often dependent on several steps, solvents, and methods which are confronted by sustainability challenges and the hazardous impacts the nanomaterials contribute to the environment (Baig et al., 2021). These steps such as carbonization, physical and chemical vapor deposition/activation, and liquid-phase synthesis, among others, produce pollutants that affect the environmental balance of the air, water, and soil and ultimately public health (Heidarinejad et al., 2020). Therefore, a lingering question is on the actual environmental sustainability of nanoparticles and the management of their environmentally hazardous contribution.

The use of life cycle assessment (LCA) as a tool in understanding the potential environmental challenges of the application of nanoparticles has been studied to ensure the environmental sustainability of nanoparticles. This involves identifying the production materials, energy, and emissions and their effects on the environment across each stage of development. LCA is an established international standard methodology (International Organization for Standardization 14,040 series), which evaluates four critical points: the goal and scope, life cycle inventory, life cycle impact assessment, and interpretation (Klöpffer & Grahl, 2014). The use of this methodology has revealed three major challenges involved in the applications of nanotechnology: unavailable or not detailed functionalities of nanoparticles, a lack of transparent life cycle inventory, and a lack of characterization methods for produced nanoparticles (Peña et al., 2021).

15.2 Goal and Scope of Life Cycle Assessment

In the delineation of the objectives and parameters governing the Life Cycle Assessment (LCA) applied to nanoparticles, a critical prerequisite is the establishment of a comprehensive functional unit (FU). The functional unit is construed to encapsulate the entirety of the nanoparticles' life cycle, encompassing a spectrum of intricate considerations (Pati et al., 2014). These encompass the operational conditions during usage, methodologies applied in the disposal phase, protocols employed in the management of nanoparticles, conformity with recommended usage practices by consumers, and a meticulous analysis of potential rebound effects affecting both environmental and health domains. The holistic exploration of these multifaceted dimensions facilitates a through evaluation of the nanoparticles' life cycle. This approach is indispensable for cultivating a nuanced comprehension of the intricate interplay between these nanomaterials and their surroundings. By comprehensively assessing environmental and health impacts across the diverse stages of nanoparticle utilization, this elucidation enables the identification of potential risks and benefits inherent in their life cycle. This systematic evaluation, guided by the principles of LCA, thereby contributes to an informed decision-making process regarding the integration of nanoparticles into various applications, ensuring a judicious and responsible approach to their utilization in contemporary scientific and industrial contexts (Curran, 2016).

15.3 Concept of Life Cycle Assessment (LCA)

Two methods are commonly used to assess materials, especially nanoparticles, risk assessment and life cycle assessment (LCA). While risk assessment focuses on the quantitative and qualitative evaluation of risks confronting the environment and public health by contaminants, LCA focuses more on the quantitative evaluation of the input, output, and possible environmental impact of a system all through its life cycle (Guinée et al., 2017). Guinée et al. (2017) stated that the LCA method is more used when assessing nanoparticles especially when both methods cannot be integrated. Guinée et al. (2017) also stated that the general framework in a specific order for LCA includes the goal and scope of definition, inventory analysis, impact assessment, interpretation, and then applications which can include product development and improvement, strategic planning, etc. These frameworks can be made specific based on particular needs or the nanomaterial in question. Broadly speaking, LCA can also involve the assessment of the nanoparticles and how they relate to climate change, photochemical ozone formation, and resource depletion (Girish et al., 2015).

LCA of nanoparticles looks at how the life cycles of nanoparticle-incorporated products differ from conventional products. It evaluates the energy efficiency at each phase and looks at disposal/recycling options. Apart from the ecological and human toxicological impact, LCA also looks at the geographical impact of nanoparticles compared to conventional products. Hence, LCA should always be conducted by manufacturing companies; however, government intervention is often needed in funding research and building databases. However, conducting LCA of nanoparticles is usually confronted with certain challenges, such as the lack of reliable inventory data, the absence of toxicological test results, and the reproducibility of results (Desai, 2012).

Earlier studies conducted on LCA of NMs encountered three main challenges in modeling nanomaterials in the LCA framework (Hischier & Walser, 2012). The first is insufficient use of an adequate functional unit that encompasses all the detailed and additional functionalities of NMs. The second challenge is the lack of transparent life cycle inventory (LCI) data in the generation of transparent life cycle inventory (LCI) data in the production of nanomaterials (NMs) is often hindered by the lack of disclosure from manufacturers regarding materials and energy inputs. This non-disclosure is frequently attributed to considerations related to commercial confidentiality. Lastly, another major challenge is the lack of characterization techniques for released NMs that are a prominent part within the LCIA context.

As a result of the rapid technological advancement in nanotechnology, the environmental toxicity pathways of the NMs require enhanced evaluation from an LCA perspective. Despite the diversity in methodological approaches, data collection methodologies, and selected characterization methods employed across various studies in the field, a lack of consistency is observed. Consequently, the resultant findings may lack robustness, rendering them inconclusive or potentially contradictory. Therefore, the present state-of-the-art LCA application in nanotechnology needs to be explored to gain insights into the present practices of LCA application in nanotechnology and its future outlook. The FU is the quantified performance of NMs used as a reference unit in the LCA study in which a fixed value is set, and the output results on the environmental impacts from the impact categories reflect on this selected FU, where the margins of error and explicitly specified data uncertainties should be incorporated (Mullen & Morris, 2021).

15.4 Life Cycle Assessment Inventory

Inventory analysis involves the collection of reliable data on the input, processes, and outputs with clear explanations for any assumptions, advantages, and disadvantages. This is to help a clean LCA alongside transparency and credibility of the product system (Miseljic & Olsen, 2014). Challenges here include the confidential processes of certain nanoparticle manufacturing which makes it unavailable for scrutiny by the government or independent agencies (Desai, 2012). LCA by the manufacturing company is usually focused on the company data but for the governmental agency could be based on sector-wide averages which will take the future consequence of an appropriate decision into consideration. This requires a complete list of inventories provided by manufacturing companies to the government and other regulatory agencies especially since nanotechnology as a field tends to get new developments which must be circulated to the necessary agencies as well. However, this is not often so, as several nanoparticles are being used by consumers with unclear understanding of the manufacturing processes, maintenance, and disposal/recycling roles (Bratovcic, 2019). Also, specific materials will be discharged during use either intentionally or otherwise that could adversely affect the health of the user or the environment. Furthermore, post-utilization, the extent of interactivity between nanoparticles and other materials, as well as their compatibility with designated disposal sites, raises pertinent concerns. The potential ramifications of improper disposal practices also warrant consideration. These aspects are of primary interest to government agencies, underscoring the essentiality of comprehensive inventory data for nanoparticles to facilitate informed decision-making processes (Purwasasmita, 2017).

To conduct a comparative LCA (such as comparing NMs with conventional materials), the FU should not be solely based on weight as it is not functionally comparable between the two products (Hischier & Walser, 2012; Nizam et al., 2021; Park et al., 2019). However, the FU needs to indicate the significance of evaluating the function and performance of these specific systems wherein all processes, as well as their inputs and outputs, are linearly scaled (Mullen & Morris, 2021; Nizam et al., 2021). A proper and adequate functional unit that considers all the additional functionalities of NMs must be addressed in future studies in this field to offer more realistic and fair potential benefits of nanomaterials in enhanced technologies. An inadequate definition of the FU leads to more uncertainty in the study (Cucurachi & Rocha, 2019; Hischier et al., 2017). Nanomaterials possess complex features and compositions, which may change their physical–chemical interactions throughout the life cycle, especially at the disposal stage, making them challenging to configure into a novel item, since the NM properties can be unpredictable. Presently, there are no international regulations on the disposal management of NMs.

The LCI phase, also referred to as the data collection phase, is crucial to any LCA study. This phase is the most work-intensive and time-consuming phase in an LCA, noting it requires a detailed data input of all the processes included in the scoping of NMs. Collections of complete and reliable data, which encapsulates clear explanations of applied assumptions, advantages and disadvantages, and transparency and credibility criteria, are limited. Different up-to-date databases are available and can be used together with LCA software; however, those data (production of electricity, coal, or packaging) are generic and can only be used for processes that are not product-specific. Inventory database such as Ecoinvent is the most widely used by the researchers in the previous studies. The inventory inputs included raw materials, energy (renewable and nonrenewable), and water, while outputs are the products and coproducts (emissions to air, water, and soil) (Mullen & Morris, 2021). The LCIA stage is purposed to assess the environmental impacts and analyze the data to evaluate the contribution to each impact category premised on the inventory analysis. The most studied impact categories included global warming potential (56%), acidification (38%), and ecotoxicity (36%). These predominant impact categories were studied frequently due to their significant impacts on the environment, disrupting the food chains by bioaccumulation and eventually harming human health and all living beings. The varied flow of toxicities of NMs in the human system is presented in Fig. 15.1. And for these reviewed studies, the average value of global warming potential is between 0.7 and 3.0 kg CO_2 eq per functional unit, while acidification releases a wide range of emissions that ranged from 1.0 to 20.0 kg SO_2 eq per functional unit. In addition, for the ecotoxicity, the range is measured between 3.0 and 8.0 CTUe/kg of which the FU is weighted in kilograms.

Sackey et al. (2019) conducted an assessment on the impacts at the midpoint level and categorized the potential impacts into eight groups. The results showed that NMs (nano-silica asphalt mixtures) performed better in global warming, ozone

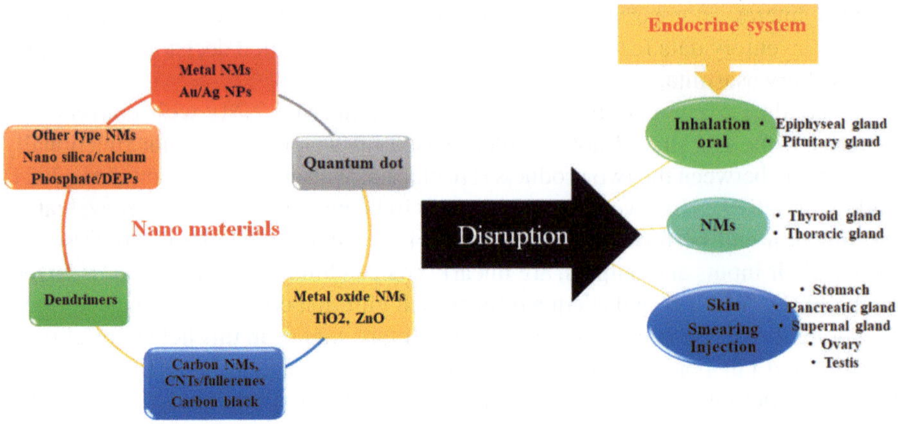

Fig. 15.1 Toxic effects of diverse types of nanoparticles on various organs. (Source: Ajdary et al., 2018)

depletion, eutrophication, photochemical oxidation, and ecotoxicity than conventional asphalt mixtures. Reviewed studies by Bartolozzi et al. (2019) and Petrakli et al. (2020) that reported nonperformance until endpoint level carried out a sum of 10–12 midpoint categories for the application of nanostructured materials in building blocks and nano-enhanced, carbon fiber-reinforced polymer prototypes, respectively. In contrast to conventional materials, the utilization of NMs has outturned reduced environmental impacts with specificity for climate change, photochemical ozone depletion, solid matter (human health and ecosystem), and acidification. Hischier et al. (2015) examined facade coating systems incorporating engineered nanomaterials (NMs), encompassing nano-titanium dioxide coatings, a comprehensive evaluation was undertaken, extending to the end-of-life stage where the disposal of these coatings was considered. The study validates that the incorporation of the investigated engineered NMs in coatings yields an enhanced environmental performance. Notably, the long-term emissions' impact on final landfilling systems has received minimal attention in the existing literature. It is imperative, however, to mitigate the disposal of unused paints containing nanomaterials to the greatest extent feasible. While life cycle impact assessment (LCIA) at the end-of-life stage holds paramount significance within the life cycle assessment (LCA) framework, the availability of pertinent data is currently limited, thereby introducing an element of uncertainty into the overall study outcomes (Fig. 15.2).

15.5 Conclusion

Nanotechnology has espoused diverse nanomaterials to facilitate the achievement of improved health, agriculture, and environment, among others. Nanochitosan is one of the nanomaterials incorporated in different sectors that encapsulate

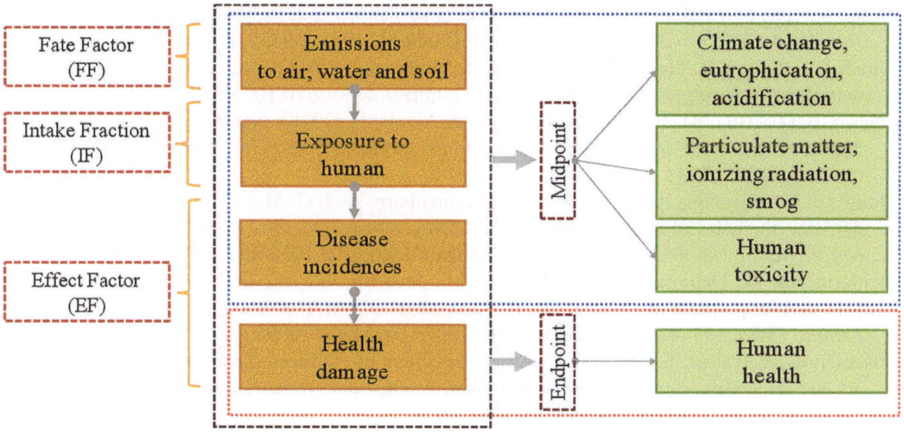

Fig. 15.2 Toxicity flow of NMs. (Source: Nizam et al., 2021)

biomedical, textile, electronics, energy, construction, and marine, among others. However, there is a need to situate the adverse conditions associated with means to address these issues through life cycle assessment methodology for enhanced outturns, all of which have received minimal focus. A comprehensive LCA inventory requires that the FU must substantially evaluate the components, functionalities, processes, and outputs involved for enhanced technologies. In all, literatures have provided results to support the better performance of NMs compared to conventional constituents with evident reduced environmental impacts.

References

Ajdary, M., Moosavi, M. A., Rahmati, M., Falahati, M., Mahboubi, M., Mandegary, A., Jangjoo, S., Mohammadinejad, R., & Varma, R. S. (2018). Health concerns of various nanoparticles: A review of their in vitro and in vivo toxicity. *Nanomaterials, 8*(9), 634. https://doi.org/10.3390/nano8090634

Baig, N., Kammakakam, I., & Falath, W. (2021). Nanomaterials: A review of synthesis methods, properties, recent progress, and challenges. *Materials Advances, 2*(6), 1821–1871. https://doi.org/10.1039/d0ma00807a

Bartolozzi, I., Daddi, T., Punta, C., Fiorati, A., & Iraldo, F. (2019). Life cycle assessment of emerging environmental technologies in the early stage of development: A case study on nanostructured materials. *Journal of Industrial Ecology, 24*, 101–115. https://doi.org/10.1111/jiec.12959

Bratovcic, A. (2019). Different applications of nanomaterials and their impact on the environment. *International Journal of Material Science and Engineering, 5*(1), 1–7. https://doi.org/10.14445/23948884/ijmse-v5i1p101

Cucurachi, S., & Rocha, C. F. B. (2019). Nanotechnology in eco-efficient construction. In *Materials, processes and application: Life-cycle assessment of engineered nanomaterials* (2nd ed., pp. 815–846). Woodhead Publishing Series in Civil and Structural Engineering.

Curran, M. A. (2016). Overview of goal and scope definition in life cycle assessment. In *LCA compendium – The complete world of life cycle assessment* (pp. 1–62). https://doi.org/10.1007/978-94-024-0855-3_1

Desai, N. (2012). Challenges in the development of nanoparticle-based therapeutics. *The AAPS Journal, 14*(2), 282–295. https://doi.org/10.1208/s12248-012-9339-4

Girish, U., Dhingra, R., Naidu, S., Atuahene, I., & Sawhney, R. (2015). Life cycle assessment of nanomaterials. In *Springer eBooks* (pp. 393–408). https://doi.org/10.1007/978-3-319-15461-9_14

Guerra, F. D., Attia, M. F., Whitehead, D. C., & Alexis, F. (2018). Nanotechnology for environmental remediation: Materials and applications. *Molecules, 23*, 1760. https://doi.org/10.3390/molecules23071760

Guinée, J. B., Heijungs, R., Vijver, M. G., & Peijnenburg, W. J. G. M. (2017). Setting the stage for debating the roles of risk assessment and life-cycle assessment of engineered nanomaterials. *Nature Nanotechnology, 12*(8), 727–733. https://doi.org/10.1038/nnano.2017.135

Heidarinejad, Z., Dehghani, M. H., Heidari, M., Javedan, G., Ali, I., & Sillanpää, M. (2020). Methods for preparation and activation of activated carbon: A review. *Environmental Chemistry Letters, 18*(2), 393–415. https://doi.org/10.1007/s10311-019-00955-0

Hischier, R., & Walser, T. (2012). Life cycle assessment of engineered nanomaterials: State of the art and strategies to overcome existing gaps. *Science and Total Environment, 425*, 271–282. https://doi.org/10.1016/j.scitotenv.2012.03.001

Hischier, R., Nowack, B., Gottschalk, F., Hincapie, I., Steinfeldt, M., & Som, C. (2015). Life cycle assessment of façade coating systems containing manufactured nanomaterials. *Journal of Nanoparticle Resources, 17*, 68. https://doi.org/10.1007/s11051-015-2881-0

Hischier, R., Salieri, B., & Pini, M. (2017). Most important factors of variability and uncertainty in an LCA study of nanomaterials—Findings from a case study with nano titanium dioxide. *NanoImpactibility, 7*, 17–26. https://doi.org/10.1016/j.impact.2017.05.001

Kim, J., Rivera, J. L., Meng, T. Y., Laratte, B., & Chen, S. (2016). Review of life cycle assessment of nanomaterials in photovoltaics. *Solar Energy, 133*, 249–258. https://doi.org/10.1016/j.solener.2016.03.060

Klöpffer, W., & Grahl, B. (2014). *Life Cycle Assessment (LCA)*. Wiley-VCH Verlag GmbH & Co. KGaA. https://doi.org/10.1002/9783527655625

Miseljic, M., & Olsen, S. I. (2014). Life-cycle assessment of engineered nanomaterials: A literature review of assessment status. *Journal of Nanoparticles Research, 16*, 2427. https://doi.org/10.1007/s11051-014-2427-x

Mullen, E., & Morris, M. (2021). Green nanofabrication opportunities in the semiconductor industry: A life cycle perspective. *Nanomaterials, 11*, 1085. https://doi.org/10.3390/nano11051085

Nizam, N. U. M., Hanafiah, M. M., & Woon, K. S. (2021). A content review of life cycle assessment of nanomaterials: Current practices, challenges, and future prospects. *Nanomaterials (Basel), 11*(12), 3324. https://doi.org/10.3390/nano11123324

Park, W.-J., Kim, T., Roh, S., & Kim, R. (2019). Analysis of life cycle environmental impact of recycled aggregate. *Applied Science, 9*, 1021. https://doi.org/10.3390/app9051021

Pati, P., McGinnis, S., & Vikesland, P. J. (2014). Life cycle assessment of 'green' nanoparticle synthesis methods. *Environmental Engineering Science, 31*(7), 410–420. https://doi.org/10.1089/ees.2013.0444

Peña, C., Civit, B., Gallego-Schmid, A., Druckman, A., Pires, A. C., Weidema, B., Mieras, E., Wang, F., Fava, J., Milà i Canals, L., Cordella, M., Arbuckle, P., Valdivia, S., Fallaha, S., & Motta, W. (2021). Using life cycle assessment to achieve a circular economy. *The International Journal of Life Cycle Assessment*. https://doi.org/10.1007/s11367-020-01856-z

Petrakli, F., Gkika, A., Bonou, A., Karayannis, P., Koumoulos, E., Semitekolos, D., Trompeta, A.-F., Rocha, N., Santos, R., Simmonds, G., et al. (2020). End-of-life recycling options of (Nano) Enhanced CFRP composite prototypes waste—A life cycle perspective. *Polymers, 12*, 2129. https://doi.org/10.3390/polym12092129

Purwasasmita, B. S. (2017). Green materials for sustainable development. *IOP Conference Series: Earth and Environmental Science, 60*, 012004. https://doi.org/10.1088/1755-1315/60/1/012004

Sackey, S., Lee, D.-E., & Kim, B.-S. (2019). Life cycle assessment for the production phase of nano-silica-modified asphalt mixtures. *Applied Science, 9*, 1315. https://doi.org/10.3390/app9071315

Chapter 16
Emerging Technology and Future Directions in Environmental Nanotoxicology

Patrick Omoregie Isibor, Ifeoluwa Ihotu Kayode-Edwards, and Olugbenga Samson Taiwo

16.1 Present State of Environmental Nanotoxicology

Environmental nanotoxicology investigates the impact of nanomaterials on ecosystems and organisms, delving into their potential risks and interactions within the environment. Present research in this field has unearthed various findings that shed light on the complexities and implications of nanomaterial exposure.

Research has demonstrated that the behavior and toxicity of nanoparticles in the environment are greatly influenced by their physicochemical characteristics (Li et al., 2015a; Goswami et al., 2017; Turan et al., 2019). The interactions of nanoparticles with living organisms and environmental matrices are determined by factors such as size, shape, surface area and chemistry, solubility and dissolution rate, aggregation and agglomeration state, and composition (Ogunkunle et al., 2021; Anastasiadis et al., 2022). Studies have indicated that these characteristics can impact organisms across trophic levels by influencing cellular uptake, bioaccumulation, and consequent biological responses (Lu et al., 2021; Sibiya et al., 2022).

Furthermore, investigations into the mechanisms of nanotoxicity have highlighted various pathways through which nanoparticles exert adverse effects. Oxidative stress, inflammation, genotoxicity, disruption of cellular processes, mitochondrial dysfunction, lysosomal damage, membrane damage, protein alterations, autophagy dysregulation, and endoplasmic reticulum stress are some of the identified mechanisms underlying nanoparticle-induced toxicity (Mohammadinejad et al., 2019; Khan et al., 2020; Xuan et al., 2023). Additionally, studies have demonstrated the capacity of nanoparticles to bioaccumulate in organisms and transfer

P. O. Isibor (✉) · I. I. Kayode-Edwards · O. S. Taiwo
Department of Biological Sciences, Covenant University, Ota, Ogun State, Nigeria
e-mail: patrick.isibor@covenantuniversity.edu.ng; kayode-edwards.ifeoluwa@stu.cu.edu.ng; gbenga.taiwo@covenantuniversity.edu.ng

P. O. Isibor et al. (eds.), *Environmental Nanotoxicology*,
https://doi.org/10.1007/978-3-031-54154-4_16

across various levels of food chains and subsequently food webs, raising concerns about their long-term impacts on ecosystems (Babaei et al., 2022).

Research efforts have also focused on understanding the fate and behavior of nanoparticles in various environmental compartments. Nanoparticles can undergo transformations in complex environmental matrices such as soil, water bodies, atmosphere, and biological systems, altering their reactivity, stability, and ultimate impact (Turan et al., 2019; Abbas et al., 2020; Ju et al., 2022). Moreover, studies have highlighted how organic matter and natural colloids alter the behavior of nanoparticles, affecting their transport, bioavailability, and ecological impacts (Abbas et al., 2020; Ju et al., 2022; Wang et al., 2022b).

However, knowledge gaps that present challenges persist, necessitating further research. Challenges include the requirement for standardized toxicity assessment techniques (Pikula et al., 2020), understanding the impact of the interactions between nanoparticles and other environmental stressors (Forest, 2021), and predicting the ecological effects of continuous or repeated exposure to nanoparticles (Hu et al., 2016). The main issues in nanotoxicology, according to Pikula et al. (2020), persist with the need to further adapt present testing guidelines to nanoparticles, develop new ones, and validate, store, systematize, and share the findings of multiple studies and projects. Hu et al. (2016) also discuss a number of intricate issues related to nanotoxicological research and the safety of nanomaterials, such as transgenerational effects, adaptation to and reversibility of nanotoxicity, and biological and environmental coronas outside of the protein corona that may arise from low-concentration, long-term exposure to nanomaterials. Addressing these gaps in knowledge requires interdisciplinary collaboration, long-term monitoring programs, advancements in analytical techniques, and a concerted effort to establish standardized protocols for assessing nanoparticle toxicity in environmental systems. Closing these gaps will be crucial for well-informed decision-making and sustainable management of nanomaterials in the environment.

Present research in environmental nanotoxicology provides valuable insights into the behavior, interactions, and potential risks that nanoparticles may pose to ecosystems. These investigations provide a basis for developing strategies to manage and mitigate the environmental impacts of nanomaterials while paving the way for future advancements in the field of environmental nanotoxicology.

16.2 Gaps in Knowledge

There are still a number of important knowledge gaps in the field of environmental nanotoxicology, which prevents a thorough understanding of the effects of nanomaterials on ecosystems.

Due to the small number of trial subjects in traditional environmental toxicology analyses, in vitro and in vivo experiments must be carried out over an extended period of time, resulting in significant time and financial costs as well as potentially cruel outcomes (Wu et al., 2022). Furthermore, according to Guimaraes (2020), ecosystems are intricate webs of interactions between living things and their surroundings, and so it is difficult to comprehend how toxic substances affect these systems because of their complexity and dynamic nature. Consequently, it is important to utilize existing data to build models for predicting environmental toxicity that are more accurate and affordable than present methods (Wu et al., 2022).

A significant challenge is the absence of widely recognized and standardized protocols for evaluating the toxicity of nanoparticles in environmental systems (Giusti et al., 2019; Boros & Ostafe, 2020; Kus-Liśkiewicz et al., 2021). Maintaining uniformity in testing techniques is essential for precise risk assessments and comparisons between various studies. Moreover, there can be unreliability when extrapolating toxicity data from one species or environment to another (Aurisano et al., 2019). Bridging these gaps in knowledge to make accurate predictions across different ecosystems and species remains a challenge. Understanding the long-term impacts of chronic nanoparticle exposure on ecosystems and organisms is another challenging task (Ale et al., 2021). The majority of research focuses on acute exposure, but more research is needed to determine the long-term effects of repeated or continuous exposure to nanoparticles.

Furthermore, it is difficult to create regulations that effectively address societal, economic, and environmental issues. It remains a battle to balance innovation and precaution in regulatory frameworks. Similarly, a lack of information about the toxicity and persistence of many nanoparticles makes it difficult to make reliable risk assessments, which in turn makes it challenging to make decisions about regulations and efficient management techniques (Ahmad et al., 2022; Cunningham et al., 2023). Additionally, nanoparticles engage in interactions with different environmental elements such as soil, sediments, and aquatic systems (Turan et al., 2019; Abbas et al., 2020; Rajput et al., 2020). There is still a lack of characterization regarding their behavior, transformations, and bioavailability in complex matrices, which makes risk assessments prone to errors. Moreover, the sensitivity of different species, and even individuals within a species, to toxic substances can vary significantly (Tortella et al., 2020). Furthermore, a variety of stressors, including pollution, climate change, invasive species, and habitat degradation, frequently affect ecosystems concurrently (Jenny et al., 2020). It is also unclear how these stressors and nanoparticles interact and how that affects ecosystems as a whole. Comprehending and forecasting these fluctuations is essential for evaluating the total effects on ecosystems.

More research is also needed to determine how nanoparticles affect ecosystem functioning, structure, and biodiversity (Ahmed et al., 2022). Evaluating the overall health of an ecosystem requires an understanding of how exposure to nanoparticles affects various species and ecological processes. Although research has demonstrated the bioaccumulation of nanoparticles in living things, it is still unclear how this accumulation could impact ecosystems and trophic levels. Developing robust ecotoxicological models and predictive tools that can reliably estimate the ecological effects of nanoparticles across different environmental scenarios and species is an ongoing challenge (Spurgeon et al., 2020; Ankley et al., 2021).

Funding, technology, and expertise are all crucial for carrying out extensive toxicological investigations (Schmeisser et al., 2023). Resource limitations impose restrictions on a lot of research projects. Additionally, communicating complex scientific findings about environmental toxicology to the public in an understandable manner is essential (Lai et al., 2022). Lack of awareness and understanding can hinder support for necessary actions and policies. Finally, constantly evolving industrial processes introduce new contaminants and nanomaterials whose toxicity and environmental behavior are not well understood (Morin-Crini et al., 2022). Keeping up with these emerging substances is a continuous challenge.

Addressing these challenges requires interdisciplinary collaboration, increased research efforts, technological advancements, improved data sharing, and effective communication between scientists, policymakers, industries, and the public. Overcoming these limitations is crucial for better environmental protection and sustainable management of ecosystems.

16.3 Emerging Technologies in Environmental Nanotoxicology

16.3.1 Advanced Analytical Techniques

Advanced imaging techniques are emerging technologies in environmental nanotoxicology. High-resolution imaging tools like transmission electron microscopy (TEM), scanning electron microscopy (SEM), and atomic force microscopy (AFM) allow for the visualization of nanoparticles in environmental matrices and biological systems (Titus et al., 2019; Zhang & Wang, 2023). Because of the high resolution of these imaging tools, the size, morphology, and aggregation of nanoparticles can be viewed which can help in understanding their interactions and potential effects at the nanoscale level. TEM and SEM provide detailed structural information, aiding in understanding nanoparticle distribution and cellular uptake (Inkson, 2016), while AFM enables accurate imaging of surface characteristics and mechanical properties at the nanoscale level (Stan & King, 2020). According to Huangfu et al. (2019), these methods make it easier to directly observe and characterize nanoparticles in a variety of environmental settings, providing insight into their interactions, changes, and bioavailability. They offer vital information for toxicity evaluations and aid in the clarification of changes in cellular structures brought about by nanoparticles. Furthermore, according to Xuan et al. (2023), advanced imaging is essential for verifying models and predictions about the behavior of nanoparticles and their effects on ecosystems.

In addition, thorough characterization of nanoparticles in biological and environmental contexts is made possible by combining various imaging modalities, such as fluorescence microscopy with electron microscopy or spectroscopic techniques (Takizawa & Robinson, 2012; Hobson & Aaron, 2022; Alonso et al., 2023). This approach provides complementary data on the effects, interactions, and localization of nanoparticles at different scales.

In the field of environmental nanotoxicology, advanced spectroscopic techniques are also a valuable technological tool. The molecular-level study of nanoparticle interactions with biomolecules or environmental elements is made possible by Raman spectroscopy, surface-enhanced Raman spectroscopy (SERS), and infrared spectroscopy (Jia et al., 2018; Kumar et al., 2019; Spedalieri & Kneipp, 2022). These methods offer details on the molecular interactions, surface chemistry, and structural alterations of nanoparticles. Raman spectroscopy, for example, provides comprehensive molecular fingerprinting, explaining nanoparticle changes and detecting biomolecular changes after exposure (Siddhanta et al., 2023). Sensitivity is increased by surface-enhanced methods such as SERS, which makes it possible to find traces of biomolecules or nanoparticles in environmental samples (Song et al., 2019; Ong et al., 2020; Terry et al., 2022). Understanding the behavior of nanoparticles in environmental matrices is facilitated by revealing of their chemical compositions and functional groups through infrared spectroscopy (Faghihzadeh et al., 2016).

In environmental nanotoxicology, synchrotron radiation methods such as X-ray fluorescence (XRF) and X-ray absorption spectroscopy (XAS) are also becoming increasingly utilized. They provide high-resolution elemental and structural analysis of nanoparticles in intricate environmental systems (Servin et al., 2012; Hernandez-Viezcas et al., 2013; Castillo-Michel et al., 2017). With the use of intense, adjustable synchrotron radiation, these methods can examine the electrical, chemical, and structural characteristics of nanoparticles. These techniques aid in clarifying the transformations, speciation, and interactions of nanoparticles with biological or environmental elements. In nanotoxicology, synchrotron-based methods are used for elemental mapping, speciation analysis, and comprehending the interactions of nanoparticles in environmental matrices (Wang et al., 2010; Li et al., 2015b). Insights into the oxidation states and chemical environment of nanoparticles are provided by X-ray absorption spectroscopy (XAS), which helps determine the fate and transformation of these particles in ecosystems (Zhang et al., 2023). With remarkable sensitivity and spatial resolution, X-ray fluorescence (XRF) allows elemental mapping, which locates and distributes nanoparticles in biological tissues and environmental samples (Vanhoof et al., 2020).

In environmental nanotoxicology, high-throughput screening (HTS) has emerged as a significant technique that makes it possible to evaluate the toxicity of nanoparticles in a variety of biological systems and environmental matrices. This technology speeds up data collection and analysis by enabling the quick, parallel testing of multiple nanoparticles or environmental samples (Nel et al., 2013; Bhusal et al., 2022). It makes it easier to identify important toxicity indicators, such as changes in biomolecules, cellular reactions, or effects on ecosystems. When combined with omics technologies, HTS platforms provide a thorough understanding of the molecular mechanisms behind the toxicity of nanoparticles (Nizamani et al., 2023). Additionally, HTS techniques speed up risk assessment, facilitate the creation of safer nanomaterials, and assist in the development of predictive models for the toxicity of nanoparticles (Tirumala et al., 2021). HTS platforms facilitate the efficient handling of large datasets, enabling researchers to make well-informed decisions about the environmental implications of nanomaterials.

Another innovative technology is nanoparticle tracking analysis (NTA), which offers high-resolution, real-time measurements of nanoparticle behavior, concentration, and size distribution in intricate environmental matrices (Zhang et al., 2019). According to Wang et al. (2019), NTA makes it possible to precisely assess the stability of nanoparticles, their aggregation kinetics, and their interactions with biological or environmental elements. It sheds light on the possible effects of nanoparticles on ecosystems and organisms by helping to understand how they behave in various environmental conditions. Due to NTA's real-time tracking of individual nanoparticles, decision-making and risk assessments can be made with greater accuracy and understanding of the behavior of these particles (Gupta et al., 2021). This analysis has a wide range of applications in environmental nanotoxicology, from determining how engineered nanomaterials behave in soil and water systems to comprehending how they interact with living things (Wright, 2020). By providing information on structure–activity relationships and assisting in the design of nanomaterials with a lower environmental impact, it supports the development of safer nanomaterials.

16.3.2 Omics Technologies

Omics technologies, encompassing genomics, transcriptomics, proteomics, metabolomics, and epigenomics, enable the study of the molecular responses of biological systems that have been exposed to nanoparticles. These technologies offer insights into how exposure to nanoparticles alters gene expression, protein levels, and metabolic pathways.

A blueprint for identifying genetic factors influencing susceptibility or resistance to nanoparticle exposure is provided by genomics, which reveals an organism's entire gene set (Kumah et al., 2023). Transcriptomics studies patterns of gene expression, reveals changed pathways, reveals dynamic cellular responses to nanoparticles, and pinpoints possible toxicity mechanisms (Alsagaby et al., 2020). Proteomics probes further, examining alterations in protein expression and modifications caused by exposure to nanoparticles, revealing complex cellular reactions (Zhang et al., 2022). A comprehensive understanding of cellular reactions to nanoparticle exposure can be obtained through the metabolomics, the study of small molecule metabolites (Awashra & Młynarz, 2023). Epigenomics clarifies changes in noncoding RNAs, histone modifications, and DNA methylation, which may reveal heritable modifications caused by nanotoxicants (Wei et al., 2020).

Together, these omics technologies make it possible to understand the complex molecular landscape of toxicity caused by nanoparticles. This understanding enables customized risk assessments for a variety of organisms and environmental conditions, as well as the identification of predictive biomarkers and the clarification of toxicity pathways. Their use helps in the creation of safer nanomaterials and accurate environmental risk assessments, ranging from early hazard identification to ecosystem-level assessments.

16.3.2.1 Genomics

Genomic analyses provide information on the genetic reactions and changes brought on by exposure to nanoparticles, revealing vulnerabilities and adaptive mechanisms. This entails determining modifications to DNA sequences, comprehending patterns of gene expression, and determining genetic variations caused by exposure to nanoparticles (Kumah et al., 2023). According to Himida et al. (2023), genomic analyses provide insight into the mechanisms that underlie biological responses to nanotoxicants, clarifying their short- and long-term effects on ecosystems.

Furthermore, the identification of biomarkers indicative of exposure to nanoparticles and associated risks is made easier by genomics. These biomarkers function as early warning signs of nanomaterial-induced environmental stress, which helps with the creation of monitoring plans to evaluate and lessen the negative effects (Kumah et al., 2023). Furthermore, developments in genomics make it possible to forecast the possible ecological effects of nanoparticles, which helps with the development of safer nanomaterials and ecologically friendly applications (Kumah et al., 2023).

16.3.2.2 Transcriptomics

This branch of omics studies focuses on RNA transcript analysis, offering a comprehensive understanding of the regulatory mechanisms and patterns of gene expression affected by exposure to nanoparticles. Deciphering the intricate dynamics of cellular reactions to nanotoxicants is made possible by transcriptomics (Alsagaby et al., 2020). It also reveals changes in the gene expression profiles, emphasizing particular genes and pathways impacted by nanoparticles. Relevant advances in knowledge of the biological processes and signaling pathways underlying nanoparticle-induced toxicity are obtained by recognizing these alterations (Wang et al., 2022a).

Furthermore, transcriptomic analyses help identify early molecular markers of nanomaterial-induced environmental stress. According to Imadi et al. (2015), these markers function as sensitive indicators of toxicity, which makes it easier to create proactive monitoring plans to identify and reduce possible threats to ecosystems. Additionally, the development of predictive models for evaluating the possible effects of nanoparticles on various organisms and ecosystems is facilitated by transcriptomics-driven research (Furxhi et al., 2019; Zhu et al., 2020). In order to guarantee the ethical and sustainable application of nanotechnology, this predictive ability informs regulatory frameworks and directs the creation of safer nanomaterials.

16.3.2.3 Proteomics

The area of omics research explores the detailed study of proteins, as well as their modifications and alterations, to help comprehend the signaling pathways and cellular reactions caused by exposure to nanoparticles. Investigating the complex protein landscapes in organisms impacted by nanomaterials is facilitated by proteomics (Zhang et al., 2022). Analyzing protein expression, interactions, and posttranslational modifications can provide significant insights into the molecular mechanisms and cellular reactions brought on by exposure to nanoparticles. These discoveries contribute to an improved comprehension of the biological processes and pathways affected by nanotoxicants (Rothen-Rutishauser et al., 2019).

Furthermore, proteomic analyses provide the chance to pinpoint particular protein biomarkers suggestive of toxicity induced by nanoparticles. Additionally, these biomarkers function as early warning signs of detrimental impacts on ecosystems and organisms, allowing for proactive risk assessment and monitoring techniques (Driessen et al., 2015). Proteomics also plays a role in clarifying the mechanism of action of nanomaterials by analyzing their interactions with biological constituents and signaling pathways (Mortimer et al., 2021). This information helps with the development of focused interventions to reduce the environmental impact of nanomaterials and the design of safer nanomaterials.

16.3.2.4 Metabolomics

In order to provide a thorough overview of the biochemical pathways impacted by nanoparticle exposure, metabolomics focuses on studying the entire set of small molecules or metabolites within biological systems (Awashra & Młynarz, 2023). The investigation of dynamic alterations in metabolite profiles induced by nanomaterials is made easier by metabolomics. Researchers are able to discover more about the disruptions and adaptations that occur within cellular metabolic pathways by analyzing these changes (Lin et al., 2022). Comprehending these alterations clarifies the influence of nanotoxicants on significant biochemical mechanisms and their possible consequences for the well-being of organisms and the stability of ecosystems (Zhu et al., 2020).

Additionally, according to Lv et al. (2015), metabolomic analyses aid in the identification of particular metabolite signatures suggestive of toxicity induced by nanoparticles. According to Zhu et al. (2020), these biomarkers function as sensitive indicators of environmental stress brought on by nanomaterials, which helps identify and evaluate possible threats to ecosystems in advance. Furthermore, research driven by metabolomics is also helpful in elucidating the mechanisms behind the toxicity of nanoparticles, offering vital information for the development of safer nanomaterials and focused environmental management strategies.

16.3.2.5 Epigenomics

This field studies changes in epigenetic modifications brought on by exposure to nanomaterials, including DNA methylation, histone modifications, and regulation of noncoding RNA. According to Turunen et al. (2018), epigenomics offers a deeper understanding of how nanomaterials affect gene expression without changing the underlying DNA sequence. Understanding how nanoparticles impact the control of genes and cellular functions can also be obtained by looking at epigenetic modifications. Understanding the long-term effects of nanotoxicants on biological systems and potential effects on future generations requires an understanding of epigenomics (Wei et al., 2020).

Moreover, epigenomic analyses aid in the identification of epigenetic markers suggestive of toxicity induced by nanoparticles (Choudhury et al., 2017; Sun et al., 2022). According to Dusinka et al. (2017), these markers function as sensitive indicators of environmental stress brought on by nanomaterials, allowing for the development of targeted interventions and early warning systems to reduce potential risks. Furthermore, epigenomics provides vital insights into the biological interactions and effects of nanomaterials by assisting in the decipherment of their mode of action within cells and organisms (Himič et al., 2023).

16.3.3 Nanoecotoxicology Testing Methods

These techniques entail creating models and assays to evaluate the toxicity of nanoparticles to diverse organisms at different trophic levels. They aid in the comprehension of the interactions and effects that nanoparticles have on ecosystems. Nanoecotoxicology testing techniques clarify the effects of nanoparticles on organisms, from cellular responses to population-level impacts, by utilizing a variety of assays and models (Savage et al., 2019). These methods show the complexity of nanoparticle toxicity pathways by allowing researchers to evaluate endpoints like oxidative stress, genotoxicity, immunotoxicity, and bioaccumulation (Hughes & Asmatulu, 2021; Oviyaa Sri et al., 2021).

The advantages of nanoecotoxicology testing techniques include their capacity to simulate actual exposure situations and evaluate the particular impacts of nanoparticles on various organisms (Forest, 2022). These techniques help in identifying vulnerable species, comprehending toxicity and its mode of action, and forecasting possible ecological repercussions (Hughes & Asmatulu, 2021). These techniques have a wide range of applications, from determining the toxicity of nanoparticles to aquatic and terrestrial organisms to determining how they affect biodiversity and ecosystem functioning. Testing techniques for nanoecotoxicology assist in risk assessments, regulatory decisions, and the creation of guidelines for the environmentally safe use of nanomaterials (Hughes & Asmatulu, 2021). Despite their benefits, standardizing protocols and applying lab results to real-world situations still present challenges (Handy et al., 2012).

16.3.4 Computational Modeling

A virtual window into the behavior, interactions, and possible risks of nanoparticles in a variety of ecosystems is provided by computational modeling (Shamsi et al., 2019). These models fill in the gaps left by experiments to improve the comprehension of the behavior of nanoparticles. They range from molecular simulations to predictive frameworks (Sahai et al., 2021). The behavior, interactions, and potential toxicity of nanoparticles can be predicted with the use of computational tools like quantitative structure–activity relationship (QSAR) models and molecular dynamics simulations (Mehta et al., 2019). These models enable more thorough risk assessments by supplementing experimental data.

The atomic-level interactions between nanoparticles and biological entities are uncovered by molecular dynamics simulations, which also clarify binding affinities, surface interactions, and possible toxicity mechanisms (Casalini et al., 2019; Fortouna et al., 2021; Yan et al., 2023). When assessing hazards and making regulatory decisions, quantitative structure–activity relationship (QSAR) models can be used to predict the toxicity of nanoparticles based on their physicochemical properties (Mansouri et al., 2018).

Computational modeling offers the advantages of cost-effective forecasts, expedited nanotoxicity evaluations, and quick screening of a variety of nanoparticles. Additionally, computational models predict the fate of the environment, which helps with risk assessments in a variety of scenarios (NIH, 2020). Applications include the assessment of nanoparticle transport in environmental matrices and the prediction of nanoparticle bioaccumulation in organisms (Gajewicz et al., 2012; Kleandrova et al., 2014). According to Hendren et al. (2013), these models support proactive environmental management, inform regulatory policies, and direct the design of safer nanomaterials. However, refining model accuracy and accounting for complex environmental variables remain challenges (Utembe et al., 2018).

16.3.5 Biosensors and Monitoring Devices

The accurate detection and quantification of nanoparticles in ecosystems is made possible by biosensors and monitoring equipment. These tools offer real-time information on the concentration and presence of nanoparticles, which supports proactive risk assessment and environmental monitoring and allows for rapid responses to potential risks (Sharma et al., 2021; Bishnoi et al., 2022). Biosensors are capable of detecting me concentrations of nanoparticles by employing diverse detection principles, including optical, electrochemical, or biological sensing (Bhomkar et al., 2014). Their sensitivity and specificity help to monitor the presence of nanoparticles and provide insight into their persistence and distribution in different environmental compartments. Gaviria-Arroyave et al. (2020), for instance, reported on the use of fluorescent biosensors based on nanomaterials for environmental monitoring.

These devices have several benefits, such as portability that makes on-site assessments possible and continuous data streams that enable dynamic monitoring (Vigneshvar et al., 2016; Peña-Bahamonde et al., 2018). Early detection of nanoparticle contamination is also made possible by biosensors, which improves risk assessment and management techniques. Their uses include on-site evaluations of organisms' exposure to nanoparticles and environmental monitoring of nanoparticle pollutants in the air, water, and soil (Ajayi et al., 2022). These technologies assist in developing strategies for minimizing environmental impact, supporting regulatory compliance, and providing information for mitigation efforts. However, there are difficulties with calibration and standardization in a variety of environmental matrices (Prabowo et al., 2021; Bollella & Katz, 2020).

16.3.6 Data Integration and Big Data Analytics

The processing of large datasets and the extraction of valuable information about the behavior and toxicity of nanoparticles in ecosystems are made possible by data integration and big data analytics (Richarz, 2019). Large-scale database-based techniques such as data mining, machine learning (ML) algorithms, and expert systems can yield more precise forecasts than conventional methods when it pertains to identifying environmental pollution, tracking the sources of pollutants, investigating environmental behavior, and toxicology (Wu et al., 2022). Drawing correlations and patterns regarding the toxicity of nanoparticles under various environmental conditions, species, and exposure scenarios is rendered easier by integrating data from multiple sources and utilizing big data analytics (Bathi et al., 2022). Comprehensive analysis is enabled through data integration, which gathers information from various studies, experiments, and environmental monitoring initiatives. This combined data is further utilized by big data analytics, which reveals relationships, trends, and models for the behavior and toxicity of nanoparticles (Sukhanova et al., 2018). For instance, as part of the NanoPUZZLES project, Richarz et al. (2017) compiled data and performed modeling of nanoparticle behavior and toxicity. After a thorough review of the compilation, the project determined what data was available and assessed it in accordance with the specifications needed to create a data collection that would be useful for nano-QSAR modeling.

The benefits of big data analytics and data integration include their capacity to manage heterogeneous datasets and their ability to detect patterns, anomalies, and new behaviors in nanoparticle interactions. These methods make it possible to find exposure–response connections, which have an impact on risk evaluations and regulatory decision-making. Applications for them include predicting the fate of nanoparticles in ecosystems and evaluating their effects at different trophic levels. Big data analytics and data integration support proactive management approaches for the sustainable use of nanomaterials, direct research priorities, and provide evidence-based policy (Sabharwal & Miah, 2021). However, there are still issues with data standardization and interoperability across various datasets and studies (Walshe, 2021).

16.4 Future Directions in Environmental Nanotoxicology

Future directions in environmental nanotoxicology are poised to address existing gaps and present limitations, particularly in nanoparticle detection, quantification, and monitoring. This inclination also seeks to improve knowledge of how nanomaterials affect ecosystems and human health. It entails combining multi-omics techniques for a thorough examination of how nanomaterials affect various species and ecosystems, as well as possible uses for risk assessment and control of nanoparticles and creative design of nanomaterials to minimize environmental impact (Hussain et al., 2023). It is crucial to conduct more research on the interactions between biological systems and nanoparticles as well as the long-term impacts of chronic nanoparticle exposure on living things. By utilizing bioinformatics and big data analytics, creating reliable predictive models, and creating risk assessment frameworks, advances in analytical tools and imaging techniques can improve the assessment of nanoparticle effects on biodiversity, ecosystem services, and trophic interactions. Innovative nanomaterial design for reduced environmental impact is a key resolution technique to be explored, keeping in mind that ethical and societal implications cannot be overlooked. Nanoparticle risk assessment and regulation are equally important.

In order to evaluate the interactions and responses in organisms exposed to nanomaterials, the multi-omics strategy integrates multiple omics disciplines, including transcriptomics, proteomics, metabolomics, and epigenomics (Prat & Degli-Esposti, 2019; Ahmad et al., 2021; Hussain et al., 2023). A comprehensive evaluation of the molecular reactions to nanomaterial exposure for predictive toxicology is made possible by the integration of genomics, transcriptomics, proteomics, metabolomics, and epigenomics (Martins et al., 2019). This thorough analysis supports predictive modeling to foresee possible nanoparticle toxicity on living things and ecosystems, allowing risk assessment prior to extensive environmental exposure. The identification of particular biomarkers suggestive of toxicity induced by nanoparticles is also made easier by multi-omics approaches (Li et al., 2023). These biomarkers function as sensitive environmental stress indicators, facilitating early identification and evaluation of the hazards connected to environmental nanomaterials (Kim & Kang, 2021). Furthermore, researchers can better understand the complex molecular mechanisms underlying nanoparticle toxicity by integrating multiple omic datasets. This enhanced comprehension of the biological processes and cellular pathways that are impacted by nanomaterials clarifies their mode of action and helps to clarify the complex effects of these materials on living things (Yu et al., 2022). Moreover, multi-omics techniques provide a thorough evaluation of the effects of nanomaterials on ecosystems (Kim et al., 2020). This entails being aware of how various trophic levels, biodiversity, and ecological interactions are affected. These understandings are essential for assessing the general well-being and adaptability of ecosystems to nanoparticle exposure. Furthermore, the application of multi-omics data facilitates the development of safer nanomaterials with lower environmental impact (Hussain et al., 2023). Through an understanding of the molecular pathways underlying the toxicity of nanoparticles, researchers can alter the characteristics of nanomaterials to reduce side effects while maintaining their functionality.

Another future opportunity in environmental nanotoxicology that is poised to fill in the gaps and overcome the constraints presently in place is more investigation into the interactions between nanoparticles and biological systems. A more accurate prediction of the behavior and possible effects of nanoparticles can be achieved by comprehending their interactions with biomolecules, cells, and tissues (Cortez-Jugo et al., 2021). To fully comprehend the ecological impact of nanoparticles, it is equally important to evaluate their effects on trophic interactions, ecosystem services, and biodiversity. It is imperative that more research be done on the long-term impacts of chronic nanoparticle exposure on ecosystems and organisms (Hartmann et al., 2019; Churilov et al., 2020). Research concentrating on cases of extended exposure will shed more light on cumulative impacts and possible ecological repercussions. A promising area of research is investigating remediation strategies based on nanotechnology to mitigate the effects of nanoparticle contamination in the environment (Bhat et al., 2022; Malik et al., 2022; Thomas et al., 2022). These remedial methods that are both efficient and environmentally friendly can help lessen their negative effects on the environment (Das et al., 2022; Mondal & Palit, 2022; Naskar et al., 2022).

Improved imaging and analytical tool development will enable more accurate nanoparticle characterization in intricate environmental matrix systems. Regulatory decision-making will be aided by the development of strong predictive models and risk assessment frameworks tailored to nanomaterials. Ecological contexts, species sensitivities, and different exposure scenarios should all be taken into account in these models. Additionally, processing and interpreting huge datasets produced by multi-omics research and environmental monitoring will be made easier by utilizing bioinformatics and big data analytics (Ahmad et al., 2021). Assessing this wealth of data can highlight trends and connections in the toxicity of nanoparticles in various ecosystems. Environmental nanotoxicology faces difficulties with big data analysis, mainly with regard to data standardization and interoperability across various studies and datasets (Walshe, 2021). Using machine learning (ML) and artificial intelligence (AI) techniques is one way to address these issues. Large volumes of toxicological data, including high-content image-based screening data and data from toxicological databases, can be analyzed and interpreted using these methods (Singh et al., 2023).

Furthermore, the behavior and toxic effects of nanomaterials can be predicted using physiologically based pharmacokinetic (PBPK) and nano-quantitative structure–activity relationship (QSAR) models, respectively (Singh et al., 2023).

Toxicogenomics is the study of the genetic basis of toxic responses in living organisms, and PBPK and nano-QSAR are well-known machine learning tools for adverse event analysis that are used to comprehend the mechanisms by which chemical compounds can cause toxic effects (Singh et al., 2023).

A combination of these approaches provides robust scientific evidence essential for developing regulatory frameworks and guidelines for the safe and responsible use of nanomaterials. This information aids policymakers in establishing standards and regulations that ensure environmental safety without hindering technological advancements. Integrating these approaches also aids in the development of targeted environmental management strategies. This includes designing effective remediation methods, monitoring tools, and mitigation strategies tailored to specific nanoparticle-induced impacts in diverse environmental settings. This integrative approach enhances predictive capabilities, enabling the identification of early biomarkers and the development of targeted interventions to mitigate nanotoxicity. It fosters a more nuanced comprehension of environmental risks associated with nanoparticles, guiding safer nanomaterial design and environmental management strategies.

16.5 Conclusion

Key advancements in environmental nanotoxicology emphasize a nuanced understanding of nanoparticle effects on ecosystems. Research delves into mechanisms, routes of exposure, and impacts on diverse organisms. It highlights the importance of predictive models and innovative assessment tools for risk evaluation. Moreover, a shift toward sustainable nanomaterial development and eco-friendly practices emerges. These advancements underscore the need for interdisciplinary collaborations and robust regulations to ensure responsible use while safeguarding environmental health. Overall, the field continually evolves, offering crucial insights into mitigating potential risks and maximizing the benefits of nanotechnology, fostering a more sustainable and eco-conscious approach to its application.

The intricate nature of nanotoxicology demands ongoing research and cross-disciplinary cooperation. By merging expertise, it becomes possible to navigate risks posed by nanomaterials to ecosystems. Continued collaboration fosters robust strategies, enabling safe implementation while preserving environmental integrity in the field of environmental nanotechnology. The call for continued research and interdisciplinary collaboration in environmental nanotoxicology aims to fortify the understanding of nanoparticle impacts on ecosystems. Through joint efforts, the development of informed strategies for responsible nanoparticle use, safeguarding environmental health and fostering sustainable practices in nanotechnology, is possible.

References

Abbas, Q., Yousaf, B., Ali, M. U., Munir, M. A. M., El-Naggar, A., Rinklebe, J., & Naushad, M. (2020). Transformation pathways and fate of engineered nanoparticles (ENPs) in distinct interactive environmental compartments: A review. *Environment International, 138*, 105646.

Ahmad, F., Mahmood, A., & Muhmood, T. (2021). Machine learning-integrated omics for the risk and safety assessment of nanomaterials. *Biomaterials Science, 9*(5), 1598–1608.

Ahmad, A., Imran, M., & Sharma, N. (2022). Precision nanotoxicology in drug development: Current trends and challenges in safety and toxicity implications of customized multifunctional nanocarriers for drug-delivery applications. *Pharmaceutics, 14*(11), 2463.

Ahmed, S. F., Kumar, P. S., Kabir, M., Zuhara, F. T., Mehjabin, A., Tasannum, N., Hoang, A. T., Kabir, Z., & Mofijur, M. (2022). Threats, challenges and sustainable conservation strategies for freshwater biodiversity. *Environmental Research, 214*, 113808.

Ajayi, R. F., Barry, S., Nkuna, M., Ndou, N., Rakgotho, T., Nqunqa, S., Ngema, N., Thipe, V., & Muluadzi, T. (2022). Nanoparticles in biosensor development for the detection of pathogenic bacteria in water. In *Emerging freshwater pollutants* (pp. 331–358). Elsevier.

Ale, A., Gutierrez, M. F., Rossi, A. S., Bacchetta, C., Desimone, M. F., & Cazenave, J. (2021). Ecotoxicity of silica nanoparticles in aquatic organisms: An updated review. *Environmental Toxicology and Pharmacology, 87*, 103689.

Alonso, D., Garcia, J., & Micó, V. (2023). Fluholoscopy—Compact and simple platform combining fluorescence and holographic microscopy. *Biosensors, 13*(2), 253.

Alsagaby, S. A., Vijayakumar, R., Premanathan, M., Mickymaray, S., Alturaiki, W., Al-Baradie, R. S., AlGhamdi, S., Aziz, M. A., Alhumaydhi, F. A., Alzahrani, F. A., & Alwashmi, A. S. (2020). Transcriptomics-based characterization of the toxicity of ZnO nanoparticles against chronic myeloid leukemia cells. *International Journal of Nanomedicine, 4*, 7901–7921.

Anastasiadis, S. H., Chrissopoulou, K., Stratakis, E., Kavatzikidou, P., Kaklamani, G., & Ranella, A. (2022). How the physicochemical properties of manufactured nanomaterials affect their performance in dispersion and their applications in biomedicine: A review. *Nanomaterials, 12*(3), 552.

Ankley, G. T., Cureton, P., Hoke, R. A., Houde, M., Kumar, A., Kurias, J., Lanno, R., McCarthy, C., Newsted, J., Salice, C. J., & Sample, B. E. (2021). Assessing the ecological risks of per- and polyfluoroalkyl substances: Current state-of-the science and a proposed path forward. *Environmental Toxicology and Chemistry, 40*(3), 564–605.

Aurisano, N., Albizzati, P. F., Hauschild, M., & Fantke, P. (2019). Extrapolation factors for characterizing freshwater ecotoxicity effects. *Environmental Toxicology and Chemistry, 38*(11), 2568–2582.

Awashra, M., & Młynarz, P. (2023). The toxicity of nanoparticles and their interaction with cells: An in vitro metabolomic perspective. *Nanoscale Advances, 5*(10), 2674–2723.

Babaei, M., Tayemeh, M. B., Jo, M. S., Yu, I. J., & Johari, S. A. (2022). Trophic transfer and toxicity of silver nanoparticles along a phytoplankton-zooplankton-fish food chain. *Science of the Total Environment, 842*, 156807.

Bathi, J. R., Wright, L., & Khan, E. (2022). Critical review of engineered nanoparticles: Environmental concentrations and toxicity. *Current Pollution Reports, 8*(4), 498–518.

Bhat, S. A., Sher, F., Hameed, M., Bashir, O., Kumar, R., Vo, D. V. N., Ahmad, P., & Lima, E. C. (2022). Sustainable nanotechnology based wastewater treatment strategies: Achievements, challenges and future perspectives. *Chemosphere, 288*, 132606.

Bhomkar, P., Goss, G., & Wishart, D. S. (2014). A simple and sensitive biosensor for rapid detection of nanoparticles in water. *Journal of Nanoparticle Research, 16*, 1–17.

Bhusal, A., Dogan, E., Nieto, D., Mousavi Shaegh, S. A., Cecen, B., & Miri, A. K. (2022). 3D bioprinted hydrogel microfluidic devices for parallel drug screening. *ACS Applied Bio Materials, 5*(9), 4480–4492.

Bishnoi, M., Deepika, M. N., & Jain, A. (2022). Biomedical applications of nano-biosensor. In *Nanotechnology for biomedical applications* (pp. 219–246). Singapore.

Bollella, P., & Katz, E. (2020). Biosensors—Recent advances and future challenges. *Sensors, 20*(22), 6645.
Boros, B. V., & Ostafe, V. (2020). Evaluation of ecotoxicology assessment methods of nanomaterials and their effects. *Nanomaterials, 10*(4), 610.
Casalini, T., Limongelli, V., Schmutz, M., Som, C., Jordan, O., Wick, P., Borchard, G., & Perale, G. (2019). Molecular modeling for nanomaterial–biology interactions: Opportunities, challenges, and perspectives. *Frontiers in Bioengineering and Biotechnology, 7*, 268.
Castillo-Michel, H. A., Larue, C., Del Real, A. E. P., Cotte, M., & Sarret, G. (2017). Practical review on the use of synchrotron based micro-and nano-X-ray fluorescence mapping and X-ray absorption spectroscopy to investigate the interactions between plants and engineered nanomaterials. *Plant Physiology and Biochemistry, 110*, 13–32.
Choudhury, S. R., Ordaz, J., Lo, C. L., Damayanti, N. P., Zhou, F., & Irudayaraj, J. (2017). From the cover: Zinc oxide nanoparticles-induced reactive oxygen species promotes multimodal cyto- and epigenetic toxicity. *Toxicological Sciences, 156*(1), 261–274.
Churilov, D. G., Polischuk, S. D., Churilov, G. I., Churilova, V. V., & Byshova, D. N. (2020). Investigation of the long-term toxic effect of nanoparticles of different physical-chemical characteristics. *Agronomy Research, 18*(3), 1973–1991.
Cortez-Jugo, C., Czuba-Wojnilowicz, E., Tan, A., & Caruso, F. (2021). A focus on "bio" in bio–nanoscience: The impact of biological factors on nanomaterial interactions. *Advanced Healthcare Materials, 10*(16), 2100574.
Cunningham, B. E., Sharpe, E. E., Brander, S. M., Landis, W. G., & Harper, S. L. (2023). Critical gaps in nanoplastics research and their connection to risk assessment. *Frontiers in Toxicology, 5*, 1154538.
Das, P. K., Mohanty, C., Purohit, G. K., Mishra, S., & Palo, S. (2022). Nanoparticle assisted environmental remediation: Applications, toxicological implications and recommendations for a sustainable environment. *Environmental Nanotechnology, Monitoring and Management, 18*, 100679.
Driessen, M. D., Mues, S., Vennemann, A., Hellack, B., Bannuscher, A., Vimalakanthan, V., Riebeling, C., Ossig, R., Wiemann, M., Schnekenburger, J., & Kuhlbusch, T. A. (2015). Proteomic analysis of protein carbonylation: A useful tool to unravel nanoparticle toxicity mechanisms. *Particle and Fibre Toxicology, 12*, 1–18.
Dusinska, M., Tulinska, J., El Yamani, N., Kuricova, M., Liskova, A., Rollerova, E., Rundén-Pran, E., & Smolkova, B. (2017). Immunotoxicity, genotoxicity and epigenetic toxicity of nanomaterials: New strategies for toxicity testing? *Food and Chemical Toxicology, 109*, 797–811.
Faghihzadeh, F., Anaya, N. M., Schifman, L. A., & Oyanedel-Craver, V. (2016). Fourier transform infrared spectroscopy to assess molecular-level changes in microorganisms exposed to nanoparticles. *Nanotechnology For Environmental Engineering, 1*, 1–16.
Forest, V. (2021). Combined effects of nanoparticles and other environmental contaminants on human health-an issue often overlooked. *NanoImpact, 23*, 100344.
Forest, V. (2022). Experimental and computational nanotoxicology—Complementary approaches for nanomaterial hazard assessment. *Nanomaterials, 12*(8), 1346.
Fortouna, Y., de Vera, P., Verkhovtsev, A. V., & Solov'Yov, A. V. (2021). Molecular dynamics simulations of sodium nanoparticle deposition on magnesium oxide. *Theoretical Chemistry Accounts, 140*(7), 84.
Furxhi, I., Murphy, F., Poland, C. A., Sheehan, B., Mullins, M., & Mantecca, P. (2019). Application of Bayesian networks in determining nanoparticle-induced cellular outcomes using transcriptomics. *Nanotoxicology, 13*(6), 827–848.
Gajewicz, A., Rasulev, B., Dinadayalane, T. C., Urbaszek, P., Puzyn, T., Leszczynska, D., & Leszczynski, J. (2012). Advancing risk assessment of engineered nanomaterials: Application of computational approaches. *Advanced Drug Delivery Reviews, 64*(15), 1663–1693.
Gaviria-Arroyave, M. I., Cano, J. B., & Peñuela, G. A. (2020). Nanomaterial-based fluorescent biosensors for monitoring environmental pollutants: A critical review. *Talanta Open, 2*, 100006.
Giusti, A., Atluri, R., Tsekovska, R., Gajewicz, A., Apostolova, M. D., Battistelli, C. L., Bleeker, E. A., Bossa, C., Bouillard, J., Dusinska, M., & Gómez-Fernández, P. (2019). Nanomaterial grouping: Existing approaches and future recommendations. *NanoImpact, 16*, 100182.

Goswami, L., Kim, K. H., Deep, A., Das, P., Bhattacharya, S. S., Kumar, S., & Adelodun, A. A. (2017). Engineered nano particles: Nature, behavior, and effect on the environment. *Journal of Environmental Management, 196*, 297–315.

Guimaraes, P. R., Jr. (2020). The structure of ecological networks across levels of organization. *Annual Review of Ecology, Evolution, and Systematics, 51*, 433–460.

Gupta, S., Aga, D., Pruden, A., Zhang, L., & Vikesland, P. (2021). Data analytics for environmental science and engineering research. *Environmental Science and Technology, 55*(16), 10895–10907.

Handy, R. D., van den Brink, N., Chappell, M., Mühling, M., Behra, R., Dušinská, M., Simpson, P., Ahtiainen, J., Jha, A. N., Seiter, J., & Bednar, A. (2012). Practical considerations for conducting ecotoxicity test methods with manufactured nanomaterials: What have we learnt so far? *Ecotoxicology, 21*, 933–972.

Hartmann, S., Louch, R., Zeumer, R., Steinhoff, B., Mozhayeva, D., Engelhard, C., Schönherr, H., Schlechtriem, C., & Witte, K. (2019). Comparative multi-generation study on long-term effects of pristine and wastewater-borne silver and titanium dioxide nanoparticles on key lifecycle parameters in Daphnia magna. *NanoImpact, 14*, 100163.

Hendren, C. O., Lowry, M., Grieger, K. D., Money, E. S., Johnston, J. M., Wiesner, M. R., & Beaulieu, S. M. (2013). Modeling approaches for characterizing and evaluating environmental exposure to engineered nanomaterials in support of risk-based decision making. *Environmental Science and Technology, 47*(3), 1190–1205.

Hernandez-Viezcas, J. A., Castillo-Michel, H., Andrews, J. C., Cotte, M., Rico, C., Peralta-Videa, J. R., Ge, Y., Priester, J. H., Holden, P. A., & Gardea-Torresdey, J. L. (2013). In situ synchrotron X-ray fluorescence mapping and speciation of CeO_2 and ZnO nanoparticles in soil cultivated soybean (Glycine max). *ACS Nano, 7*(2), 1415–1423.

Himič, V., Syrmos, N., Ligarotti, G. K., & Ganau, M. (2023). Latest insights on genomic and epigenomic mechanisms of nanotoxicity. In *Impact of engineered nanomaterials in genomics and epigenomics* (pp. 397–417). Wiley.

Hobson, C. M., & Aaron, J. S. (2022). Combining multiple fluorescence imaging techniques in biology: When one microscope is not enough. *Molecular Biology of the Cell, 33*(6), tp1.

Hu, X., Li, D., Gao, Y., Mu, L., & Zhou, Q. (2016). Knowledge gaps between nanotoxicological research and nanomaterial safety. *Environment International, 94*, 8–23.

Huangfu, X., Xu, Y., Liu, C., He, Q., Ma, J., Ma, C., & Huang, R. (2019). A review on the interactions between engineered nanoparticles with extracellular and intracellular polymeric substances from wastewater treatment aggregates. *Chemosphere, 219*, 766–783.

Hughes, S., & Asmatulu, E. (2021). Nanotoxicity and nanoecotoxicity: Introduction, principles, and concepts. *Nanotoxicology and Nanoecotoxicology, 1*, 1–19.

Hussain, M., Zahra, N., Lang, T., Zain, M., Raza, M., Shakoor, N., Adeel, M., & Zhou, H. (2023). Integrating nanotechnology with plant microbiome for next-generation crop health. *Plant Physiology and Biochemistry, 196*, 703.

Imadi, S. R., Kazi, A. G., Ahanger, M. A., Gucel, S., & Ahmad, P. (2015). Plant transcriptomics and responses to environmental stress: An overview. *Journal of Genetics, 94*, 525–537.

Inkson, B. J. (2016). Scanning electron microscopy (SEM) and transmission electron microscopy (TEM) for materials characterization. In *Materials characterization using nondestructive evaluation (NDE) methods* (pp. 17–43). Woodhead publishing.

Jenny, J. P., Anneville, O., Arnaud, F., Baulaz, Y., Bouffard, D., Domaizon, I., Bocaniov, S. A., Chèvre, N., Dittrich, M., Dorioz, J. M., & Dunlop, E. S. (2020). Scientists' warning to humanity: Rapid degradation of the world's large lakes. *Journal of Great Lakes Research, 46*(4), 686–702.

Jia, M., Li, S., Zang, L., Lu, X., & Zhang, H. (2018). Analysis of biomolecules based on the surface enhanced Raman spectroscopy. *Nanomaterials, 8*(9), 730.

Ju, Y., Li, X., Ju, L., Feng, H., Tan, F., Cui, Y., Yang, Y., Wang, X., Cao, J., Qiao, P., & Xiao, L. (2022). Nanoparticles in the earth surface systems and their effects on the environment and resource. *Gondwana Research, 110*, 370–392.

Khan, A. A., Allemailem, K. S., Almatroudi, A., Almatroodi, S. A., Mahzari, A., Alsahli, M. A., & Rahmani, A. H. (2020). Endoplasmic reticulum stress provocation by different nanoparticles: An innovative approach to manage the cancer and other common diseases. *Molecules, 25*(22), 5336.

Kim, H. M., & Kang, J. S. (2021). Metabolomic studies for the evaluation of toxicity induced by environmental toxicants on model organisms. *Metabolites, 11*(8), 485.

Kim, H. M., Long, N. P., Min, J. E., Anh, N. H., Kim, S. J., Yoon, S. J., & Kwon, S. W. (2020). Comprehensive phenotyping and multi-omic profiling in the toxicity assessment of nanopolystyrene with different surface properties. *Journal of Hazardous Materials, 399*, 123005.

Kleandrova, V. V., Luan, F., Gonzalez-Diaz, H., Ruso, J. M., Speck-Planche, A., & Cordeiro, M. N. D. (2014). Computational tool for risk assessment of nanomaterials: Novel QSTR-perturbation model for simultaneous prediction of ecotoxicity and cytotoxicity of uncoated and coated nanoparticles under multiple experimental conditions. *Environmental Science and Technology, 48*(24), 14686–14694.

Kumah, E. A., Fopa, R. D., Harati, S., Boadu, P., Zohoori, F. V., & Pak, T. (2023). Human and environmental impacts of nanoparticles: A scoping review of the current literature. *BMC Public Health, 23*(1), 1–28.

Kumar, A., Khandelwal, M., Gupta, S. K., Kumar, V., & Rani, R. (2019). Fourier transform infrared spectroscopy: Data interpretation and applications in structure elucidation and analysis of small molecules and nanostructures. In *Data processing handbook for complex biological data sources* (pp. 77–96). Academic Press.

Kus-Liśkiewicz, M., Fickers, P., & Ben Tahar, I. (2021). Biocompatibility and cytotoxicity of gold nanoparticles: Recent advances in methodologies and regulations. *International Journal of Molecular Sciences, 22*(20), 10952.

Lai, A., Clark, A. M., Escher, B. I., Fernandez, M., McEwen, L. R., Tian, Z., Wang, Z., & Schymanski, E. L. (2022). The next frontier of environmental unknowns: Substances of unknown or variable composition, complex reaction products, or biological materials (UVCBs). *Environmental Science and Technology, 56*(12), 7448–7466.

Li, X., Liu, W., Sun, L., Aifantis, K. E., Yu, B., Fan, Y., Feng, Q., Cui, F., & Watari, F. (2015a). Effects of physicochemical properties of nanomaterials on their toxicity. *Journal of Biomedical Materials Research Part A, 103*(7), 2499–2507.

Li, Y. F., Zhao, J., Qu, Y., Gao, Y., Guo, Z., Liu, Z., Zhao, Y., & Chen, C. (2015b). Synchrotron radiation techniques for nanotoxicology. *Nanomedicine: Nanotechnology, Biology and Medicine, 11*(6), 1531–1549.

Li, Y., Zhang, Z., Jiang, S., Xu, F., Tulum, L., Li, K., Liu, S., Li, S., Chang, L., Liddell, M., & Tu, F. (2023). Using transcriptomics, proteomics and phosphoproteomics as new approach methodology (NAM) to define biological responses for chemical safety assessment. *Chemosphere, 313*, 137359.

Lin, S., Zhang, H., Wang, C., Su, X. L., Song, Y., Wu, P., Yang, Z., Wong, M. H., Cai, Z., & Zheng, C. (2022). Metabolomics reveal nanoplastic-induced mitochondrial damage in human liver and lung cells. *Environmental Science and Technology, 56*(17), 12483–12493.

Lu, J., Wang, P., Tian, S., Qian, W., Huang, Y., Wang, Z., Zhu, X., & Cai, Z. (2021). TiO_2 nanoparticles enhanced bioaccumulation and toxic performance of PAHs via trophic transfer. *Journal of Hazardous Materials, 407*, 124834.

Lv, M., Huang, W., Chen, Z., Jiang, H., Chen, J., Tian, Y., Zhang, Z., & Xu, F. (2015). Metabolomics techniques for nanotoxicity investigations. *Bioanalysis, 7*(12), 1527–1544.

Malik, S., Dhasmana, A., Preetam, S., Mishra, Y. K., Chaudhary, V., Bera, S. P., Ranjan, A., Bora, J., Kaushik, A., Minkina, T., & Jatav, H. S. (2022). Exploring microbial-based green nanobiotechnology for wastewater remediation: A sustainable strategy. *Nanomaterials, 12*(23), 4187.

Mansouri, K., Grulke, C. M., Judson, R. S., & Williams, A. J. (2018). OPERA models for predicting physicochemical properties and environmental fate endpoints. *Journal of Cheminformatics, 10*(1), 1–19.

Martins, C., Dreij, K., & Costa, P. M. (2019). The state-of-the art of environmental toxicogenomics: Challenges and perspectives of "omics" approaches directed to toxicant mixtures. *International Journal of Environmental Research and Public Health, 16*(23), 4718.

Mehta, C. H., Narayan, R., & Nayak, U. Y. (2019). Computational modeling for formulation design. *Drug Discovery Today, 24*(3), 781–788.

Mohammadinejad, R., Moosavi, M. A., Tavakol, S., Vardar, D. Ö., Hosseini, A., Rahmati, M., Dini, L., Hussain, S., Mandegary, A., & Klionsky, D. J. (2019). Necrotic, apoptotic and autophagic cell fates triggered by nanoparticles. *Autophagy, 15*(1), 4–33.

Mondal, S., & Palit, D. (2022). Prospects and implementation of nanotechnology in environmental remediation and clean up. In *Natural resources conservation and advances for sustainability* (pp. 271–287). Elsevier.

Morin-Crini, N., Lichtfouse, E., Fourmentin, M., Ribeiro, A. R. L., Noutsopoulos, C., Mapelli, F., Fenyvesi, É., Vieira, M. G. A., Picos-Corrales, L. A., Moreno-Piraján, J. C., & Giraldo, L. (2022). Removal of emerging contaminants from wastewater using advanced treatments: A review. *Environmental Chemistry Letters, 20*(2), 1333–1375.

Mortimer, M., Wang, Y., & Holden, P. A. (2021). Molecular mechanisms of nanomaterial-bacterial interactions revealed by omics—The role of nanomaterial effect level. *Frontiers in Bioengineering and Biotechnology, 9*, 683520.

Naskar, J., Boatemaa, M. A., Rumjit, N. P., Thomas, G., George, P. J., Lai, C. W., Mousavi, S. M., & Wong, Y. H. (2022). Recent advances of nanotechnology in mitigating emerging pollutants in water and wastewater: Status, challenges, and opportunities. *Water, Air, and Soil Pollution, 233*(5), 156.

National Institute of Health (NIH). (2020). Computational modeling. In *National institute of biomedical imaging and bioengineering*. U.S. Department of Health and Human Services. Available at: https://www.nibib.nih.gov/science-education/science-topics/computational-modeling. Accessed: 28 Nov, 2023

Nel, A. E., Nasser, E., Godwin, H., Avery, D., Bahadori, T., Bergeson, L., Beryt, E., Bonner, J. C., Boverhof, D., Carter, J., & Castranova, V. (2013). A multi-stakeholder perspective on the use of alternative test strategies for nanomaterial safety assessment. *ACS Nano, 7*(8), 6422–6433.

Nizamani, M. M., Zhang, Q., Muhae-Ud-Din, G., & Wang, Y. (2023). High-throughput sequencing in plant disease management: A comprehensive review of benefits, challenges, and future perspectives. *Phytopathology Research, 5*(1), 44.

Ogunkunle, C. O., Oyedeji, S., Okoro, H. K., & Adimula, V. (2021). Interaction of nanoparticles with soil. In *Nanomaterials for soil remediation* (pp. 101–132). Elsevier.

Ong, T. T., Blanch, E. W., & Jones, O. A. (2020). Surface enhanced Raman spectroscopy in environmental analysis, monitoring and assessment. *Science of the Total Environment, 720*, 137601.

Oviyaa Sri, M., Ilangovan, S. S., Srisugamathi, G., Nilofar Nisha, J., Akshhayya, C., Sounder, S. S., & Srilakkshmi, K. M. (2021). Techniques, methods, procedures and protocols in nanotoxicology. In *Nanotoxicology and nanoecotoxicology* (Vol. 2, pp. 267–302). Springer International Publishing.

Peña-Bahamonde, J., Nguyen, H. N., Fanourakis, S. K., & Rodrigues, D. F. (2018). Recent advances in graphene-based biosensor technology with applications in life sciences. *Journal of Nanobiotechnology, 16*, 1–17.

Pikula, K., Zakharenko, A., Chaika, V., Kirichenko, K., Tsatsakis, A., & Golokhvast, K. (2020). Risk assessments in nanotoxicology: Bioinformatics and computational approaches. *Current Opinion in Toxicology, 19*, 1–6.

Prabowo, B. A., Cabral, P. D., Freitas, P., & Fernandes, E. (2021). The challenges of developing biosensors for clinical assessment: A review. *Chem, 9*(11), 299.

Prat, O., & Degli-Esposti, D. (2019). New challenges: Omics technologies in ecotoxicology. In *Ecotoxicology* (pp. 181–208). Elsevier.

Rajput, V., Minkina, T., Ahmed, B., Sushkova, S., Singh, R., Soldatov, M., Laratte, B., Fedorenko, A., Mandzhieva, S., Blicharska, E., & Musarrat, J. (2020). Interaction of copper-based nanoparticles to soil, terrestrial, and aquatic systems: Critical review of the state of the science and future perspectives. *Reviews of Environmental Contamination and Toxicology, 252*, 51–96.

Richarz, A. N. (2019). Big data in predictive toxicology: Challenges, opportunities and perspectives. In *Big data in predictive toxicology* (pp. 1–37). Royal Society of Chemistry.

Richarz, A. N., Avramopoulos, A., Benfenati, E., Gajewicz, A., Golbamaki Bakhtyari, N., Leonis, G., Marchese Robinson, R. L., Papadopoulos, M. G., Cronin, M. T., & Puzyn, T. (2017). Compilation of data and modelling of nanoparticle interactions and toxicity in the NanoPUZZLES project. *Modelling the Toxicity of Nanoparticles, 947*, 303–324.

Rothen-Rutishauser, B., Bourquin, J., & Petri-Fink, A. (2019). Nanoparticle-cell interactions: Overview of uptake, intracellular fate and induction of cell responses. In *Biological responses to nanoscale particles: Molecular and cellular aspects and methodological approaches* (pp. 153–170). Springer.

Sabharwal, R., & Miah, S. J. (2021). A new theoretical understanding of big data analytics capabilities in organizations: A thematic analysis. *Journal of Big Data, 8*(1), 1–17.

Sahai, N., Gogoi, M., & Ahmad, N. (2021). Mathematical modeling and simulations for developing nanoparticle-based cancer drug delivery systems: A review. *Current Pathobiology Reports, 9*, 1–8.

Savage, D. T., Hilt, J. Z., & Dziubla, T. D. (2019). In vitro methods for assessing nanoparticle toxicity. In *Nanotoxicity: Methods and protocols* (pp. 1–29). Springer.

Schmeisser, S., Miccoli, A., von Bergen, M., Berggren, E., Braeuning, A., Busch, W., Desaintes, C., Gourmelon, A., Grafström, R., Harrill, J., & Hartung, T. (2023). New approach methodologies in human regulatory toxicology–Not if, but how and when! *Environment International, 178*, 108082.

Servin, A. D., Castillo-Michel, H., Hernandez-Viezcas, J. A., Diaz, B. C., Peralta-Videa, J. R., & Gardea-Torresdey, J. L. (2012). Synchrotron micro-XRF and micro-XANES confirmation of the uptake and translocation of TiO_2 nanoparticles in cucumber (Cucumis sativus) plants. *Environmental Science and Technology, 46*(14), 7637–7643.

Shamsi, M., Mohammadi, A., Manshadi, M. K., & Sanati-Nezhad, A. (2019). Mathematical and computational modeling of nano-engineered drug delivery systems. *Journal of Controlled Release, 307*, 150–165.

Sharma, P., Pandey, V., Sharma, M. M. M., Patra, A., Singh, B., Mehta, S., & Husen, A. (2021). A review on biosensors and nanosensors application in agroecosystems. *Nanoscale Research Letters, 16*, 1–24.

Sibiya, A., Jeyavani, J., Santhanam, P., Preetham, E., Freitas, R., & Vaseeharan, B. (2022). Comparative evaluation on the toxic effect of silver (ag) and zinc oxide (ZnO) nanoparticles on different trophic levels in aquatic ecosystems: A review. *Journal of Applied Toxicology, 42*(12), 1890–1900.

Siddhanta, S., Kuzmin, A. N., Pliss, A., Baev, A. S., Khare, S. K., Chowdhury, P. K., Ganguli, A. K., & Prasad, P. N. (2023). Advances in Raman spectroscopy and imaging for biomedical research. *Advances in Optics and Photonics, 15*(2), 318–384.

Singh, A. V., Varma, M., Laux, P., Choudhary, S., Datusalia, A. K., Gupta, N., Luch, A., Gandhi, A., Kulkarni, P., & Nath, B. (2023). Artificial intelligence and machine learning disciplines with the potential to improve the nanotoxicology and nanomedicine fields: A comprehensive review. *Archives of Toxicology, 97*(4), 963–979.

Song, D., Yang, R., Long, F., & Zhu, A. (2019). Applications of magnetic nanoparticles in surface-enhanced Raman scattering (SERS) detection of environmental pollutants. *Journal of Environmental Sciences, 80*, 14–34.

Spedalieri, C., & Kneipp, J. (2022). Surface enhanced Raman scattering for probing cellular biochemistry. *Nanoscale, 14*(14), 5314–5328.

Spurgeon, D., Lahive, E., Robinson, A., Short, S., & Kille, P. (2020). Species sensitivity to toxic substances: Evolution, ecology and applications. *Frontiers in Environmental Science, 8*, 588380.

Stan, G., & King, S. W. (2020). Atomic force microscopy for nanoscale mechanical property characterization. *Journal of Vacuum Science and Technology B, 38*(6), 21.

Sukhanova, A., Bozrova, S., Sokolov, P., Berestovoy, M., Karaulov, A., & Nabiev, I. (2018). Dependence of nanoparticle toxicity on their physical and chemical properties. *Nanoscale Research Letters, 13*, 1–21.

Sun, Q., Li, T., Yu, Y., Li, Y., Sun, Z., & Duan, J. (2022). The critical role of epigenetic mechanisms involved in nanotoxicology. *Wiley Interdisciplinary Reviews: Nanomedicine and Nanobiotechnology, 14*(6), e1789.

Takizawa, T., & Robinson, J. M. (2012). Correlative fluorescence and transmission electron microscopy in tissues. In *Methods in cell biology* (Vol. 111, pp. 37–57). Academic Press.

Terry, L. R., Sanders, S., Potoff, R. H., Kruel, J. W., Jain, M., & Guo, H. (2022). Applications of surface-enhanced Raman spectroscopy in environmental detection. *Analytical Science Advances, 3*(3–4), 113–145.

Thomas, S., Thomas, M. S., & Pothen, L. A. (Eds.). (2022). *Nanotechnology for environmental remediation*. Wiley.

Tirumala, M. G., Anchi, P., Raja, S., Rachamalla, M., & Godugu, C. (2021). Novel methods and approaches for safety evaluation of nanoparticle formulations: A focus towards in vitro models and adverse outcome pathways. *Frontiers in Pharmacology, 12*, 612659.

Titus, D., Samuel, E. J. J., & Roopan, S. M. (2019). Nanoparticle characterization techniques. In *Green synthesis, characterization and applications of nanoparticles* (pp. 303–319). Elsevier.

Tortella, G. R., Rubilar, O., Durán, N., Diez, M. C., Martínez, M., Parada, J., & Seabra, A. B. (2020). Silver nanoparticles: Toxicity in model organisms as an overview of its hazard for human health and the environment. *Journal of Hazardous Materials, 390*, 121974.

Turan, N. B., Erkan, H. S., Engin, G. O., & Bilgili, M. S. (2019). Nanoparticles in the aquatic environment: Usage, properties, transformation and toxicity—A review. *Process Safety and Environmental Protection, 130*, 238–249.

Turunen, T. A., Väänänen, M.-A., & Ylä-Herttuala, S. (2018). Epigenomics. In *Encyclopedia of cardiovascular research and medicine* (pp. 258–265). Elsevier.

Utembe, W., Wepener, V., Yu, I. J., & Gulumian, M. (2018). An assessment of applicability of existing approaches to predicting the bioaccumulation of conventional substances in nanomaterials. *Environmental Toxicology and Chemistry, 37*(12), 2972–2988.

Vanhoof, C., Bacon, J. R., Fittschen, U. E., & Vincze, L. (2020). 2020 atomic spectrometry update–a review of advances in X-ray fluorescence spectrometry and its special applications. *Journal of Analytical Atomic Spectrometry, 35*(9), 1704–1719.

Vigneshvar, S., Sudhakumari, C. C., Senthilkumaran, B., & Prakash, H. (2016). Recent advances in biosensor technology for potential applications–an overview. *Frontiers in Bioengineering and Biotechnology, 4*, 11.

Walshe, R. (2021). The road to big data standardisation. In *The elements of big data value* (p. 333). Springer.

Wang, B., Wang, Z., Feng, W., Wang, M., Hu, Z., Chai, Z., & Zhao, Y. (2010). New methods for nanotoxicology: Synchrotron radiation-based techniques. *Analytical and Bioanalytical Chemistry, 398*, 667–676.

Wang, C., Lv, B., Hou, J., Wang, P., Miao, L., & Ci, H. (2019). Quantitative measurement of aggregation kinetics process of nanoparticles using nanoparticle tracking analysis and dynamic light scattering. *Journal of Nanoparticle Research, 21*, 1–15.

Wang, S., Alenius, H., El-Nezami, H., & Karisola, P. (2022a). A new look at the effects of engineered ZnO and TiO_2 nanoparticles: Evidence from transcriptomics studies. *Nanomaterials, 12*(8), 1247.

Wang, X., Liang, D., Wang, Y., Peijnenburg, W. J., Monikh, F. A., Zhao, X., Dong, Z., & Fan, W. (2022b). A critical review on the biological impact of natural organic matter on nanomaterials in the aquatic environment. *Carbon Research, 1*(1), 13.

Wei, C. C., Yen, P. L., Chaikritsadakarn, A., Huang, C. W., Chang, C. H., & Liao, V. H. C. (2020). Parental CuO nanoparticles exposure results in transgenerational toxicity in Caenorhabditis elegans associated with possible epigenetic regulation. *Ecotoxicology and Environmental Safety, 203*, 111001.

Wright, J.S. (2020). *Colloidal characterisations for environmental exposure assessment in support of the risk assessment of nano-biomaterials for biomedical applications*. pp 79. http://dspace.unive.it/bitstream/handle/10579/16325/865297-1223485.pdf?sequence=2

Wu, X., Zhou, Q., Mu, L., & Hu, X. (2022). Machine learning in the identification, prediction and exploration of environmental toxicology: Challenges and perspectives. *Journal of Hazardous Materials, 438*, 129487.

Xuan, L., Ju, Z., Skonieczna, M., Zhou, P. K., & Huang, R. (2023). Nanoparticles-induced potential toxicity on human health: Applications, toxicity mechanisms, and evaluation models. *MedComm, 4*(4), e327.

Yan, X., Yue, T., Winkler, D. A., Yin, Y., Zhu, H., Jiang, G., & Yan, B. (2023). Converting nanotoxicity data to information using artificial intelligence and simulation. *Chemical Reviews, 123*, 8575.

Yu, C., Zeng, H., Wang, Q., Chen, W., Chen, W., Yu, W., Lou, H., & Wu, J. (2022). Multi-omics analysis reveals the molecular responses of Torreya grandis shoots to nanoplastic pollutant. *Journal of Hazardous Materials, 436*, 129181.

Zhang, S., & Wang, C. (2023). Precise analysis of nanoparticle size distribution in TEM image. *Methods and Protocols, 6*(4), 63.

Zhang, M., Yang, J., Cai, Z., Feng, Y., Wang, Y., Zhang, D., & Pan, X. (2019). Detection of engineered nanoparticles in aquatic environments: Current status and challenges in enrichment, separation, and analysis. *Environmental Science: Nano, 6*(3), 709–735.

Zhang, Y., Yang, H., Yu, Y., & Zhang, Y. (2022). Application of nanomaterials in proteomics-driven precision medicine. *Theranostics, 12*(6), 2674.

Zhang, G., LiBretto, N., Purdy, S., Cesar, L., & Miller, J. (2023). X-ray absorption spectroscopy (XAS): Surface structural determination of alloy nanoparticles. In *Springer handbook of advanced catalyst characterization* (pp. 659–669). Springer International Publishing.

Zhu, Y., Wu, X., Liu, Y., Zhang, J., & Lin, D. (2020). Integration of transcriptomics and metabolomics reveals the responses of earthworms to the long-term exposure of TiO_2 nanoparticles in soil. *Science of the Total Environment, 719*, 137492.

Correction to: Environmental Nanotoxicology

Patrick Omoregie Isibor, Geetha Devi, and Alex Ajeh Enuneku

Correction to:
P. O. Isibor et al. (eds.), *Environmental Nanotoxicology*, https://doi.org/10.1007/978-3-031-54154-4

The book was inadvertently published with incorrect author name arrangement and affiliation in Chapter 11 as:

Oyegbade Samuel Adeniyi
Department of Biochemistry, Federal University of Technology, Minna, Nigeria

Oni Gbotemi Jerry
Department of Biological Sciences, College of Science and Technology, Covenant University, Ota, Nigeria

Ahmed Wopa Wurie
Department of Biological Sciences, College of Science and Technology, Covenant University, Ota, Nigeria

Abiodun Eniola Opeyemi
Department of Microbiology, University of Ibadan, Oyo, Nigeria

---

The updated versions of the chapters can be found at
https://doi.org/10.1007/978-3-031-54154-4_11
https://doi.org/10.1007/978-3-031-54154-4_12

P. O. Isibor et al. (eds.), *Environmental Nanotoxicology*,
https://doi.org/10.1007/978-3-031-54154-4_17

The author name arrangement and affiliation details have been changed as follows:

Samuel Adeniyi Oyegbade
Department of Biological Sciences, College of Science and Technology, Covenant University, Ota, Nigeria

Jerry Gbotemi Oni
Department of Biochemistry, Federal University of Technology, Minna, Nigeria

Wopa Wurie Ahmed
Department of Biochemistry, Federal University of Technology, Minna, Nigeria

Eniola Opeyemi Abiodun
Department of Microbiology, University of Ibadan, Oyo, Nigeria

In addition to this the author name arrangement and spelling in Chapter 12 was inadvertently published as:

Oyegbade Samuel Adeniyi

Oni Gbotemi Jerry

Ahmed Wopa Wurie

Abiodun Eniola Opeyem

The author name arrangement and spelling have been changed as follows:

Samuel Adeniyi Oyegbade

Jerry Gbotemi Oni

Wopa Wurie Ahmed

Eniola Opeyemi Abiodun

Index

P. O. Isibor et al. (eds.), *Environmental Nanotoxicology*,
https://doi.org/10.1007/978-3-031-54154-4

MIX
Papier aus verantwortungsvollen Quellen
Paper from responsible sources
FSC® C105338

If you have any concerns about our products,
you can contact us on
ProductSafety@springernature.com

In case Publisher is established outside the EU,
the EU authorized representative is:
Springer Nature Customer Service Center GmbH
Europaplatz 3, 69115 Heidelberg, Germany

Printed by Libri Plureos GmbH
in Hamburg, Germany